NACAN PRODUCTS LIMITED
Jan 86.

Reactive Oligomers

ACS SYMPOSIUM SERIES **282**

Reactive Oligomers

Frank W. Harris, EDITOR
The University of Akron

Harry J. Spinelli, EDITOR
E. I. du Pont de Nemours and Company

Based on a symposium sponsored by
the Divisions of Polymer Chemistry and
Polymeric Materials Science and Engineering
at the 187th Meeting
of the American Chemical Society,
St. Louis, Missouri,
April 8–13, 1984

American Chemical Society, Washington, D.C. 1985

Library of Congress Cataloging in Publication Data
Reactive oligomers.
(ACS symposium series, ISSN 0097-6156; 282)

"Based on a symposium sponsored by the Divisions of Polymer Chemistry and Polymeric Materials Science and Engineering at the 187th Meeting of the American Chemical Society, St. Louis, Missouri, April 8-13, 1984."

Bibliography: p.
Includes indexes.

1. Polymers and polymerization—Congresses.
2. Chemistry, Physical organic—Congresses.

I. Harris, Frank Wayne, 1942- . II. Spinelli, Harry J., 1949- . III. American Chemical Society. Division of Polymer Chemistry. IV. American Chemical Society. Division of Polymeric Materials: Science and Engineering. V. American Chemical Society. Meeting (187th: 1984: St. Louis, Mo.) VI. Title: Oligomers. VII. Series.

QD380.R418   1985   668.9   85-9215
ISBN 0-8412-0922-7

Copyright © 1985

American Chemical Society

All Rights Reserved. The appearance of the code at the bottom of the first page of each chapter in this volume indicates the copyright owner's consent that reprographic copies of the chapter may be made for personal or internal use or for the personal or internal use of specific clients. This consent is given on the condition, however, that the copier pay the stated per copy fee through the Copyright Clearance Center, Inc., 27 Congress Street, Salem, MA 01970, for copying beyond that permitted by Sections 107 or 108 of the U.S. Copyright Law. This consent does not extend to copying or transmission by any means—graphic or electronic—for any other purpose, such as for general distribution, for advertising or promotional purposes, for creating a new collective work, for resale, or for information storage and retrieval systems. The copying fee for each chapter is indicated in the code at the bottom of the first page of the chapter.

The citation of trade names and/or names of manufacturers in this publication is not to be construed as an endorsement or as approval by ACS of the commercial products or services referenced herein; nor should the mere reference herein to any drawing, specification, chemical process, or other data be regarded as a license or as a conveyance of any right or permission, to the holder, reader, or any other person or corporation, to manufacture, reproduce, use, or sell any patented invention or copyrighted work that may in any way be related thereto. Registered names, trademarks, etc., used in this publication, even without specific indication thereof, are not to be considered unprotected by law.

PRINTED IN THE UNITED STATES OF AMERICA

ACS Symposium Series

M. Joan Comstock, *Series Editor*

Advisory Board

Robert Baker
U.S. Geological Survey

Martin L. Gorbaty
Exxon Research and Engineering Co.

Roland F. Hirsch
U.S. Department of Energy

Herbert D. Kaesz
University of California—Los Angeles

Rudolph J. Marcus
Office of Naval Research

Vincent D. McGinniss
Battelle Columbus Laboratories

Donald E. Moreland
USDA, Agricultural Research Service

W. H. Norton
J. T. Baker Chemical Company

Robert Ory
USDA, Southern Regional
Research Center

Geoffrey D. Parfitt
Carnegie-Mellon University

James C. Randall
Phillips Petroleum Company

Charles N. Satterfield
Massachusetts Institute of Technology

W. D. Shults
Oak Ridge National Laboratory

Charles S. Tuesday
General Motors Research Laboratory

Douglas B. Walters
National Institute of
Environmental Health

C. Grant Willson
IBM Research Department

FOREWORD

The ACS SYMPOSIUM SERIES was founded in 1974 to provide a medium for publishing symposia quickly in book form. The format of the Series parallels that of the continuing ADVANCES IN CHEMISTRY SERIES except that, in order to save time, the papers are not typeset but are reproduced as they are submitted by the authors in camera-ready form. Papers are reviewed under the supervision of the Editors with the assistance of the Series Advisory Board and are selected to maintain the integrity of the symposia; however, verbatim reproductions of previously published papers are not accepted. Both reviews and reports of research are acceptable, because symposia may embrace both types of presentation.

CONTENTS

Preface ix

1. **High Temperature Polymers from Thermally Curable Oligomers** 1
   Paul M. Hergenrother

2. **Synthesis of Bisphenol-Based Acetylene-Terminated Thermosetting Resins** 17
   J. S. Wallace, F. E. Arnold, and W. A. Feld

3. **Arylether Sulfone Oligomers with Acetylene Termination from the Ullman Ether Reaction** 31
   P. M. Lindley, L. G. Picklesimer, B. Evans, F. E. Arnold, and J. J. Kane

4. **Network Structure in Polymers from Bisphthalonitriles** 43
   Jeffrey A. Hinkley

5. **Intermolecular and Intramolecular Reactions of Substituted Norbornenyl Imides** 53
   Chaim N. Sukenik, Vinay Malhotra, and Uday Varde

6. **Thermally Stable Polymers for Electronic Applications** 63
   D. J. Dawson, W. W. Fleming, J. R. Lyerla, and J. Economy

7. **Ethynyl End-Capped Polyimide Oligomers Containing Oxyethylene Linkages: Synthesis and Characterization** 81
   Frank W. Harris and K. Sridhar

8. **Polyaromatics with Terminal or Pendant Styrene Groups: A New Class of Thermally Reactive Oligomers** 91
   Virgil Percec and Brian C. Auman

9. **N-Cyanourea-Terminated Resins** 105
   S. C. Lin

10. **Chain Extendable Oligomers for High-Solids Coatings** 117
    J. W. Holubka

11. **New Telechelic Polymers and Sequential Copolymers by Polyfunctional Initiator–Transfer Agents (Inifers): End Reactive Polyisobutylenes by Semicontinuous Polymerization** 125
    Rudolf Faust, Agota Fehervari, and Joseph P. Kennedy

12. **Functionalization of Polymeric Organolithium Compounds: Synthesis of Macromolecular Ketones and Telechelic Amines** 139
    Roderic P. Quirk, Wei-Chih Chen, and Pao-Luo Cheng

13. **Free-Radical Ring-Opening Polymerization: Use in Synthesis of Reactive Oligomers** 147
    William J. Bailey, Benjamin Gapud, Yin-Nian Lin, Zhende Ni, and Shang-Ren Wu

14. **Reactive Difunctional Siloxane Oligomers: Synthesis and Characterization** 161
    Iskender Yilgör, Judy S. Riffle, and James E. McGrath

15. Synthesis of Poly(phenylene Oxides) by Electrooxidative Polymerization of Phenols...175
    Eishun Tsuchida, Hiroyuki Nishide, and Toshihiko Maekawa

16. Coupling and Capping Reactions on Poly(2,6-dimethyl-1,4-phenylene Oxide)..187
    Dwain M. White and George R. Loucks

17. Characterization of Hydroxyl-Terminated Liquid Polymers of Epichlorohydrin...199
    Simon H. Yu

18. Thermal and Photo Stability of Polyarylates: Styrylpyridine-Based Polymers...209
    Hoh-Jiear Yan and Eli M. Pearce

19. Glycol Bis(allyl Phthalates) as Cocross-linkers for Diallyl Phthalate Resins...225
    Akira Matsumoto and Masayoshi Oiwa

20. Special Functional Triglyceride Oils as Reactive Oligomers for Simultaneous Interpenetrating Networks............................237
    L. H. Sperling, J. A. Manson, and G. M. Jordhamo

Author Index...251

Subject Index..251

PREFACE

INTEREST IN THE CHEMISTRY OF REACTIVE OLIGOMERS is growing rapidly, and this book is a reflection of this interest. In fact, the symposium upon which this book is based generated so much attention that it was teleconferenced to several different sites throughout the country! This attraction is easy to understand when one considers the wide variety of applications that can be addressed with reactive oligomers. For example, several different oligomeric systems for composite, electronic, adhesive, and high-solid coating applications are described in this volume. Syntheses of several telechelic oligomers that should prove extremely useful in the preparation of block and graft copolymers are also included. In addition, for comparison purposes, several authors describe syntheses that employ cationic, anionic, free-radical, and step-growth polymerization techniques.

Although the authors discuss a diversity of interests and chemical approaches, the chapters are connected by a common thread: All are concerned with the synthesis and/or reactions of well-defined low molecular weight compounds that are terminated with reactive functional groups. We believe that the information presented not only serves as an excellent introduction to the field, but also is of considerable value to those already working with these systems.

FRANK W. HARRIS
The University of Akron
Akron, OH 44325

HARRY J. SPINELLI
E. I. du Pont de Nemours and Company
Wilmington, DE 19898

High Temperature Polymers from Thermally Curable Oligomers

PAUL M. HERGENROTHER

Langley Research Center, National Aeronautics and Space Administration, Hampton, VA 23665

> The synthesis, physical and mechanical properties of thermally curable oligomers primarily for use as high temperature composite matrices are reviewed. High temperature in this context is defined as having usable mechanical properties at 177°C and higher.

In the late 1950's research was initiated on high temperature polymers, primarily to meet the demand for functional and structural resins for advanced aircraft and weapon systems and the electronics industry. Many different polymers with remarkable thermal stability evolved from this effort. However, most of these high temperature polymers for structural applications exhibited processing problems due to the evolution of volatiles and/or poor flow and wetting even at temperatures of ~371°C and pressures ~1.38 MPa (200 psi). In the late 1960's, research was directed towards improving the processability of high temperature polymers. The approaches primarily involved incorporating more flexible moieties within the polymer backbone to permit processing as thermoplastics and placing reactive groups on the ends of oligomers which could be thermally reacted to chain extend.

The reactive oligomer approach has served as an attractive route to high performance/high temperature materials for a variety of applications. To place reactive oligomers in proper prospective, the advantages and disadvantages are briefly reviewed. The reactive oligomer route alleviates many shortcomings of the high molecular weight linear polymer route by providing processable materials with better solubility and wettability, lower melt or softening temperature and minimal volatile evolution during cure. It also provides cured materials with better solvent and moisture resistance and generally higher initial elevated temperature mechanical properties. The disadvantages are a processable form with shorter shelf-life and a cured material with less toughness and lower long term thermooxidative stability. This paper reviews thermally curable oligomers that afford high performance/high temperature polymers primarily as laminating resins and is not intended to be

This chapter not subject to U.S. copyright.
Published 1985, American Chemical Society

comprehensive. High temperature in this context is defined as having usable mechanical properties at 177°C and higher.

Results and Discussion

Epoxies - The term "reactive oligomer" is relatively new but the concept is fairly old. This concept has been used for many years with systems such as epoxies, phenolics, unsaturated esters, cyanates, isocyanates and many other crosslinked systems. An example of a 177°C curing epoxy system (Narmco's 5208) which was introduced into the marketplace about 1971 is shown in Eq. 1.

$$CH_2 \overbrace{\left[\underset{}{\bigcirc}-N \left(CH_2-\underset{O}{CH}-CH_2 \right)_2 \right]_2} + SO_2 \left(\underset{}{\bigcirc}-NH_2 \right)_2 + \begin{pmatrix} CH_2-\underset{O}{CH}-CH_2-O-\underset{CH_3}{\overset{CH_3}{\bigcirc-C-\bigcirc}}-O-CH_2-\underset{O}{CH}-CH_2 \\ \begin{pmatrix} CH_2 \\ CH_2-\underset{O}{CH}-CH_2-O-\underset{CH_3}{\overset{CH_3}{\bigcirc-C-\bigcirc}}-O-CH_2-\underset{O}{CH}-CH_2 \end{pmatrix}_{\sim 3} \end{pmatrix}$$

MY720 (100 PARTS)   DDS (28 PARTS)

SU-8 (8.2 PARTS)   (1)

↓ Δ

SOLUBLE TACKY OLIGOMER

↓ 177-204°C

THERMOSET

The tetraglycidyl derivative of 4,4'-methylenedianiline (Ciba-Geigy's MY-720) and an epoxy novolac (Celanese's SU-8) are partially reacted with 4,4'-diaminodiphenylsulfone (DDS) to yield a soluble tacky oligomer (1). This oligomer is used to melt impregnate a reinforcement such as glass or carbon/graphite. The resulting prepreg is tacky, drapable and has nominal shelflife at ambient temperature of two weeks. Composites fabricated with a final cure of 2 hr at 204°C under 0.34 MPa (50 psi) exhibit good strength properties from -54°C to 177°C but poor ambient temperature damage tolerance (ability to carry a load after sustaining impact damage). There are many other 177°C epoxy systems available such as Fiberite's 934 and 976, Hexcel's 263, and Hercules 3501 and 3502. The composition of these systems vary. Some employ a catalyst such as boron trifluoride monoethylamine while others use various other components (e.g. glycidyl esters) to modify the formulation.

Cyanates - Cyanate chemistry serves as another example where a reactive oligomer is initially formed and subsequently cured to a highly crosslinked resin. The initial report on cyanates as laminating resins in 1968 (2) involved the dicyanate of 2,2'-bis(4-hydroxyphenyl)propane (bisphenol-A). More recent work has involved other dicyanates including 1,3-dicyanatobenzene (resorcinol dicyanate), a white crystalline solid melting at 80°C (3), which is transformed by heating to a tacky oligomer. This form has been used for melt impregnation of glass or carbon/graphite reinforcement to yield a prepreg with room temperature tack and drape. Carbon/graphite reinforced composites exhibited excellent mechanical properties at

26 to 177°C (4, 5). Cured neat resin properties of resorcinol dicyanate are given in Table I.

Table I. Resorcinol Dicyanate Cured Resin Properties

Test Temp., °C	26	177	232
Tensile Strength, MPa (Ksi)	124 (18)	88.2 (12.8)	43.4 (6.3)
Tensile Modulus, GPa (Ksi)	4.76 (690)	3.93 (570)	3.03 (440)
Elongation, %	3.2	4.2	9.2
Flexural Strength, MPa (Ksi)	224 (32.5)	158 (22.9)	68.9 (10)
Flexural Modulus, GPa (Ksi)	5.03 (730)	3.72 (540)	2.69 (390)
Compressive Yield Strength, MPa (Ksi)	227 (33)	111.7 (16.2)	75.8 (11)
Compressive Modulus, GPa (Ksi)	2.90 (420)	1.93 (280)	1.52 (220)
Izod Impact-Notched, Ft. lb./in	0.35	--	--

Source: Reference 4.

Cyanamides - Cyanamides also represent a class of materials where reactive oligomers have been prepared. A representative example of the type of modification done to cyanamides to moderate the initial reaction to obtain linear soluble melt-processable oligomers is shown in Eq. 2. A bis(aryl sulfonyl cyanamide) was initially reacted with two moles of a bis(cyanamide) to yield an oligomeric mixture (ideally represented in Eq. 2 as a simple compound). These fire-resistant materials have shown promising properties as composite resin matrices (6).

Bismaleimides - Bismaleimides resins were first introduced into the market in the early 1970's. As with other resin systems, there are many variations of bismaleimides. The Kermid and Kinel bismaleimide products as marketed in the U. S. by Rhodia are representative examples. Bismaleimide chemistry is represented in Eq. 3 where curing can be accomplished thermally through the unsaturation in the maleimide or by way of the Michael Reaction where an appropriate curing agent such as aromatic diamine adds across the activated double bond. In most instances, a combination of curing thermally through the double bond and via an aromatic diamine is used in actual practice. Bismaleimides are frequently formulated with other materials to obtain the proper combination of properties. As an example, materials known as the BT resins (B stands for bismaleimide and T means triazine) are blends of various amounts of a bismaleimide and a dicyanate (bisphenol A dicyanate) (7). One form of the BT resin is thought to be the principal component in a laminating resin system known as 5245-C (8) which has displayed excellent translation of fiber properties in composites. In general, bismaleimide formulations offer a higher use temperature than epoxy systems.

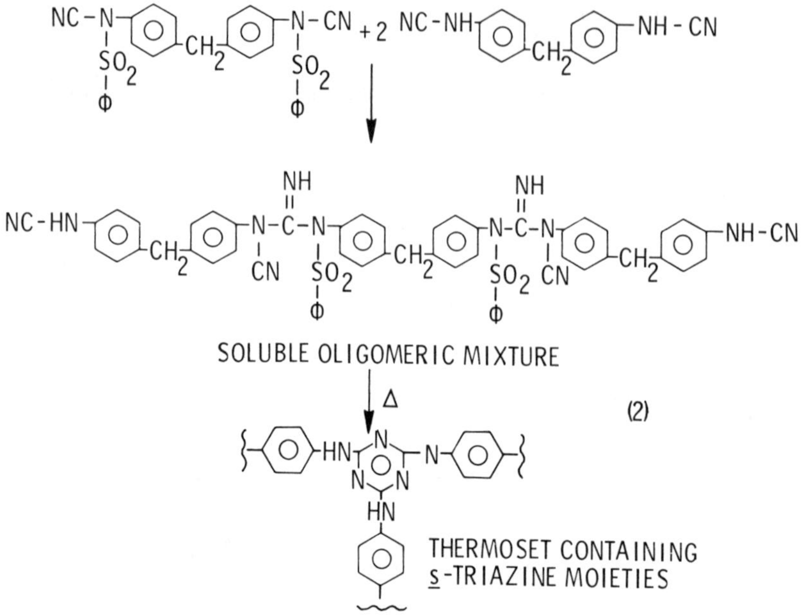

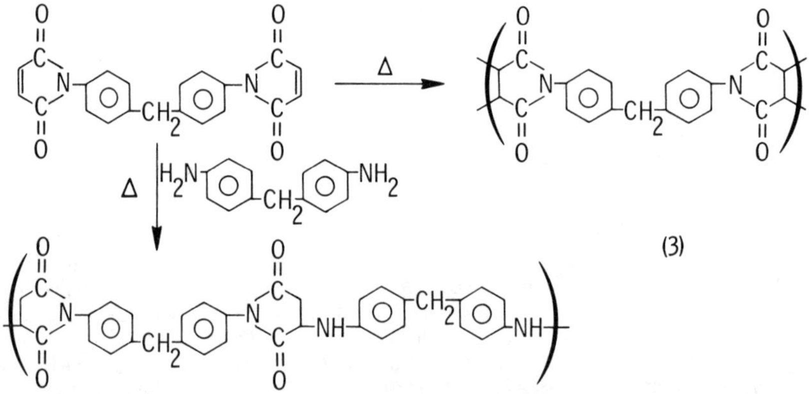

Sulfones - Although thermosets as exemplified by the cured epoxies, cyanates and bismaleimides are high modulus materials, they are brittle when void of toughening agents (e.g. rubbers or thermoplastics). In addition, these materials absorb moisture which lowers their use temperature and stiffness. This is especially apparent in the compression strength at elevated temperature of composites which have been exposed to hot/humid environments. In an attempt to develop a composite matrix resin with improved moisture resistance, a significant effort has been devoted to the synthesis and polymerization of bis(3-ethynylphenoxy-4-phenyl)sulfone, commonly referred to as acetylene-terminated sulfone (ATS). Several different forms of ATS, containing various amounts of oligomers, have been evaluated (9-12). The presence of oligomers is advantageous since they depress the melting point to provide a tacky material at room temperature (desirable for melt impregnation and tack and drape in prepreg). The most recent synthetic route (12) to ATS is shown in Eq. 4. Neat resin moldings cured for 1 hr each at

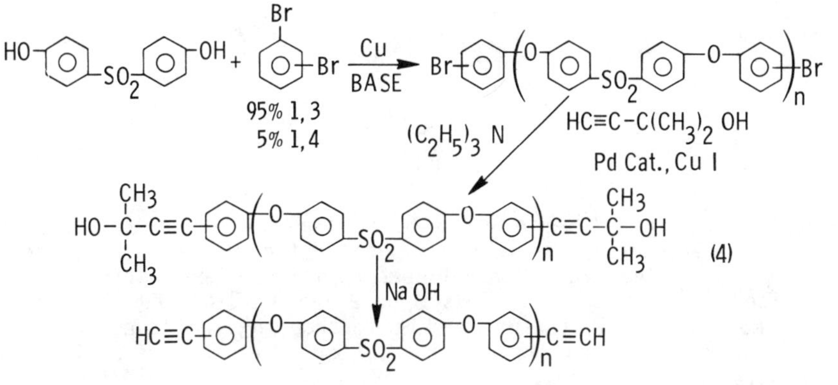

177 and 204°C and 18 hr at 218°C gave the properties in Table II. Room temperature properties of unidirectional AS-4 graphite fiber laminates (cured 2 hr each at 177 and 250°C) included flexural strength of 1.56 GPa (226,000 psi), flexural modulus of 117.2 GPa (17,000,000 psi) and short beam shear strength of 79.3 MPa (11,500 psi) (13). Laminate properties were also reported to be good at 177°C after exposure to hot/humid environment (9). Unfortunately, the brittleness of cured ATS as indicated by fracture energy of 35 J/M² limits its use in composite applications. Materials with low toughness and only moderate stiffness exhibit low levels of impact

Table II. Neat Resin Properties of ATS[1]

Tensile Strength, MPa (Ksi)	48.9 (7.1)
Tensile Modulus, MPa (Ksi)	3619.4 (525)
Elongation, %	1.4
Glass Transition Temperature (Tg), °C	300
Fracture Energy (G_{Ic}), J/M²	35[2]

1. Source: Reference 13.
2. Source: Reference 14.

resistance. In addition, microcracking of the resin in a composite is frequently encountered.

Other acetylene-terminated resins (ATR) similar to ATS as shown in Eq. 5 have also been prepared (13, 15). The cured

$$HO-Ar-OH + Br-\bigcirc-Br \xrightarrow[BASE]{Cu} Br-\bigcirc-(O-Ar-O-\bigcirc)_n-Br$$

$$\downarrow HC\equiv C-C(CH_3)_2 OH, \text{ Pd cat.}$$

$$HC\equiv C-\bigcirc-(O-Ar-O-\bigcirc)_n-C\equiv CH \xleftarrow{NaOH} HO-\underset{CH_3}{\overset{CH_3}{\underset{|}{C}}}-C\equiv C-\bigcirc-(O-Ar-O-\bigcirc)_n-C\equiv C-\underset{CH_3}{\overset{CH_3}{\underset{|}{C}}}-OH$$

WHERE

$$Ar = \bigcirc,\ \bigcirc-SO_2-\bigcirc,\ \bigcirc-S-\bigcirc,\ \bigcirc-\overset{O}{\underset{||}{C}}-\bigcirc,\ \bigcirc-\bigcirc,\ \bigcirc-\underset{CF_3}{\overset{CF_3}{\underset{|}{C}}}-\bigcirc$$

AND $\bigcirc-\underset{CH_3}{\overset{CH_3}{\underset{|}{C}}}-\bigcirc$ n = 1 to 3

(5)

acetylene-terminated resin based upon bisphenol A was reported to absorb significantly less moisture under hot/humid environment than ATS (13). The cured hexafluoropropylidene containing resin exhibited the lowest weight loss upon aging at 316°C for 200 hr in air. More work will undoubtedly be forthcoming on these ATR which will include toughening and more comprehensive evaluation as composite resin matrices.

Prior to the ATS work, maleimide-terminated sulfones (16) were prepared as depicted in Eq. 6. After curing the maleimide-terminated sulfone structure where n = 1 at 275°C for 0.5 hr, the shear modulus was 1.2×10^{10} dyne/cm² (173,880 psi) and the T_g = 239°C. When n = 2, the shear modulus and Tg dropped to 1.0×10^{10} dyne/cm² (144,900 psi) and 185°C respectively. A resin to be used as a composite matrix should have a high shear modulus to prevent microbuckling of the fibers under a compression load. The shear modulus is related to the Young's modulus by the equation G = E/2 (1 + ν) where G = shear modulus, E = Young's modulus and ν = Poisson's ratio. A cured resin with a Young's modulus of ~3.44 GPa (500,000 psi) and a fracture energy of ~5000 J/M² is desired for use as a composite resin matrix (17). Unidirectional carbon/graphite fiber (e.g. Hercules AS-4) reinforced composites should have a minimum compressive strength at room temperature of ~1.38 GPa (200,000 psi) and at 82°C after moisture saturation of ~1.03 GPa (150,000 psi) for use in commercial airplanes.

Nadimide-terminated polysulfones (norbornene-terminated), as shown in Eq. 7 where the linear sulfone segment had a number average molecular weight (M_n) of ~20,000g/mole were first reported in 1980 (18). This work was performed in an attempt to develop a tough solvent resistant resin for use in composites. The thermally induced reaction of the nadimide terminal groups involves a combination of reactions of the nadimide and maleimide/cyclopentadiene formed from a retro Diels-Adler reaction. Graphite fabric (T-300, 8-Harness satin) reinforced composites gave RT flexural strength of 0.63 GPa (91,900 psi) and RT flexural strength after 28 days in methylene chloride of 0.12 GPa (17,000 psi) (19). It was not surprising that a crosslinked polymer with a linear sulfone segment of M_n of ~20,000g/mole was found to be sensitive to methylene chloride, a principal component in paint strippers, and an extremely aggressive solvent for polysulfones.

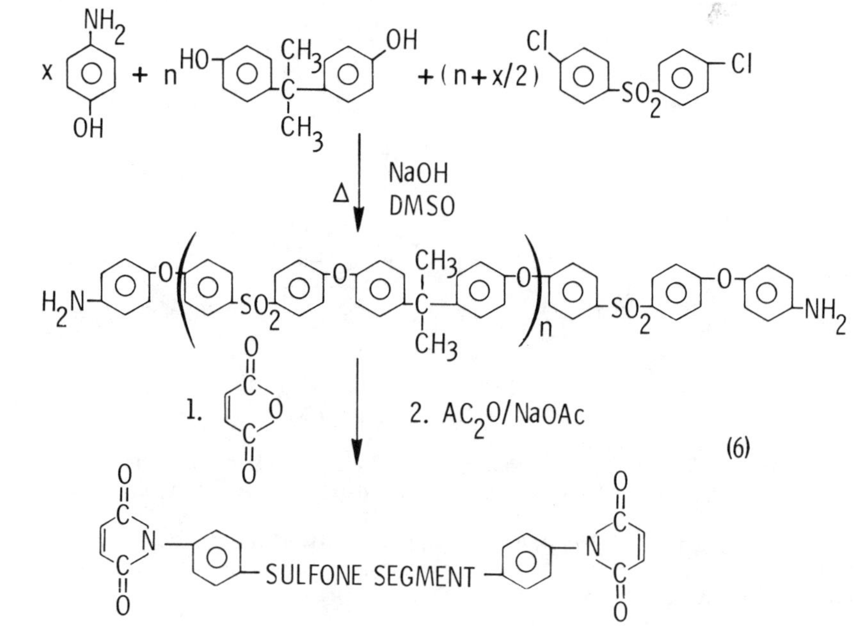

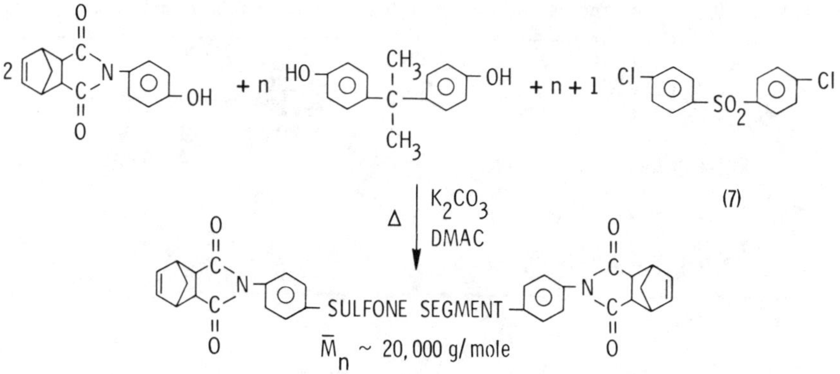

Ethynyl-terminated sulfone oligomers (ETS) were prepared by end-capping hydroxy-terminated sulfone oligomers with 4-ethynylbenzoyl chloride as indicated in Eq. 8 (20). The

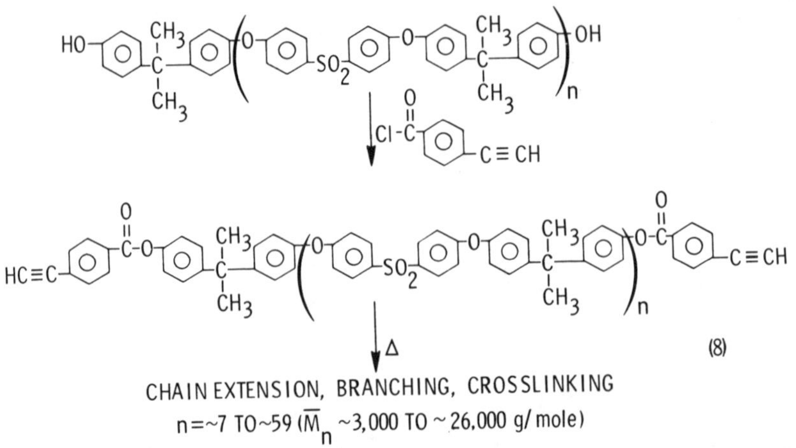

$$HO-\underset{CH_3}{\underset{|}{\overset{CH_3}{\overset{|}{C}}}}\text{—Ar—}(\text{O—Ar—SO}_2\text{—Ar—O—Ar—}\underset{CH_3}{\underset{|}{\overset{CH_3}{\overset{|}{C}}}}\text{—Ar})_n\text{—OH}$$

$$\downarrow \text{Cl—}\overset{O}{\overset{\|}{C}}\text{—Ar—C}\equiv\text{CH}$$

$$HC\equiv C\text{—Ar—}\overset{O}{\overset{\|}{C}}\text{—O—Ar—}\underset{CH_3}{\underset{|}{\overset{CH_3}{\overset{|}{C}}}}\text{—Ar—}(\text{O—Ar—SO}_2\text{—Ar—O—Ar—}\underset{CH_3}{\underset{|}{\overset{CH_3}{\overset{|}{C}}}}\text{—Ar})_n\text{—O—}\overset{O}{\overset{\|}{C}}\text{—Ar—C}\equiv\text{CH} \qquad (8)$$

$$\downarrow \Delta$$

CHAIN EXTENSION, BRANCHING, CROSSLINKING

n = ~7 TO ~59 ($\overline{M}_n$ ~3,000 TO ~26,000 g/mole)

thermally cured ETS exhibited higher T_g's, better solvent resistance, and slightly better adhesive and composite properties at 177°C than a comparable linear polysulfone (Union Carbide's UDEL-P1700, from the reaction of bisphenol-A and 4,4'-dichlorodiphenyl sulfone) (21). The properties of ETS where the sulfone segment had M_n of 4000, 8000 and 12000 g/mole are presented in Table III.

Table III. Properties of Cured Ethynyl-Terminated Sulfones

M_n, g/mole of sulfone segment	4000	8000	12000
Tg, °C (250°C cure)	202	200	196
Film Swelling in chloroform, %	< 10	~ 20	~ 55
G_{Ic}, J/M²	790	1300	2100
RT Thin Film Tensile Strength, MPa (Ksi)	--	--	83.4 (12.1)
RT Thin Film Tensile Modulus, GPa (Ksi)	--	--	2.45 (355)
RT Thin Film Elongation, %	--	--	4.6

Source: Reference 22.

Imides - Polyimides (PI) have been conventionally prepared by the chemical or thermal cyclodehydration of polyamic acids formed from the solution reaction of aromatic tetracarboxylic dianhydrides and aromatic diamines. The early PI were insoluble and relatively intractable. The polyamic acid was the processable intermediate. However, the polyamic acid precursor has two major shortcomings, hydrolytic instability and the evolution of volatiles during the thermal conversion to PI. In addition, residual solvent was left in adhesive tapes and prepregs to obtain tack, drape and flow. During the fabrication of components, the evolution of volatiles caused processing problems and led to porosity in the part. As work progressed on PI, other synthetic routes were investigated (e.g. reaction of esters of aromatic tetracarboxylic acids with diamines and reaction of aromatic diisocyanates with aromatic tetracarboxylic dianhydrides). Information on structure/property relationships permitted the design of thermoplastic PI. However, until recently, little work was done to optimize the molecular weight/processability/mechanical properties of thermoplastic PI. Generally, temperatures 80 to 120°C higher than the glass transition temperature (Tg) and pressure > 1.38 MPa were required in compression molding. In most instances, processable PI were obtained by compromising the use temperature.

To improve the processability of PI and to alleviate or eliminate volatiles, work began in the late 1960's under NASA Lewis sponsorship (23), to end-cap oligomers with reactive groups. The first report involved the use of 3,6-endomethylene-1,2,3,6-tetrahydrophthalic anhydride (5-norbornene-2,3-dicarboxylic anhydride, nadic anhydride) and alkyl derivatives thereof (e.g. citraconic anhydride) and 1,2,3,4-tetrahydrophthalic anhydride (4-cyclohexene-1, 2-dicarboxylic anhydride) as the reactive end group on imide oligomers (23-25). This initial effort led to the development of P13N [with P for polyimide, 13 for M_n of ~1300g/mole and N for nadic end cap] whose structure is shown below.

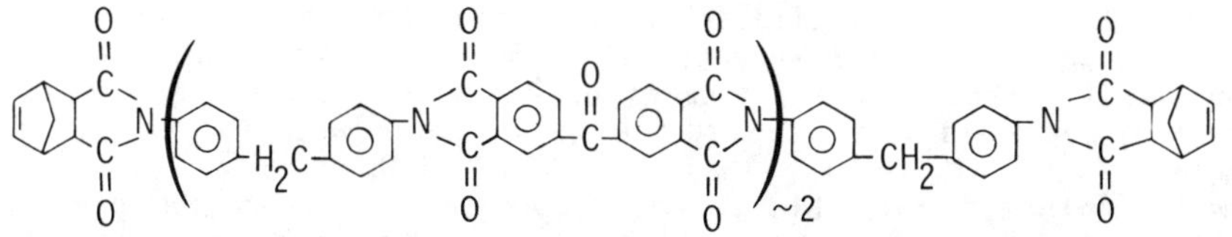

The amic acid oligomer could be thermally cyclodehydrated to yield the nadimide-terminated oligomer which exhibited poor flow at temperature < 275°C. The end-groups begin to react at a moderate rate at ~275°C with the evolution of a small amount (1-2%) of cyclopentadiene. The thermal chain extension, referred to as pyrolytic polymerization, occurs through a combination of addition reactions involving the unsaturation of the nadimide moiety and maleimide/cyclopentadiene formed from a reverse Diels-Alder reaction. This route provided an improvement in processing over the linear polyamic acids and ester/amine derived polyimides but the cured P13N resin was brittle. Early work reported good high temperature properties for small dense glass reinforced composites (25). Problems (e.g. porosity and microcracking) were encountered in the fabrication of large composites and filled moldings apparently due to the evolution of water (from incomplete imidization), residual solvent (N,N-dimethylformamide, DMF) and cyclopentadiene, poor resin flow and resin shrinkage. In addition, variability in the stability of the amic acid solution was a problem.

To resolve these shortcomings, work was initiated at NASA Lewis Research Center which led to the development of PMR-15 (with PMR meaning the in-situ polymerization of monomeric reactants and 15 for an average M_n linear segment of ~1500 g/mole) (26, 27). The early work involved the dimethyl ester of 3,3'4,4'-benzophenonetetracarboxylic acid, 4,4'-methylenedianiline and the monomethyl ester of 5-norbornene-2,3-dicarboxylic acid where the stoichiometry was adjusted to provide an oligomer with M_n of ~1500g/mole. A methanol solution of these monomers was used to impregnate the reinforcement. Residual solvent (~11%) was required to provide tack and drape to the prepreg. In the fabrication of composites, the cure cycle was programmed to first remove the residual solvent and then most of the volatiles (water and methanol) from thermal cyclization to the imide. This cycle is commonly called B-staging. As the temperature was increased, the B-staged resin still had sufficient flow under pressure to compact further and thereby provide dense (low void content) composites. In actual practice, complete conversion to imide prior to polymerization through the norbornenyl end-group may not occur. This is only of academic concern since dense composites with excellent mechanical properties have been fabricated (28, 29). PMR-15 composites have a significant data base and are being used in a variety of high performance, high temperature applications. PMR-15 prepreg is commercially available from several major prepreg suppliers.

To obtain tack and drape in the prepreg without the use of residual solvent, the P13N and PMR-15 technology was further modified at NASA Langley Research Center. The half-ethyl esters of 3,3',4,4'-benzophenonetetracarboxylic acid and nadic acid were blended with a liquid eutectic amine mixture, Jeffamine AP-22 or Ancamine-DL (homologs and isomers of methylenedianiline), with the stoichiometry adjusted to provide an oligomer with M_n of ~1600g/mole (30). This material was called LARC-160. The viscous monomeric mixture is amenable to melt impregnation to provide prepreg with tack and drape without excessive residual solvent. As with PMR-15, the prepreg is B-staged to remove condensation volatiles and subsequently cured to yield dense composites with good high temperature properties. LARC-160 is also available from several prepreggers.

Nadic anhydride was also used to terminate an amic acid oligomer prepared from the reaction of 3,3',4,4'-benzophenonetetracarboxylic dianhydride and 3,3'-methylenedianiline where the stoichiometry was adjusted to yield a nadimide-terminated imide oligomer with M_n ~1300g/mole. This material has been referred to as LARC-13 and has been evaluated as a high temperature adhesive (31). A high temperature adhesive (BXR10214-151C) similar to PMR-15 is commercially available (32).

About 1970, research was initiated under Air Force funding on acetylene-terminated imide oligomers (ATI) which could be thermally chain extended through the acetylenic end-groups (33, 34). This effort resulted in the development of HR-600 (Eq. 9) and subsequent commercialization by Gulf Oil Chemicals Company in the form of Thermid -600. Neat resin properties of HR-600 are presented in Table IV while preliminary composite properties are given in Table V.

Although HR-600/Thermid-600 provided promising neat resin and composite properties, major processing problems have plagued these as well as other acetylene-terminated oligomers. Resin flow and wetting is inhibited due to the reaction of the terminal ethynyl groups prior to the formation of a complete melt or soft state. This becomes even more severe due to heat transfer problems as

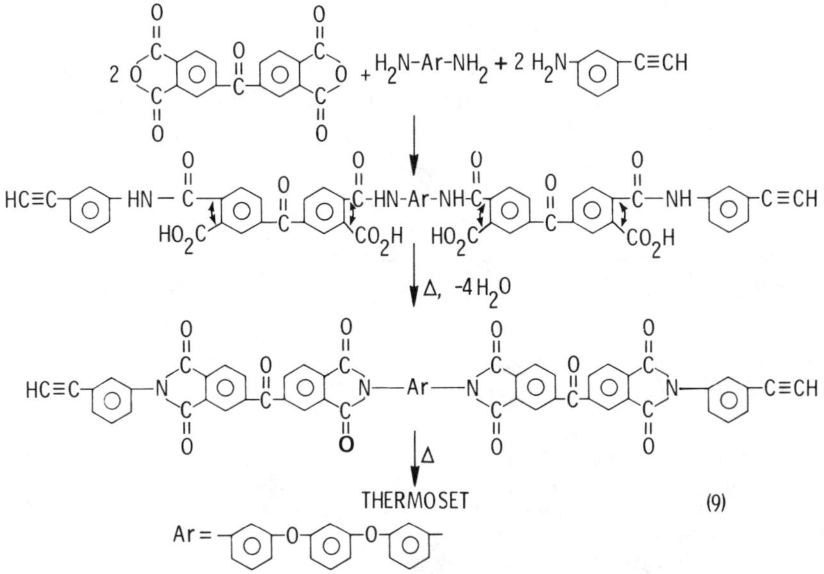

Table IV. Cured Acetylene-Terminated Imide (HR-600) Properties[1,2]

Tensile Strength, MPa (Ksi)	96.5 (14)
Tensile Modulus, GPa (Ksi)	3.79 (550)
Elongation, %	2.6
Flexural Strength, MPa (Ksi)	124.1 (18)
Flexural Modulus, GPa (Ksi)	4.48 (650)
Compressive Strength, MPa (Ksi)	213.7 (31)
Barcol Hardness	45
Tg, °C (after 48 hr @ 370°C)	350

[1] Cured 2 hr @ 250°C under 13.8 MPa (2000 psi) and 16 hr at 316°C in air

[2] Source: Reference 35.

larger components are fabricated. Because of the poor solubility and processability of HR-600/Thermid-600 materials, work shifted to the chemical conversion of the acetylene-terminated amic acid oligomer to the acetylene-terminated isoimide oligomer (structure shown below). The isoimide oligomer was reported to have better

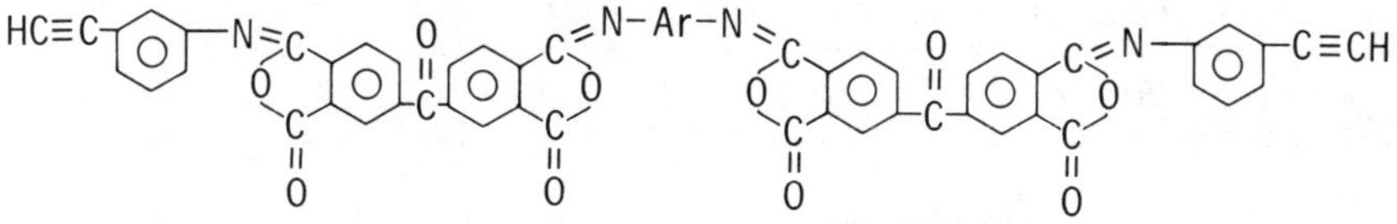

AND ISOMERS

Table V. - Thermid -600/Unidirectional HT-S Graphite
Fiber Laminate Properties

Test Condition	Flexural Strength, GPa (Ksi)	Short Beam Shear Strength, MPa (Ksi)
26°C	1.28 (185)	83.4 (12.1)
260°C after 500 hr @ 260°C	1.17 (169)	60.0 (8.7)
288°C after 500 hr @ 288°C	0.93 (144)	51.0 (7.4)
316°C after 500 hr @ 316°C	1.04 (151)	41.4 (6.0)

Source: Reference 36.

solubility and processability than ATI (37). At high temperatures (e.g. 300°C), the isoimide rearranges to imide. Various forms of the acetylene-terminated imides and isoimides are available from National Starch Corporation.

Imide oligomers containing other reactive end groups such as vinyl (38), cyano (39, 40) phenylacetylene, phenylvinylacetylene (H_5C_6-CH=CH-C≡C-) (41), phenyldiacetylene (H_5C_6-C≡C-C≡C-) (41) and propargyl (42) have been reported.

Phenylquinoxalines - Polyphenylquinoxalines (PPQ) prepared from the reaction of aromatic bis(o-diamines) and aromatic bis(phenyl-α-diketones) are high temperature thermoplastics. They are processable with little or no volatile evolution at relatively high temperatures (> 316°C) and pressure (~1.38 MPa) by virtue of their thermoplasticity. Like other thermoplastics, the processability is governed primarily by the chemical structure, molecular weight and molecular weight distribution.

To improve the processability of PPQ, appropriate phenylquinoxaline oligomers were end-capped with acetylenic groups using 3-(3,4-diaminophenoxy)phenylacetylene (43) or 4-(3- and 4-ethynyl-phenoxy)benzil (44, 45) (Eq. 10). The processability was improved but at the sacrifice of the thermooxidative stability. In general, cured acetylene-terminated heterocyclic polymers are less stable in a thermooxidative environment than the parent linear polymer.

A thin molding of an acetylene-terminated phenylquinoxaline, fabricated by compression molding at 316°C for 26 hr and at 371°C for 5 hr, gave tensile strength of 103 MPa (15,000 psi), tensile modulus of 2.62 GPa (380,000 psi) and elongation of 5% (46). Preliminary unidirectional graphite fiber laminate properties are reported in Table VI.

Phenylquinoxaline oligomers have also been end-capped with 4,4'-divinylbenzil (48). The terminal vinyl groups underwent a thermally induced crosslinking reaction to yield materials with high T_gs (~390°C) but poor thermooxidative stability.

Other Reactive Oligomers - In the mid 1960's, a reactive oligomeric precursor to a polybenzimidazole (PBI) was available as

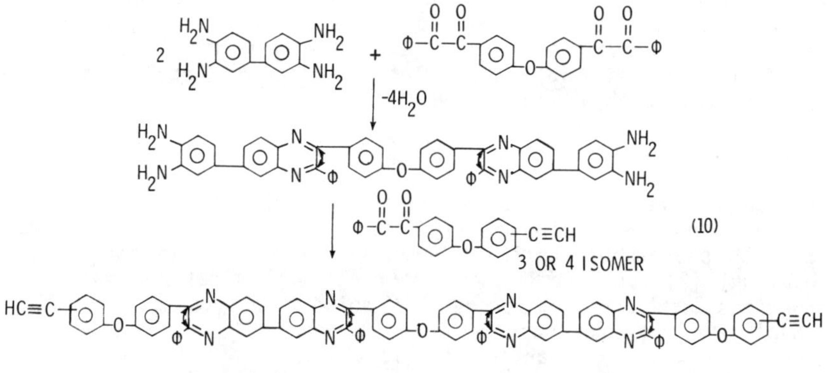

$$\text{(10)}$$

Table VI - Unidirectional HT-S Graphite Fiber Laminate Properties of Acetylene-Terminated Phenylquinoxaline Resin[1,2]

Test Condition	Flexural St., GPa (Ksi)	Flexural Mod., GPa (msi)	Short Beam Shear St., MPa (Ksi)
26°C	1.65 (239)	133.1 (19.3)	100.7 (14.6)
232°C after 2000 hr at 232°C	1.37 (199)	126.8 (18.4)	51.7 (7.5)
260°C after 500 hr 260°C	1.30 (188)	126.2 (18.3)	42.1 (6.1)

[1] Cured at 316°C under 1.38 MPa (200 psi) for 4 hr
[2] Source: Reference 47.

Imidite 850 from Narmco Materials Inc. The oligomeric form was prepared from the partial reaction of 3,3'4,4'-tetraaminobiphenyl and diphenyl isophthalate. Reactive oligomeric forms of PBI are still under evaluation as laminating resins for use in high temperature composite applications. Experimental quantities of acetylene-terminated phenylene oligomers were introduced by Hercules in the early 1970's under the name of H-Resins (49). Cure takes place in the 200-300°C range to yield highly crosslinked resins. These materials were contemplated for use as coatings, composites, and moldings. Apparently due to an unfavorable combination of price, processability and performance, H-resins were not commercialized.

Cured ethynyl-terminated ester oligomers prepared from the reaction of hydroxy-terminated ester oligomers with 4-ethynylbenzoyl chloride exhibited higher T_gs and better solvent resistance than comparable unendcapped polymers (50). Biphenylene end-caps have been placed on imide (40, 51), quinoline (52, 53) and quinoxaline (53) oligomers. High temperatures (> 316°C) are required to cure oligomers with biphenylene end-caps even in the presence of catalysts (nickel and rhodium compounds). Preliminary composite properties of biphenylene end-capped imide and quinoline oligomers have been reported (53). Acetylene-terminated phenyl-as-triazine oligomers have also been prepared (54).

Conclusions

Research on thermally curable oligomers during the past 15 years have produced the following significant advances: (1) oligomers with good processability that thermally cured with minimal volatile evolution, (2) cured resins with attractive physical and mechanical properties with good moisture and solvent resistance and (3) a technology which can custom-tailor molecules to provide a broad scope of resins for various applications. The review presented herein elucidates the types of chemistry which have been studied. Considerable research and development is underway. If a favorable combination of price, processability and performance can be attained, thermally curable oligomers as precursors to high temperature polymers will find wide acceptance in industry.

The use of trade names of manufacturers does not constitute an official endorsement of such products or manufacturers, either expressed or implied, by the National Aeronautics and Space Administration.

Literature Cited

1. Cizmecioglu, M.; Gupta, A. SAMPE Quart. 1982, 13, 16.
2. Kubens, R.; Schultheis, H.; Wolf, R; Grigat, E. Kunstoffe 1968, 58(12), 827.
3. Grigat, E.; Putter, R. Chem. Ber.1964, 97, 3012.
4. Delano, C. B.; Harrison, E. S. Soc. Adv. Matl. Proc. Eng. Series 1975, 20, 243.
5. Hergenrother, P. M.; Harrison, E. S.; Gosnell, R. B. Soc. Adv. Matl. Proc. Eng. Series 1978, 23, 506.
6. R. J. Kray Soc. Adv. Matl. Proc. Eng. Ser. 1975, 20, 227.
7. Mitsubishi Gas Chemical Co., Inc. 277 Park Avenue, New York, NY 10017.
8. Narmco Materials, Inc. 1440 N. Kraemer Blvd., Anaheim, CA 92806.
9. Maximovich, M. G.; Lockerby, S. C.; Arnold, F. E.; Longhran, G. A. Sci. Adv. Proc. Eng. Ser. 1978, 23, 490.
10. Loughran, G. A.; Wereta, A; Arnold, F. E. U. S. Patent 4 131 625, 1978.
11. Loughran, G. A.; Arnold, F. E. Polym. Prepr. 1980, 21 (1), 199.
12. Harrison, J. J.; Sabourin, E. T.; Selwitz, C. M.; Feld, W. A.; Unroe, M. R.; Hedberg, F. L. Ibid, 1982, 23(2), 189 (1982).
13. Abrams, F. L.; Browning, C. E. Org. Coat. Appld. Polym. Sci. Prepr. 1983, 48, 909.

14. Browning, C. E.; Air Force Wright Aeronautical Laboratories, Dayton, OH, private communication.
15. Loughran, G. A.; Reinhart, B. A.; Arnold, F. E.; Soloski, E. J. Org. Coat. Plast. Chem. Prepr. 1980, 43, 777.
16. Kwiatkowski, G. T.; Robeson, L. M.; Brode, G. L.; Bedwin, A. W. J. Polym. Sci. Polym. Chem. Ed. 1975, 13, 961.
17. Johnston, N. J. NASA Langley Research Center, private communication.
18. Sheppard, C. H.; House, E. E.; Stander, M. in Soc. Plast. Industry Inc., 36th Ann. Conf. Prepr., Session 17-B, 1980, p. 1.
19. Sheppard, C. H.; House, E. E.; Stander, M. Natl. SAMPE Tech. Conf. Ser. 1982, 14, 70.
20. Hergenrother, P. M. J. Polym. Sci. Polym. Chem. Ed. 1982, 20, 3131.
21. Hergenrother, P. M.; Jensen, B. J. Org. Coat. Appld. Polym. Sci. Prepr. 1983, 48, 914.
22. Hergenrother, P. M.; Jensen, B. J.; Havens, S. J. Sci. Adv. Matl. Proc. Eng. Ser. 1984, 29, 1060.
23. Burns, E. A.; Lubowitz, H. R.; Jones, J. F. NASA CR-72460, 1968.
24. Lubowitz, H. R. U. S. Patent 3 528 950, 1970.
25. Lubowitz, H. R. Polym. Prepr., 1971, 12(1), 329.
26. Serafini, T. T.; Delvigs, P.; Lightsey, G. R. J. Appl. Polym. Sci. 1972, 16, 905.
27. Serafini, T. T.; Delvigs, P.; Lightsey, G. R. U. S. Patent 3 745 149, 1973.
28. Serafini, T. T. "Status Review of PMR Polyimides" in Resins for Aerospace (Clayton A. May, ed.) ACS Symposium Series 1980, 132, p. 15.
29. Freeman, W. T., Jr.; Baucom, R. M. Soc. Adv. Mat'l Proc. Eng. Ser. 1980, 25, 418.
30. St. Clair, T. L.; Jewell, R. A. Natl. SAMPE Tech. Conf. Ser. 1976, 8, 82; Sci. Adv. Matl. and Proc. Eng. Ser. 1978, 23, 520.
31. St. Clair, T. L.; Progar, D. J. Ibid, 1979, 24(2), 1081.
32. American Cyanamid Co., Aerospace Products Dept., Havre de Grace, MD.
33. Bilow, N.; Landis, A. L.; Miller, L. J. U. S. Patent 3 845 018, 1974.
34. Landis, A. L.; Bilow, N.; Boschan, R. H.; Lawrence, R. E.; Aponji, T. J., Polym. Prepr. 1974, 15(2), 537.
35. Bilow, N.; Landis, A. L. Natl. SAMPE Tech. Conf. Ser., 1976, 8, 94.
36. Hergenrother, P. M.; Johnston, N. J. in Resins for Aerospace (Ed. C. A. May), ACS Symposium Series 1980, 132, p. 3.
37. Landis, A. L.; Naselow, A. B. Natl. SAMPE Tech. Conf. Ser. 1982, 14, 236.
38. Serafini, T. T.; Delvigs, P.; Vannucci, R. D. Paper presented at 36th Ann. Conf. of Rein. Plast./Comp. Inst. of Soc. of Plast. Ind., Wash., D. C., Feb. 1981.
39. Hsu, L. C. U. S. Patent 4 061 856, 1977.
40. Hsu, L. C. Polym. Prepr. 1980, 21(1), 86.
41. Harris, F. W.; Pamidimurkkla; Gupta, R.; Das, S.; Wu, T.; Mock, G. Ibid. 1983, 24(2), 324.

42. D'Alelio, G. F.; U. S. Patent 4 166 168, 1978.
43. Kovar, R. F.; Ehlers, G. F. L.; Arnold, F. E. Polym. Prepr. 1976, 16(2), 246; J. Polym. Sci. Polym. Chem. Ed. 1977, 15, 1081.
44. Hergenrother, P. M. Org. Coat. Plast. Chem. Prepr. 1976, 36(2), 264; in Chemistry and Properties of Crosslinked Polymers (S. S. Labana, ed). Academic Press, NY, 1977, p. 107.
45. Hedberg, F. L.; Arnold, F. E. Polym. Prepr. 1977, 18(1), 826; J. Appld. Polym. Sci. 1979, 24, 763.
46. Browning, C. E.; Wereta, A. Jr.,; Hartness, J. T.; Kovar, R. F. Sci. Adv. Mat'l Proc. Eng. Ser. 1976, 21, 83.
47. Hergenrother, P. M. Polym. Eng. Sci. 1981, 21(16), 1072.
48. Wentworth, S. E.; Macaione, D. P. J. Polym. Sci. Polym. Chem. Ed. 1976, 14, 101.
49. Jabloner, H.; Cessna, L. C. J. Elastomers Plast. 1974, 6, 103.
50. Havens, S. J.; Hergenrother, P. M. Polym. Prepr. 1983, 24(2), 16; J. Polym. Sci. Polym. Chem. Ed. (in press).
51. Droske, J. P.; Gaik, U. M.; Stille, J. K. Macromolecules, 1984, 17, 10.
52. Droske, J. P.; Stille, J. K. Ibid. 1984, 17, 1.
53. Droske, J. P.; Stille, J. K.; Alston, W. B. Ibid. 1984, 17, 14.
54. Hergenrother, P. M. Macromolecules 1978, 11, 332.

RECEIVED February 4, 1985

Synthesis of Bisphenol-Based Acetylene Terminated Thermosetting Resins

J. S. WALLACE[1], F. E. ARNOLD[1], and W. A. FELD[2]

[1]Air Force Wright Aeronautical Laboratories, AFWAL/MLBP, Wright-Patterson Air Force Base, OH 45433
[2]Department of Chemistry, Wright State University, Dayton, OH 45435

> A series of high molecular weight (750-950 amu range) bis-phenol based acetylene terminated (AT) resins were synthesized by reacting four moles of 4,4'-dihalodiphenylsulfone (chloro and fluoro) with one mole of a bis-phenol (4,4'-isopropylidinediphenol, 4,4'-thiodiphenol, p,p'-biphenol, and resorcinol were used), end-capping the resulting halo-terminated products with 4-(m-hydroxyphenyl)-2-methyl-3-butyn-2-ol, and caustically cleaving the terminal acetone protecting groups to give free ethynyl functionalities. This synthesis produces a mixture of monomer and oligomer AT-products which were separated by column chromatography. Pure AT-monomers and the monomer/oligomer mixtures produced by the outlined stoichiometry were cured at 288°C (550°F) for 8 h in air. Glass transition temperatures (Tgs) of the cured (by thermomechanical analysis) and uncured (by differential scanning calorimetry) AT-systems were measured. Thermo-oxidative stability of the resins was evaluated by isothermal aging (ITA) in air at 315°C (600°F) for 200 h.

In recent years acetylene terminated resins (AT-resins) have shown promise in the area of high temperature applications and are possible replacements for state-of-the-art epoxy resin systems where humidity exposure is expected. Acetylene terminated resins have two important advantages over previously developed systems: 1) cured products have high thermal stability and exhibit good mechanical properties after exposure to humidity, and 2) the resins are cured through an addition reaction so no volatile by-products are evolved giving a voidless final matrix.(1)

These properties make AT-resins especially attractive for advanced aircraft and aerospace vehicles where material weight is a critical factor and high temperatures as well as humidity are likely to be encountered.

In an effort to optimize mechanical and processing properties, many different molecular systems have been inserted between terminal acetylene groups, with each system imparting its own unique properties to the final resin. The first of these systems to demonstrate good mechanical properties were the quinoxalines. These systems, with their large flexible molecular structures, did not however meet epoxy processing standards (i.e., melt processability and room temperature tack and drape)(2) because of a high initial Tg.(3) Two approaches were taken to resolve this problem. The first approach, which is presently still under investigation, was to incorporate reactive plasticizers into the quinoxaline resins. Preliminary data shows that this has the effect of lowering the initial Tg, but further study is required before effects on mechanical properties can be completely evaluated.(4) The second approach involves the substitution of aromatic diol and bis-diol groups (resulting in a group of compounds called phenylene Rs) into the backbone. The first group tried, the diphenyl sulfone linkage, led to resins (commonly known as ATS) with good processing properties but which were brittle in nature.(1) A factor which has been associated with resin toughness is crosslink density of the matrix. This density can be controlled by increasing or decreasing the molecular weight of the monomer and oligomer molecules which make up the resin. Reducing the crosslink density in a matrix has been shown to yield improved toughness.(5) In an effort to improve the mechanical properties of AT resins, monomers and oligomers of higher molecular weight, with various phenylene R backbones, are being evaluated. This work is centered at the Materials Laboratory of the Air Force Wright Aeronautical Laboratories and is the basis for the research presented here. The combination of a semiflexible backbone and lower crosslink density should provide an AT-resin with superior mechanical properties. The objectives of this research therefore were:

1. The synthesis of higher molecular weight (750-950 amu range), semi-rigid AT-resins, with the general formula 1 (Figure 1) with the potential of improved mechanical properties. The diol and bis-diols used in these systems were chosen because of their low cost and ready availability. Meta- rather than para-AT end capped systems were synthesized because the former generally have lower uncured Tgs which are desirable for easy processing.

2. Development of a new acetylene terminating procedure to facilitate the synthesis of the aforementioned systems.

Experimental

Synthesis of the End-Capping Agent
4-(m-Hydroxyphenyl)-2-methyl-3-butyn-2-ol I

A three-necked 500mL round-bottom flask was fitted with a reflux condenser, magnetic stir bar, stopper, and gas inlet/outlet adapters. Under dry nitrogen the flask was charged with m-bromophenol (10.0g, 57.8 mmol), 2-methyl-3-butyn-2-ol (5.0g, 59.4 mmol) and 250 mL of distilled triethylamine resulting in a pale yellow solution. The mixture was heated at reflux for 15 min while a nitrogen atmosphere was maintained. After the reflux period the catalyst system consisting of dichlorobis(triphenylphosphine) palladium II (0.1g), triphenylphosphine (0.2g) and cuprous iodide (0.1g) was added. Addition of the catalyst system caused the reaction mixture color to deepen to yellow-orange. After heating at reflux for 25 h, light-colored salts formed. Reaction progress was followed by gas chromatography (GC) using phenol as a standard. After cooling to 25°C, the reaction mixture was filtered under nitrogen through a glass frit packed with celite. The remaining light gray precipitate was rinsed with additional triethylamine and the filtrates combined, concentrated (rotary evaporator) and the resulting yellow oil dissolved in 200 mL of toluene and then washed with 120 mL of 8% hydrochloric acid. The hydrochloric acid washing was extracted with three 50 mL portions of ethyl acetate which were combined, concentrated (rotary evaporator) and the resulting residue dissolved in 25 mL of toluene. The toluene fractions were combined and dried (magnesium sulfate). The dry toluene solution was treated with ethylene diamine (brown solution) and stirred with heating (50-60°C) under nitrogen for 15 min which resulted in a blue precipitate. After cooling to room temperature, the solution was filtered to remove the precipitate and the filtrate was extracted with 250 mL of distilled water and three 125 mL portions of 10% potassium carbonate. The base extract was stirred in an ice/water bath and neutralized with 6N hydrochloric acid (addition was continued until the pH was slightly acidic). The aqueous solution was extracted with one 250 mL and three 125 mL portions of ethyl acetate which were combined, dried and evaporated. The resulting dark yellow-orange oil was induced to crystallize by dissolving it in methylene chloride, adding n-hexane until slightly cloudy, seeding and cooling by refrigeration for 24 h. The resulting solid was recrystallized from n-hexane/methylene chloride (2/3) to yield 4.1g (41%) of a white solid: mp 94-95°C; IR (KBr) 3380, 3160 cm^{-1} (O-H), 3000, 1575 cm^{-1} (aromatic), 2930 cm^{-1} (aliphatic), 2210 cm^{-1} (C C), 1215, 940 cm^{-1} (C-OH).

Anal. Calcd. for $C_{11}H_{12}O_2$: C, 74.91; H, 6.81.
Found: C, 74.52; H, 6.96.
Cu and Pd Analysis: Cu, 3ppm; Pd, 3ppm.

Example Procedure for Preparation of Halo-Terminated Intermediate Monomer/Oligomer Mixtures

1,1'-(1-Methylethylidene)bis[4-[(4-fluorophenyl)sulfonyl] phenoxy]benzene] **II**

and

-[4-[1-[4-[4-[(4-Fluorophenyl)sulfonyl]phenoxy]phenyl]-1-methylethyl]phenyl]- -[4-[(4-fluorophenyl)sulfonyl]phenoxy]poly[oxy-1, 4-phenylenesulfonyl-1,4-phenyleneoxy-1,4-phenylene(1-methylethylidene)-1,4-phenylene] **III**

A three-necked 250 mL round-bottom flask, was equipped with a magnetic stir bar, reflux condenser, Dean-Stark trap, and a gas inlet/outlet adapter. The following mixture was added:

4,4'-isopropylidenediphenol (3g, 13.2 mmol), difluorodiphenyl sulfone (13.43g, 52.8 mmol), anhydrous potassium carbonate (1.91g, 13.8 mmol), 25 mL of freshly distilled N-methyl-2-pyrrolidone and 25 mL of benzene. The reaction mixture was heated at reflux (100°C) and stirred rapidly under dry nitrogen until all water (a reaction side product) was removed by azeotropic distillation with benzene. The reaction (dark purple color) temperature was then raised and maintained at 135°C for 5h. The mixture darkened during this period. After cooling to room temperature, the reaction mixture was diluted with 100 mL of methylene chloride and washed with three 200 mL portions of 10% hydrochloric acid, one 200 mL portion of distilled water, dried (magnesium sulfate) and filtered. The filtrate was concentrated (rotary evaporator) and chromatographed on a quartz column filled with activated silica gel (460g). Unreacted difluorodiphenyl sulfone was eluted with methylene chloride/petroleum ether (¼), monomer $\underline{II}$ was eluted with methylene chloride/petroleum ether (½) to yield 6.61g of a white crystalline solid: mp 120-122°C; IR(KBr) 3070, 1575, 1480 cm^{-1} (aromatic), 1230 cm^{-1} (ArF), 1320, 1145 cm^{-1} (SO_2), 1095 cm^{-1} (ArOAr);
Anal. Calcd. for $C_{39}H_{30}F_2S_2O_6$: C,67.23; H,4.34; S,9.20; F,5.45. Found: C,67.21; H,4.58; S,9.05; F,5.44.
Oligomer $\underline{III}$ was eluted with methylene chloride to yield 1.39g of a clear viscous oil: IR (NaCl film) 3070, 1570, 1475 cm^{-1} (ArF), 1320, 1150 cm^{-1} (SO_2), 1100 cm^{-1} (ArOAr).
Total yield was 8.00g (87.3%, if based on pure monomer theoretical yield).

Example Procedure for Preparation of Acetone Protected AT-Products

Monomer Product

4,4'-[(1-Methylethylidene)bis(4,1-phenyleneoxy-4,1-phenylene-sulfonyl-4,1-phenyleneoxy-3,1-phenylene)]bis[2-methyl-3-butyn-2-ol] $\underline{IV}$

To a 50 mL, three-necked, round-bottom flask equipped with magnetic stir bar and nitrogen inlet/outlet was added 40 mL of dry DMSO. While stirring under nitrogen, 4-(m-hydroxyphenyl)-2-methyl-3-butyn-2-ol $\underline{I}$ (1.67g, 9.48 mmol) and potassium methoxide (0.67g, 9.48 mmol) were added. The mixture was stirred at 40°C for 1 h to complete generation of the potassium salt. A 250 mL three-necked round-bottom flask, equipped with a gas inlet/outlet, addition funnel, and magnetic stir bar was charged with $\underline{II}$ (3.0g, 4.31 mmol) dissolved in 40 mL of dry DMSO. Under nitrogen, the solution was heated to 90°C with stirring. The potassium salt of 4-(m-hydroxyphenyl)-2-methyl-3-butyn-2-ol $\underline{I}$ was transferred (under nitrogen) to the addition funnel and added to the solution of $\underline{II}$ over a period of 1h. After addition was complete, the reaction mixture was maintained at 90°C for an additional 3 h, cooled to room temperature and diluted with 150 mL of methylene chloride and was washed with three 200 mL portions of 10% hydrochloric acid and one 300 mL portion of distilled water. The methylene chloride solution was dried (magnesium sulfate), filtered, and concentrated (rotary evaporator) to yield a pale brown product. The crude product was chromatographed on a quartz column filled with activated silica gel (160g). The product was eluted with hexanes/ethyl acetate (3/1) to yield 3.82g (87.9%) of white solid: mp 179-181°C; IR(KBr) 3480 cm^{-1} (COH), 3060, 1570, 1475 cm^{-1} (aromatic), 1380, 1360 cm^{-1} (gem dimethyl), 1315, 1145 cm^{-1} (SO_2), 1100 cm^{-1} (ArOAr);

Anal. Calcd. for $C_{61}H_{52}S_2O_{10}$: C, 72.70; H, 5.21; S, 6.35.
Found: C, 72.14; H, 5.49; S. 6.01.

Example Procedure for Cleavage of AT-Product Acetone Protecting Groups

Monomer Product

1,1'-(1-Methylethylidene)bis[4-[4-[[4-(3-ethynylphenoxy)phenyl]sulfonyl]phenoxy]benzene] **V**

A solution of **IV** (2.0g, 2.0 mmol) in 150 mL of dry toluene was formed with heating under nitrogen in a three-necked, 250 mL, round-bottom flask which was equipped with a Dean-Stark trap, reflux condenser, magnetic stir bar and a gas inlet/outlet adapter. Dry, powdered potassium hydroxide (2.0g) was added and the mixture was stirred and heated at reflux for 1 h. Acetone formed as the reaction progressed was removed by azeotropic distillation with toluene. The toluene was replaced and the distillation procedure repeated two more times. The progress of the reaction was followed by TLC on silica gel (methylene chloride). After 3 h the reaction mixture was cooled to room temperature, filtered through celite and washed with three 200 mL portions of distilled water, dried (magnesium sulfate), and filtered. The filtrate was concentrated (rotary evaporator) and chromatographed on a quartz column filled with activated silica gel (160g). The product was eluted with hexanes/methylene chloride (3/1) to yield 1.56g (83%) of a white solid: mp 80-82°C; IR(KBr) 3300 cm^{-1} (C CH), 3050, 1570, 1475 (aromatic), 1380, 1360 (gem di-methyl), 1315, 1340 (SO_2), 1100 cm^{-1} (ArOAr) ^{1}H NMR 6.86-8.20 (m, aromatic, 32H), 3.15 (s, acetylene, 2H), 1.71 (s, methyl, 6H);

Anal. Calcd. for $C_{55}H_{40}S_2O_8$: C, 74.00; H, 4.53; S. 7.18.
Found: C, 74.36; H, 4.83; S, 7.02.

Results and Discussion

End-Capping Agent I

In previously employed end-capping schemes a palladium catalyst was used to directly connect a protected ethynyl group (2-methyl-3-butyn-2-ol) to a bulky intermediate. This procedure results in the entrapment of catalyst metals which must be laboriously removed from the final AT-product. If these metals are not removed, premature curing of the system occurs which narrows the processing window. In extreme cases where large quantities (>100ppm) of palladium remain, curing in the reaction vessel during the cleavage of the acetone protecting group has been observed.(6) The end-capping agent 4-(m-hydroxyphenyl)-2-methyl-3-butyn-2-ol I eliminates this problem. The synthesis of I is accomplished by using a palladium catalyst to replace the bromine atom of m-bromophenol with 2-methyl-3-butyn-2-ol, a protected ethynyl group. This route results in the preparation of a substituted phenolic acetylene which can be completely isolated from the catalyst metals by an ethylenediamine treatment (Figure 2).

The utility of a palladium catalyst in the synthesis of substituted aryl acetylenes is well established.(7,8,9,10) The end-capping agent I was produced by using a standard catalyst system, dichlorobis(triphenylphosphine)palladium (II)/copper (I) iodide/triphenylphosphine mixture, which has been employed in previously developed ethynylation procedures.(10) The copper (I) iodide is believed to act as a cocatalyst, reducing the palladium (II) complex to the active palladium (0) catalyst. The scheme is shown in Figure 3 (diethylamine is the solvent).(11)
Sonogashira has proposed a catalytic cycle (Figure 4) which shows: 1) the reduction of the palladium complex, 2) coordination of the aryl halide and acetylene with the palladium (0) complex and 3) the reductive elimination of the substituted aryl acetylene and regeneration of the active catalyst.(10)
The use of a basic solvent (in this case diethylamine) is important to stabilize acetylenic anions.(9) The third catalyst system component, triphenyl phosphine, is presumably added to help replace lost triphenyl phosphine ligands on the palladium complex and thus prevent metal agglomeration.

Triethylamine, because of its basic properties was chosen as the solvent for the end-capper I synthesis. Reactions were run at reflux for times ranging from 2 to 30 hours. Reaction progress was followed by gas chromatography (GC) using phenol as a standard. All reactions were run under a dry nitrogen atmosphere. Yields rapidly increased with time up to 12 hours, slowly increased with time up to 24 hours, and showed a slow decline thereafter. The reaction never appears to go to completion regardless of reaction time, and there is always residual m-bromophenol remaining (detected by GC and TLC). Dieck and Heck state that the major limitation of substituted aryl acetylene preparation is that aryl halides with strongly electron-donating substitutents have relatively low reactivity toward oxidative addition.(9) The failure of the end-capper reaction to go to completion is most likely associated with the inductive electron-donating effect of the hydroxy group of m-bromophenol. This leads to the unproven hypothesis that the palladium catalyst is involved with a competing deactivation reaction with a relatively fast rate. Thus, a portion of the catalyst is deactivated or destroyed before it can react with m-bromophenol. The acetylenic starting material, 2-methyl-3-butyn-2-ol was used in slight excess because it was known that some would be lost to side reaction products.(9) Increasing the excess had no observable effect on reaction yield (consistently 40%).

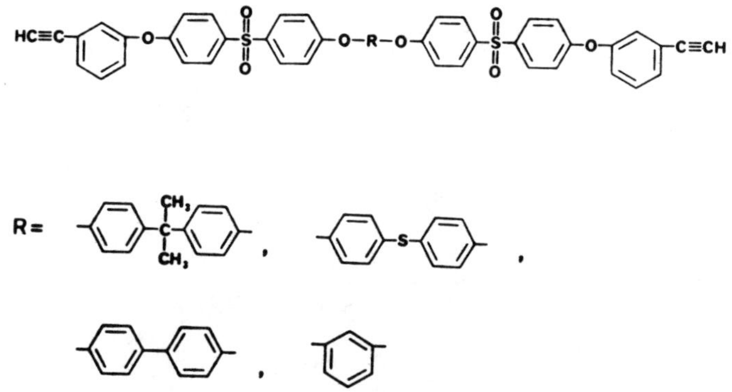

Figure 1. General Structure of Synthesized High Molecular Weight AT-Resins.

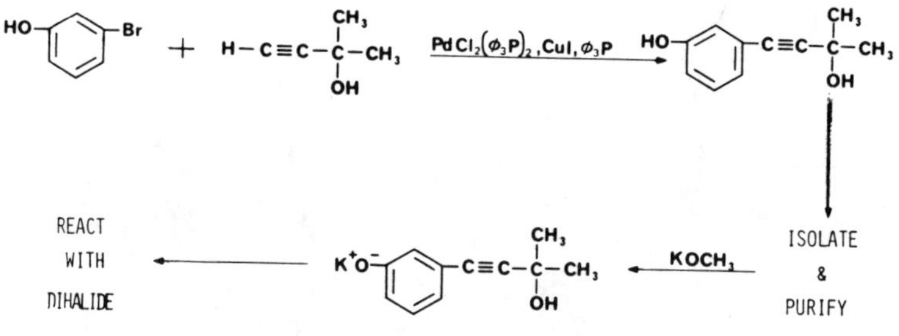

Figure 2. Synthesis of Endcapping Agent.

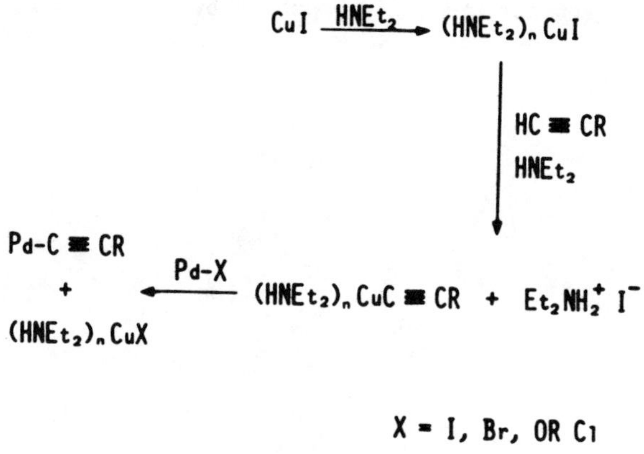

X = I, Br, OR Cl

Figure 3. Proposed Role of Copper (I) Iodide.

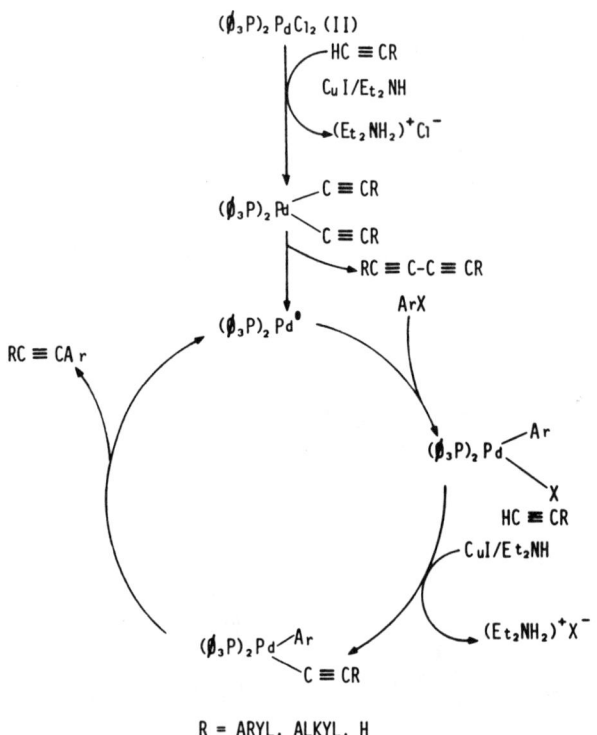

Figure 4. Catalytic Cycle Proposed by Sonogashira.

AT-Monomer and AT-Monomer/Oligomer Systems

The general approach used to synthesize the AT-monomer and AT-monomer/oligomer products consists of three synthetic steps: 1) synthesis of halo-terminated intermediates, 2) end-capping of the intermediates with end-capping agent $\underline{I}$, and 3) cleavage of the acetone protecting groups to produce the final AT-product. Production of both monomeric and oligomeric products is an inherent part of this synthetic route (Figure 5). Purification by column chromatography is required after each synthetic step.

A procedure similar to that developed by McGrath, et.al., was used to produce halo-terminated products.(<u>12</u>) The use of both 4,4'-difluoro and 4,4'-dichlorodiphenyl sulfone (chosen for the sulfone's ability to activate the displacement of the terminal halo groups) to end-cap several bis-phenols were evaluated. In an effort to produce primarily monomer products, dihalodiphenyl sulfone was used in a 4:1 molar excess with respect to the bis-phenols (the four different bis-phenols used were bisphenol-A; 4,4'-thiodiphenol; p,p'-biphenol; and resorcinol). The stoichiometry outlined above produced monomer/oligomer ratios ranging from 100/0 to 60/40 depending on the bis-phenol reactant used. Yields of the products ranged between 75 and 90% with the fluoro-terminated products generally about 10% higher than chloro-terminated products (only two of the bis-phenols were reacted with 4,4'-dichlorodiphenyl sulfone). All reactions were run in N-methyl-2-pyrrolidone (NMP) under nitrogen using potassium carbonate as a base. Typical reaction time was 4 hours with reaction progress followed by TLC. A higher reaction temperature was required for chloro displacement (150°C for chloro vs. 100°C for fluoro). Before end-capping, monomer was separated from oligomeric products by column chromatography. Monomer/oligomer ratios were determined by weighing each component after separation.

In the next step, acetylene end-capping in DMSO, an unexpected problem was encountered with the chloro terminated intermediates. As in the first step a high temperature (150°C) was required for chloro displacement. The combination of high temperature and basic conditions (potassium methoxide was used) produced interestingly, the final unprotected AT-monomer and AT-monomer/oligomer systems in low yield (20-40%). Evaluation of the stability of the potassium salt of $\underline{I}$ (using sodium nitroprusside as an indicator for acetone) showed it was losing its acetone protecting group at approximately 130°C in DMSO. This loss occurred below the 150°C temperature required for chloro displacement. Weaker bases such as potassium carbonate were tried, but gave similar results. Reaction of the fluoro terminated systems with $\underline{I}$ went smoothly and gave good yields (70-80%) of the protected AT-systems. Displacement of the fluoro atoms was accomplished at 85-90°C and $\underline{I}$ remained completely intact. Caustic cleavage of the protecting groups gave AT-monomer and AT-monomer/oligomer mixtures in yields ranging from 80 to 85%. Overall synthetic yields for this procedure ranged between 50 and 60% depending on the bis-phenol used.

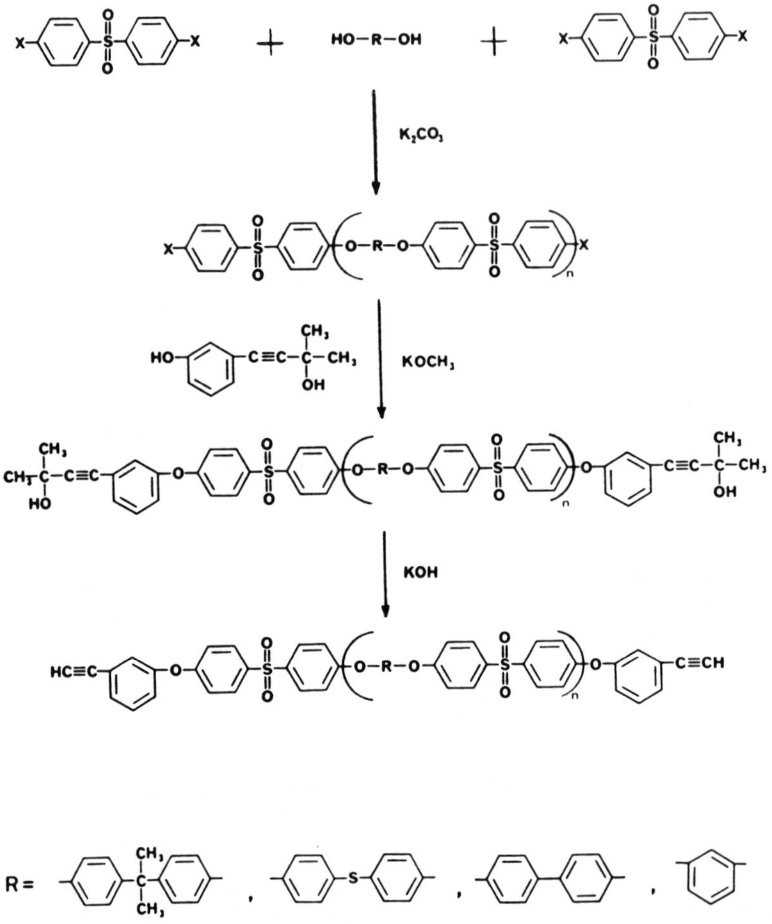

X = F, Cl

Figure 5. AT-Monomer/Oligomer Synthetic Route.

AT-Resin Thermomechanical Properties

After final chromatographic purification, samples of the AT-systems were cured in air at 288°C (550°F) for eight hours. Samples chosen for curing included pure monomers, monomer/oligomer mixtures produced by the stoichiometry outlined in the previous section, and in one case (the bisphenol-A based resin) pure oligomer. This set of samples was selected to provide data showing the effect of oligomer concentration on thermomechanical properties.

Differential scanning calorimetry (DSC) and thermomechanical analysis (TMA) were used to measure the glass transition temperatures (Tgs) of the uncured and cured AT-resins respectively (Figure 6).

For composite applications and a 350°F use temperature, cured resins must exhibit a minimum Tg of 220°C. A review of the cured Tg values shows that only the bisphenol-A based resin does not meet this criteria. For easy processing, an initial Tg of <20°C is required; this gives the uncured resin the necessary tack and drape for practical handling. Initial Tgs for the synthesized resins fall between 45 and 55°C which is too high for conventional processing techniques. The p,p'-biphenol based resin has, in addition to a high initial Tg, a Tm (melt temperature) of 180°C (Figure 7). Most AT-systems cure rapidly at temperatures over 175°C. As soon as the p,p'-biphenol system melts, rapid curing begins and no time is available for processing. The lack of a processing window makes this system almost impossible to handle by standard methods. The most attractive AT-systems seem to be the 4,4'-thiodiphenol and resorcinol based resins. Even though these systems have initial Tgs of ~50°C, they all show adequate cured Tgs and have wide processing windows (Figures 6 and 7).

Thermo-oxidative stability of the resins was evaluated by isothermal aging (ITA) in air at 315°C for 200 hours. All the systems showed good thermal stability with weight losses ranging from 20 to 38% after 200 hours. The comparable monomer/oligomer mixtures of the resins generally showed slightly greater weight losses.

Conclusions

The use of 4-(m-hydroxyphenyl)-2-methyl-3-butyn-2-ol $\underline{I}$ as an acetylene end-capping agent has the significant advantage of simplifying the task of catalyst metals removal which is a difficulty encountered with other end-capping approaches. Further study is required to improve the synthetic yield of $\underline{I}$. An effort directed at achieving a better yield would be best focused on the development of a recoverable polymer supported palladium catalyst for ethynylation of bromophenol. In addition to improving the yield, such a system would lower the synthetic cost and make large scale reactions possible.

All the AT-resins synthesized by the new procedure showed good thermal stability and several exhibited excellent cured Tgs. The high initial Tgs of the systems remain a problem. Recent investigation of an AT-reactive diluent which can be added to the

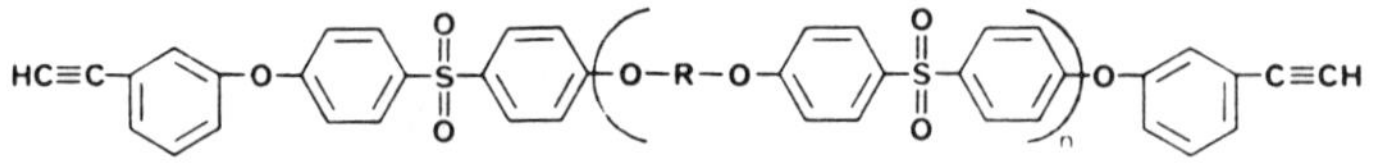

HC≡C—⟨⟩—O—⟨⟩—S(O)₂—⟨⟩—(O—R—O—⟨⟩—S(O)₂—⟨⟩)ₙ—O—⟨⟩—C≡CH

R	Ratio Monomer/Oligomer	Tg Initial (°C) [a]	Tg Cured (°C) [b,c]
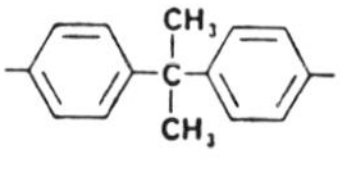 —⟨⟩—C(CH₃)₂—⟨⟩—	100/0	55	191
	80/20	55	168
	0/100	46	140
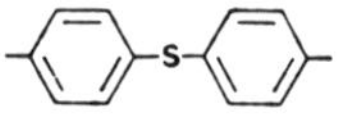 —⟨⟩—S—⟨⟩—	100/0	54	232
	60/40	50	240
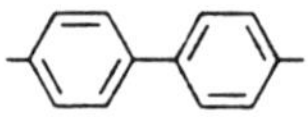 —⟨⟩—⟨⟩—	100/0	50	275
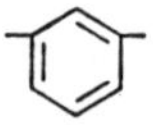 —⟨⟩—	100/0	53	269
	70/30	45	246

a) Determined by DSC (10°C/min).
b) Determined by TMA (10°C/min).
c) Cured at 550°F, 8H, in air.

Figure 6. AT-Resin Initial and Cured Glass Transition Temperatures (Tgs).

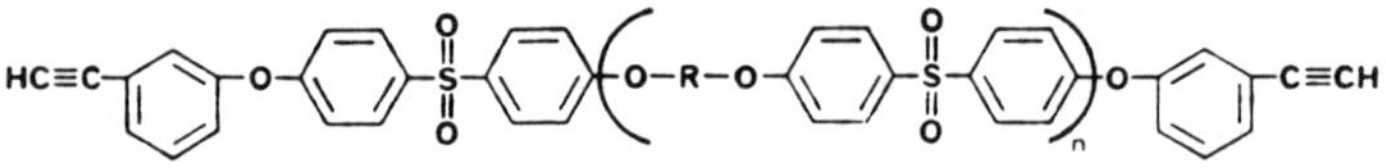

HC≡C—⟨⟩—O—⟨⟩—S(O)₂—⟨⟩—(O—R—O—⟨⟩—S(O)₂—⟨⟩)ₙ—O—⟨⟩—C≡CH

R	Ratio of Mono./Olig.	Tm (°C) [a]	Tpoly Onset(°C) [a]	Tpoly Max (°C) [a]
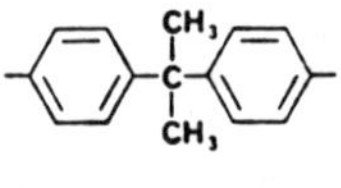 —⟨⟩—C(CH₃)₂—⟨⟩—	100/0	80	160	250
	80/20	77	160	245
	0/100	--	110	270
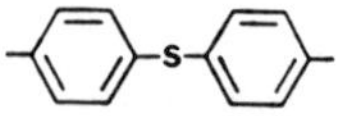 —⟨⟩—S—⟨⟩—	100/0	--	145	240
	60/40	--	150	230
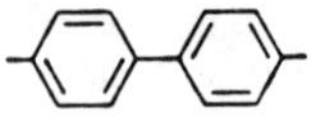 —⟨⟩—⟨⟩—	100/0	180	185	230
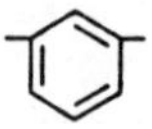 —⟨⟩—	100/0	70	130	230
	70/30	--	135	235

a) Determined by DSC (10°C/min).

Figure 7. Onset and Peak of AT-Resin Polymerization.

new systems gave promising results and might provide a solution to this problem. What effect the addition of this diluent might have on the synthesized resins' mechanical properties remains to be seen. Scale-up of the resins to the 50g level would provide material for mechanical properties characterization and determine if these increased molecular weight systems have improved toughness.

Literature Cited

1. C. Y-C. Lee, I. J. Goldfarb, T. E. Helminiak, F. E. Arnold, Soc. Adv. Matr. Proc. Eng. Ser., 28, 699 (1983).
2. J. Gamble, "The Effects on Ridgidity as a Function of Para Ethynylphenoxy Linkages in an Acetylene Terminated Thermosetting System," (M.S. Dissertation), Wright State University, 1981.
3. C. Y-C. Lee, I. J. Goldfarb, F. E. Arnold, T. E. Helminiak, Org. Coat. Appl'd Polym. Sci. Preprints, 48, (1983).
4. S. Sikka and I. J. Goldfarb, Org. Coat. Plast. Preprints, 45, 138, (1981).
5. T. E. Helminiak and W. B. Jones, Org. Coat. and Plast. Preprints, 48, 221, (1983).
6. E. T. Sobourin, The Synthesis of Polymer Precursors and Exploratory Research Based on Acetylene Displacement Reaction, AFWAL-TR-80-4151 (1980).
7. E. T. Sobourin, Preprints Div. Petr. Chem., 24, 233 (1979).
8. L. Cassar, J. Organomet. Chem., 93, 253 (1975).
9. H. A. Dieck and R. E. Heck, J. Organomet. Chem., 93, 259 (1975).
10. K. Sonagashira, Y. Tohda, N. Hagihara, Tet. Letter, 4467 (1975).
11. K. Sonagashira, T. Yatake, Y. Tohda, S. Takahashi, and N. Hagihara, J.C.S. Chem. Comm., 291 (1977).
12. D. Mohanty, J. Hedrick, K. Gobetz, B. Johnson, K. Yilgor, E. Yilgore, R. Yang, and J. McGrath, Polym. Preprints, 23(1), 284 (1982).

RECEIVED February 26, 1985

Arylether Sulfone Oligomers with Acetylene Termination from the Ullman Ether Reaction

P. M. LINDLEY[1], L. G. PICKLESIMER[1], B. EVANS[1], F. E. ARNOLD[1], and J. J. KANE[2]

[1] Air Force Wright Aeronautical Laboratories, AFWAL/MLBP, Wright-Patterson Air Force Base, OH 45433
[2] Department of Chemistry, Wright State University, Dayton, OH 45435

>   Acetylene terminated (AT) oligomeric arylether
>   sulfones which upon curing possess higher molecular
>   weight between crosslinks than earlier AT sulfones
>   have been synthesized via a four step reaction
>   sequence. The increase in chain length between
>   reactive sites was obtained by synthesis of higher
>   molecular weight diols using the nucleophilic
>   aromatic substitution reaction of
>   4,4'-dichlorodiphenyl sulfone with various diols,
>   including resorcinol, hydroquinone, bisphenol A,
>   4,4'-dihydroxybiphenyl and 4,4'-thiodiphenol. The
>   higher molecular weight diols were reacted with
>   excess dibromobenzene through the Ullmann ether
>   reaction to produce bromo endcapped arylether
>   sulfones. The final stages of the sequence involve
>   replacement of bromine with acetylene groups to get
>   AT oligomers which were evaluated to determine Tg's
>   before and after cure.

A substantial effort in our laboratory has been directed toward the synthesis and characterization of acetylene-terminated (AT) matrix resins. The most significant feature and driving force for the effort is that the thermal induced addition reaction provides a moisture insensitive cured product. This technology offers a wide variety of thermoset resins for various high temperature applications. Backbone structural design for use temperature capabilities, processing characteristics and mechanical performance has demonstrated the versatility of the AT type systems.

A variety of aromatic and aromatic heterocyclic oligomers with terminal acetylene units have been synthesized and reported (1-3) in recent years. Long term use temperatures for these materials are in the range of 250-550°F with short term usage at 600-650°F depending on the specific molecular structure between acetylene cure sites. These materials can be classified into rigid (high Tg) and flexible (low Tg) systems. For higher use temperatures, the rigid aromatic heterocyclic backbones are employed to exhibit

higher Tg's after cure. Materials which process similarly to the
state-of-the-art epoxides require the more flexible **phenylene R**
type systems, where R refers to the functional group which imparts
the flexibility to the oligomeric backbone.

The rigid high Tg systems are prepared by the formation of a
heterocyclic low molecular weight oligomeric species followed by
endcapping with an appropriate aromatic acetylene. These
heterocyclic systems which have been prepared include the
quinoxalines (4-6), imides (7) and N-phenylbenzimidazoles (8). The
formation of the heterocycle is generally a polymer forming
polycondensation reaction in which the molecular weight of the
oligomer can be controlled by adjusting the stoichiometry of the
reactants.

The most convenient method of preparing the flexible (low Tg)
system is to employ the Ullmann ether reaction of dibromobenzene
and aromatic bis-diols followed by catalytic replacement of the
bromine atoms by terminal acetylene groups. A host of commercially
available bis-diols have been used in the synthesis with both meta
and para dibromobenzene. Low Tg arylether oligomers have been
prepared containing sulfone, sulfide, carbonyl, isopropyl and
perfluoroisopropyl groups in the backbone (9).

These flexible systems do offer the advantages of epoxy-type
processing; however, the mechanical properties, in particular the
lack of toughness of the resulting resins, were drawbacks. Work
with the quinoxaline series (10) had shown that the toughness of AT
systems could be affected by decreasing the crosslink density in
the polymer network. This was done by increasing the chain length
of the units between terminal acetylenes. In an effort to
determine whether this finding with the rigid quinoxalines would
also apply to the flexible systems, work was undertaken to
synthesize higher molecular weight monomer/oligomer systems of the
Phenylene R type shown below.

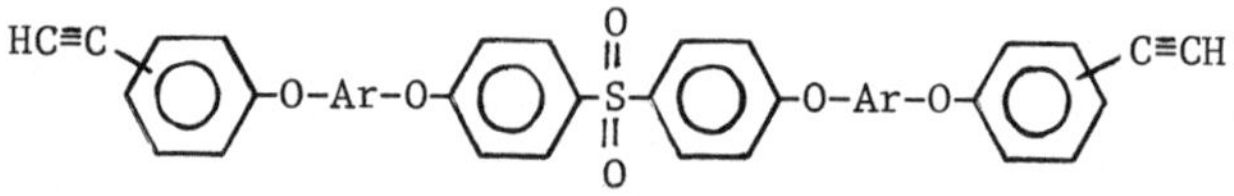

$$HC{\equiv}C{-}\text{Ph}{-}O{-}Ar{-}O{-}\text{Ph}{-}\underset{\underset{O}{\|}}{\overset{\overset{O}{\|}}{S}}{-}\text{Ph}{-}O{-}Ar{-}O{-}\text{Ph}{-}C{\equiv}CH$$

The approach used involved a four step sequence to obtain the
desired product. A number of low cost diols and bis-diols were
studied, including **bisphenol A,** resorcinol and hydroquinone.
An Ullmann ether condensation was used in the reaction sequence in
an effort to obtain very flexible systems which would have low
initial Tg's for ease of processing.

The objectives of this work were to synthesize a series of
arylether sulfone oligomers with an increased chain length between
reactive sites by the proposed synthetic route. In addition, it
was to be determined if the products obtained by this route would
have initial glass transition temperatures near or below room
temperature while having cured Tg's in the 300-350°F range of
epoxides.

Results and Discussion

The reaction sequence used to synthesize these flexible systems involved four steps which are outlined in Figure 1. The first of these was an aromatic nucleophilic substitution, a polymer forming reaction in which 4,4'-dichlorodiphenyl sulfone reacts with various diols. The second step, an Ullmann ether reaction, gives bromine terminated products in which the bromines can be replaced by ethynyl end groups in the final stages.

Both the substitution and Ullmann reactions provide sources of oligomers, making the final product a mixture of monomeric and oligomeric species. While this was desirable for the overall objective of the work, increasing chain length between crosslink sites, the presence of oligomers did complicate characterization of the products obtained from the various reactions. For this reason, a reaction scheme which would give a pure monomeric product was formulated and used for all of the diol systems to do preliminary evaluations.

Monomer Systems. The first step in the synthesis of the model monomeric systems required formation of a mono(bromophenoxy)phenol (I) system by the Ullmann reaction outlined in Figure 2. Here, meta or para dibromobenzene was reacted with the appropriate diol under either of two sets of Ullmann conditions. The first method involved using pyridine, potassium carbonate and cuprous iodide, while the second entailed using 2,4,6-collidine and cuprous oxide. Both of these syntheses will give a small amount of dibrominated material as well as the desired product since two hydroxy linkages are present in the diol, but by controlling the stoichiometry of the reaction the predominant product can be directed toward the mono-bromo material. This Ullmann reaction was done without regard to yield or to optimize conditions, but rather to get enough of the mono(bromophenoxy)phenol product to continue on in the sequence.

Through the use of one of the methods outlined above, all five of the diols shown in Figure 2 were reacted with an isomer of dibromobenzene. The products obtained from this reaction were carried through the sequence shown in Figure 3. The mono(bromophenoxy)phenol was reacted with 4,4'-dichlorodiphenyl sulfone in a nucleophilic aromatic substitution using 1-methyl-2-pyrrolidinone (NMP) as a solvent and potassium carbonate as base. The reaction was run using a 2:1 ratio of bromo compound to sulfone, and since the substitution is a polymer forming reaction under normal circumstances, good conversion to the dibromo product was observed with no oligomer formation.

The second and third steps in the monomer synthesis involve the replacement of bromine with an acetylene protected by an acetone adduct, followed by cleavage of the adduct. These steps will be discussed in more depth later as they are the same for systems containing only monomer or a monomer/oligomer mixture.

Evaluation of Monomers. Following the completion of this sequence of reactions, the monomeric models (III) for all of the diol systems were formed. The products were then evaluated for initial and final Tg's and the results compiled in Figure 4.

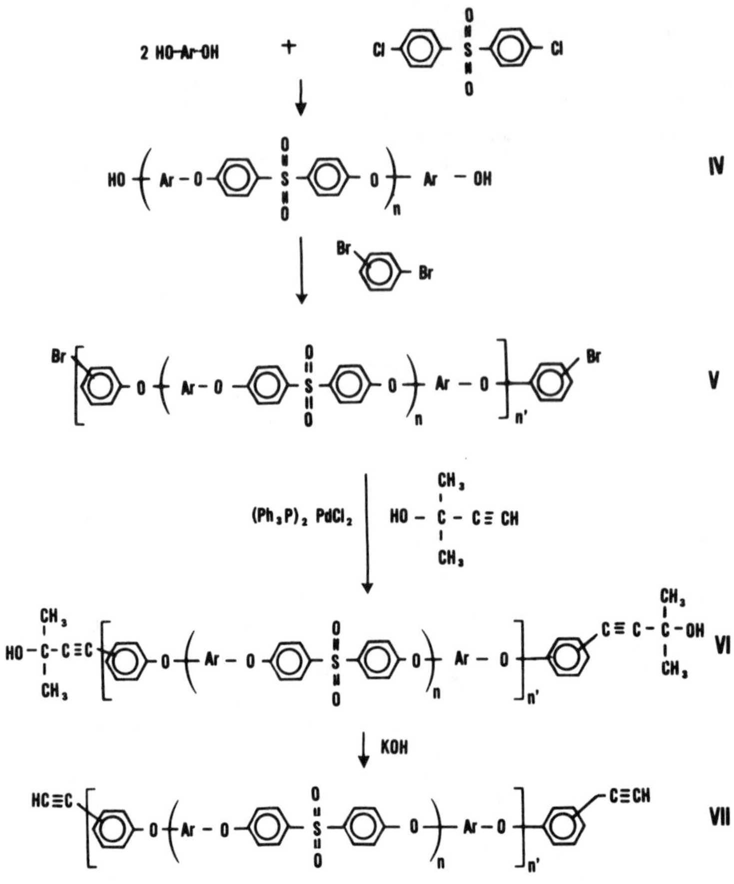

Figure 1. Reaction sequence for AT arylether sulfone oligomers.

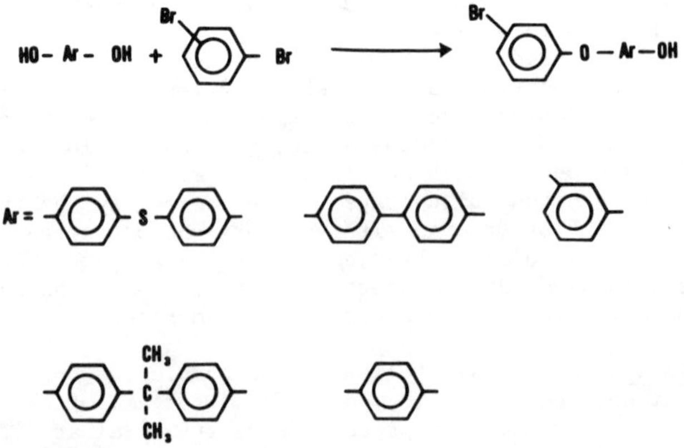

Figure 2. Synthesis of mono(bromophenoxy)phenols.

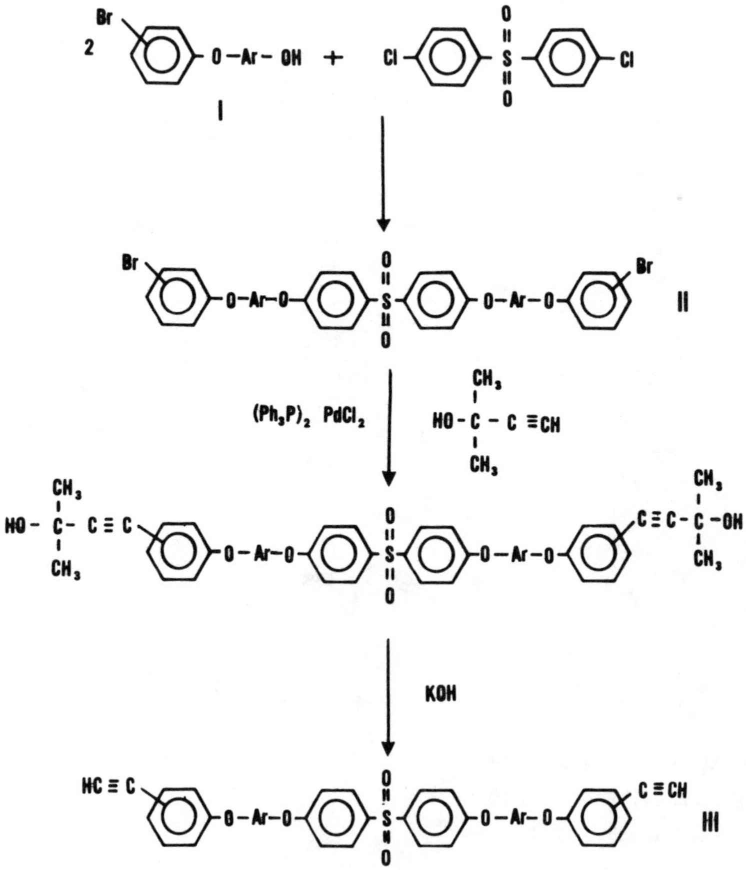

Figure 3. Synthesis of model monomer systems.

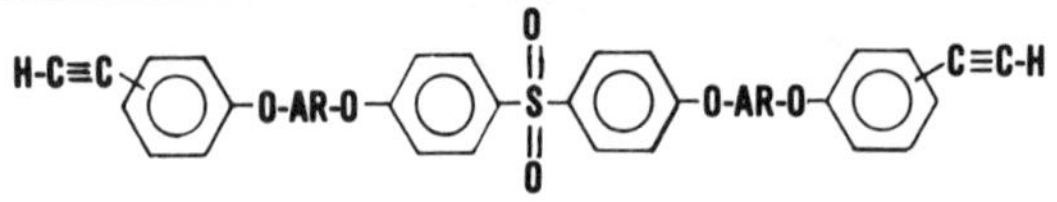

Figure 4. Evaluation of monomeric AT arylether sulfones.

AR	m/p AT-ISOMER	a Tg INITIAL	b Tg CURED	c
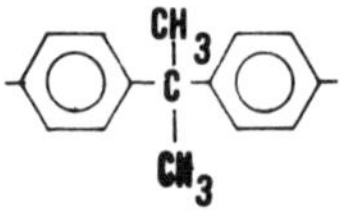 –⟨O⟩–C(CH₃)₂–⟨O⟩–	p	58°C	271°C	
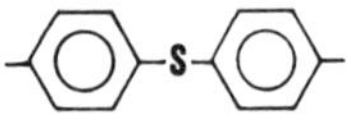 –⟨O⟩–S–⟨O⟩–	p	33° (Tm=74°C)	174°C	
 –⟨O⟩–	p	20°C	240°C	
 –⟨O⟩–	m	12°C	185°C	
 –⟨O⟩–	p	59°C	230°C	
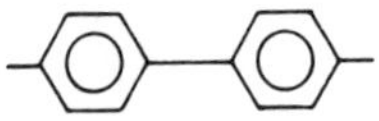 –⟨O⟩–⟨O⟩–	m	59° (Tm=186°C)	241°C	

a. DETERMINED BY DSC (10°C/MIN)
b. DETERMINED BY TMA (10°C/MIN)
c. CURED AT 550°F, 8HR

The initial Tg's found in column three of the table were determined by differential scanning calorimetry at a scan rate of 10°C/min. Final Tg's were determined by TMA on samples which had been cured in air at 550°F for 8 h. No residual exotherm was found in any of the samples when scanned by DSC following this cure cycle.

The second column in the table indicates whether the dibromobenzene used was the meta or the para isomer. In order to provide more flexibility for the rigid diols, 4,4'-dihydroxybiphenyl and hydroquinone, the meta isomer of dibromobenzene was used in preference to para. It was postulated that the meta isomer would increase the free volume of the chain thus lowering the initial Tg. The hydroquinone system was synthesized with both para and meta dibromobenzene isomers to provide a comparison between the effects of the variation in structure.

The variation in the initial Tg's show the effect on processability of the structures of the diol and the dibromobenzene. Two of the systems, those based on thiodiphenol and dihydroxybiphenyl, gave crystalline products which could not be made amorphous upon heat treatment. The effect on chain flexibility of dibromobenzene structure could be seen with the two hydroquinone systems. In this case, the meta isomer gave an initial Tg of 12°C while the more rigid para system had a Tg of 59°C. Variation of diol structure shows a parallel effect as shown with the resorcinol/p-dibromobenzene system. This material has a softening temperature of 20°C.

A similar trend was postulated for final Tg's of the cured materials. This result was observed in some cases, as with the hydroquinone systems where the more flexible meta product gave a significantly lower final Tg. The highest cured Tg's were seen in the bisphenol A, dihydroxybiphenyl and resorcinol systems. The lowest final Tg was observed with the thiodiphenol product, at 174°C, less than 350°F.

The fact that the incorporation of thiodiphenol into the backbone produced both undesired crystallinity and a low final Tg led to a decision not to synthesize the monomer/oligomer mixture for this system.

Monomer/Oligomer Synthesis. The first two steps in the four step reaction sequence of Figure 1 are capable of producing both monomer and oligomer. The first step, aromatic nucleophilic substitution, is a polymer forming reaction under the correct stoichiometric conditions. In order to favor the formation of monomer with a small amount of oligomer, the substitution was carried out at a 4:1 ratio of diol to dichlorodiphenyl sulfone. This led to a predominantly monomeric product (IV) with only the requirement that the excess diol be removed from the product to eliminate the potential presence of low molecular weight species in later reactions.

It should be noted that one of these diols, the hydroquinone, did not provide any oligomer in the first step. This was due to the formation of the quinone structure which made it impossible to use hydroquinone directly in the substitution reaction. An alternate method was used to overcome this problem which involved the use of 4-methoxyphenol to obtain the sulfone product, followed by cleavage of the methyl ether to the diol (VIII) with boron tribromide. This set of reactions is outlined in Figure 5.

All of the sulfone diols were able to form oligomers in the second step of the reaction sequence, the Ullmann ether synthesis. As with the synthesis of the mono(bromophenoxy)phenol products, two methods were used to form the dibromo materials. Method A used pyridine, potassium carbonate and cuprous iodide, while Method B employed collidine and cuprous oxide with the dibromobenzene and higher molecular weight diol (IV). The major difference between the syntheses of the mono(bromophenoxy)phenols described earlier and these lies in the stoichiometry of the reactions. In order to obtain bromo endcapped product and to allow formation of predominantly monomeric species, the reactions were run at a 10:1 ratio of dibromobenzene to diol. This ratio has been found (11) to form monomer as the major product, while also affording a reasonable amount of higher molecular weight products (V).

The monomer/oligomer mixtures were used in the third step of the reaction sequence, the replacement of bromine with 2-methyl-3-butyn-2-ol by use of the bis(triphenylphosphine) palladium chloride catalyst system. This reaction used a triethylamine/pyridine solvent system to replace the bromines on the ether sulfone with ethynyl groups protected by acetone adducts. The acetone protecting groups were then removed in a toluene/methanol/potassium hydroxide solvent system.

Evaluation of Oligomeric Systems. Completion of the cleavage reactions gave the five acetylene terminated systems (VII) evaluated in Figure 6. The third column of the table contains monomer/oligomer ratios for the various systems. These ratios were determined by a column chromatography separation of products, and were based on a weight percent ratio of the amount of monomer to oligomer. The variation in these ratios in the five systems is extreme, ranging from 60% monomer with dihydroxybiphenyl to 95% monomer with hydroquinone/p-dibromobenzene. Some of the variation is attributable to the use of 4-methoxyphenol in the aromatic nucleophilic substitution reaction of the hydroquinone products which precludes the formation of oligomers in this step. The remaining diols are capable of forming oligomers in both the first and second steps of the synthesis. Though the ratios of reactants used in these steps are the same for each system, a variation of 60-85% monomer can still be observed from dihydroxybiphenyl to resorcinol. The structure of the diol used obviously influences the formation of monomer and oligomer in the first steps of the reaction sequence. The effect of either meta or para dibromobenzene isomers is seen with the two hydroquinone products, where the meta isomer forms more oligomer than the para.

The difference between the two dibromobenzene isomers which was noted in the monomeric systems with regards to initial Tg also holds true for the oligomers. The meta system still has a lower initial Tg with the resorcinol system as an intermediate between the two hydroquinone products. The other initial glass transitions are higher than the hydroquinone systems. The dihydroxybiphenyl material exhibits a melting point at 135°C, and though lower than the pure monomer it is still high for the desired room temperature processability. Cured Tg's of all of the systems studied were above 350°F.

Isothermal aging of the samples at was done at 600°F in air. A weight loss of 20-30% was observed for all of the systems over 200 h. There is little difference between the various products since the controlling factor of the thermoxidative stability, the sulfone linkage, is common to each of them.

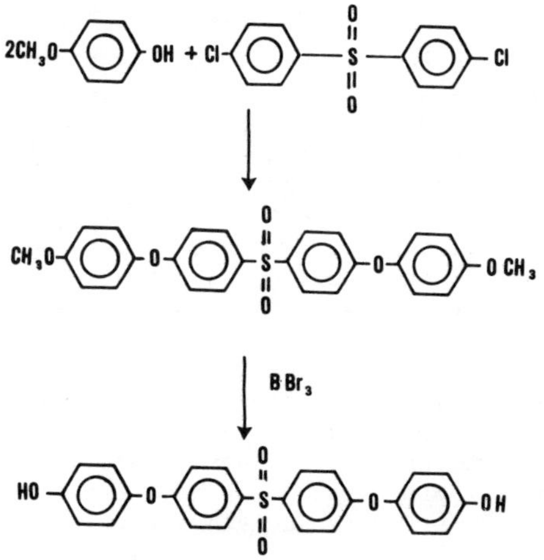

Figure 5. Diol synthesis from 4-methoxyphenol.

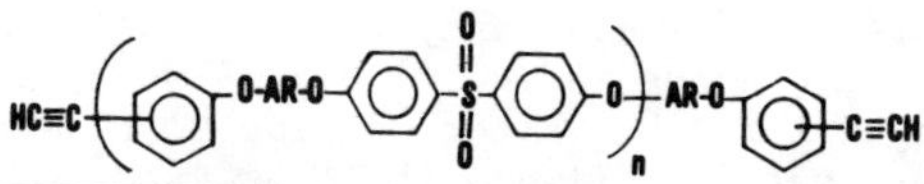

AR	m/p AT-ISOMER	MONO/OLIG RATIO	a Tg INITIAL	b c Tg CURED
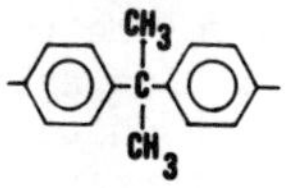 –⌬–C(CH₃)₂–⌬–	p	75/25	55°C	241°C
 ⌬	p	85/15	12°C	200°C
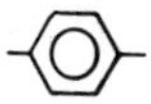 –⌬–	m	85/15	2°C	215°C
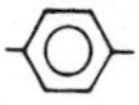 –⌬–	p	95/5	35°C	220°C
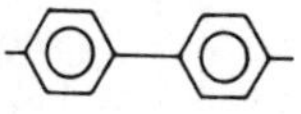 –⌬–⌬–	m	60/40	81°C (TM=135°C)	241°C

a. DETERMINED BY DSC (10°C/MIN)
b. DETERMINED BY TMA (10°C/MIN)
c. CURED AT 550°F, 8 HR

Figure 6. Evaluation of oligomeric AT arylether sulfones.

Conclusions

The synthesis of **phenylene R** type systems via the proposed route afforded two materials which had both final Tg's above the 350°F limit and initial Tg's which were at or below room temperature. The two systems which offered the most desirable properties were those based on resorcinol/p-dibromobenzene and hydroquinone/m-dibromobenzene. Of the two, the hydroquinone system offered a lower initial softening temperature while also providing a cured Tg above 200°C. For this reason the product from hydroquinone and m-dibromobenzene was chosen as the candidate system for future work.

A scale-up of the reaction sequence to obtain enough of the material for mechanical testing is in progress. Following completion of the synthetic work, evaluation of the mechanical properties of the chosen system will be carried out to determine whether increasing chain length between reactive sites will indeed bring an improvement in toughness of acetylene terminated Phenylene R systems.

Experimental

Mono(bromophenoxy)phenol (I). The mono(bromophenoxy)phenols required for the monomeric models were synthesized by two methods. Method A: A mixture of pyridine (90mL), diol (60 mmol), dibromobenzene (56.4g, 240 mmol), anhydrous potassium carbonate (33.3g, 250 mmol) and cuprous iodide (1.62g, 9 mmol) was heated at reflux under nitrogen for 20h. After cooling to room temperature, the reaction mixture was acidified with 1N HCl and the product extracted with chloroform. The chloroform was evaporated and the residue extracted with 10% aq. NaOH. The aqueous phase was acidified, extracted with chloroform, and reduced in volume to an oil that was stirred with 20% aq NaOH to afford the sodium salt of the product. The salt was isolated and dried to give a white solid. The solid was acidified to pH 1 in water, and the freed mono(bromophenoxy)phenol was washed with water and dried. C-H analysis for products was satisfactory.

Method B: A mixture of diol (50 mmol), dibromobenzene (47.0g, 200 mmol), cuprous oxide (1.97g, 15 mmol) and 2,4,6-collidine (100mL) was heated to reflux under nitrogen for 20h. The product was isolated as in Method A.

1,1'-Sulfonylbis[4-[mono(bromophenoxy)phenoxy]benzene] Monomer (II). A mixture of dry 1-methyl-2-pyrrolidinone (45mL), benzene (25mL), I (52.8 mmol), 4,4'-dichlorodiphenyl sulfone (7.58g, 26.4 mmol) and anhydrous potassium carbonate (3.83g, 27.7 mmol) was purged with nitrogen for 15 minutes. The mixture was heated to reflux and the benzene azeotrope collected between 80°C and 150°C. When most of the benzene had been removed, the reaction was heated at 165°C for 7h. After cooling to room temperature, the mixture was poured into 350mL 10% sulfuric acid and was extracted with chloroform. The organic phase was washed first with 25% sulfuric acid, then with distilled water. After drying over $MgSO_4$, the chloroform solution was evaporated to dryness affording a 75-85% yield of the monomeric sulfone.

1,1'-Sulfonylbis[4-(phenolphenoxy)benzene] Monomer and oligomer (IV). A mixture of dry 1-methyl-2-pyrrolidinone (40mL), benzene (20mL), diol (95 mmol), 4,4'-dichlorodiphenyl sulfone (6.79g, 23.7 mmol) and anhydrous potassium carbonate (3.43 g, 24.8 mmol) was purged with nitrogen for 15 minutes. The mixture was heated to reflux and the benzene azeotrope collected until the reaction temperature reached 190°C. When most of the benzene was removed, the reaction was heated at 190°C for 7h. After cooling to room temperature, the reaction was poured into 400mL 10% HCl. A dark, gummy solid separated and the aqueous solution was decanted from the gum. This solid was dissolved in 100mL glacial acetic acid, and the acid solution was precipitated into 1200mL ice water to give a light solid. The precipitate was dissolved in diethyl ether and filtered over a bed of silica gel to remove a dark impurity. The ether solution was dried giving a light amorphous solid in 75-85% yield.

Brominated Monomer/Oligomer Mixtures from IV (V). The brominated sulfone monomer/oligomer mixtures were prepared by two different methods. Method A: A mixture of pyridine (70mL), IV (11.5 mmol), dibromobenzene (26.96g, 115 mmol), anhydrous potassium carbonate (7.94g, 57.5 mmol) and cuprous iodide (0.13g, 0.7 mmol) was heated at reflux under nitrogen for 24h. After cooling to room temperature, the reaction mixture was acidified with 1N HCl and the aqueous solution extracted with ether. The organic phase was reduced in volume to a brown gum which was washed several times with hexane and then dried to give a 75-95% yield of the dibromo product.

Method B: After combining 2,4,6-collidine (9mL), IV (10.25 mmol), dibromobenzene (24.2g, 102.5 mmol) and cuprous oxide (2.95g, 20.51 mmol), the reaction mixture was heated at reflux under nitrogen for 24h. When the reaction had cooled to approx. 80°C, the mixture was filtered, then diluted with chloroform. The organic phase was warmed to 60°C with 100mL conc. HCl, washed with water and then was filtered through a short bed of silica gel. The product, a light amorphous solid, was obtained in 75-90% yield.

Acetone Protected Acetylene Monomer/Oligomer From V (VI). A solution of V (5.5 mmol), 2-methyl-3-butyn-2-ol (2.76g, 32.9mmol), pyridine (15mL), triethylamine (30mL), bis(triphenylphosphine) palladium chloride (0.10g), cuprous iodide (0.10g) and triphenylphosphine (0.20g) was degassed with nitrogen for 15 minutes. The reaction was heated at reflux for 24h, cooled and filtered, then acidified with 1N HCl. The product was extracted with toluene. The organic phase was washed with 1N HCl and then was heated for ½h at 60°C with ethylene diamine. After washing well with water, the toluene solution was dried over $MgSO_4$ and chromatographed on silica gel using chloroform as eluent. The chloroform was removed to give a light, amorphous solid in 70-85% yield.

Ethynyl Terminated Monomer/Oligomer Mixtures from VI (VII). A mixture of the bis-butynol adduct, VI (3.9 mmol), toluene (40mL) and 10% methanolic KOH (40mL) was heated to reflux under nitrogen. The methanol and toluene were removed by distillation, adding more toluene as needed to maintain the reaction volume at 40mL. After

4h, the reaction was cooled to room temperature and filtered over a bed of packed Celite. The toluene solution was washed with water, dried over $MgSO_4$, and reduced in volume to give the ethynyl product (VII) a light colored solid, in 80-95% yield. Monomer/oligomer ratios were determined by column chromatography as weight % monomer: weight % oligomer.

4,4'-[Sulfonylbis(4,1-phenyleneoxy)]bis[phenol] (VIII). A dry flask containing 1,1'-Sulfonylbis[4-(4-methoxyphenoxy)benzene] (10.0g, 22 mmol) was cooled in a dry ice - acetone bath for 15 minutes. After cooling, a 1.0M solution of Boron tribromide in methylene chloride (16.5g, 66 mmol) was added and the reaction was allowed to reach room temperature. After stirring for 18h, the solution was slowly added to 300mL water and stirred for several hours until a light solid could be filtered. The solid was washed with water and dried to give the product in 95% yield.

Characterization. Differential scanning calorimetry and thermal mechanical analysis data were obtained on a DuPont 990 thermal analyzer coupled with a DuPont DSC or TMA cell. Isothermal aging studies were carried out with an automatic multisample apparatus.

Literature Cited

1. Hedberg, F. L.; Kovar, R. F.; Arnold, F. E. In "Contemporary Topics in Polymer Science"; Pearce, E. M., Ed.; Plenum Publishing: New York, 1977; Vol. 2, p. 235.
2. Hergenrother, P. M. Macromol. Sci. Rev. Macromol. Chem. 1980, C19, 1.
3. Bilow, N. In "Resins for Aerospace"; ACS SYMPOSIUM SERIES No. 132, American Chemical Society: Washington, D.C., 1982; p.139.
4. Kovar, R. F.; Ehlers, G. F.; Arnold, F. E. J. Polym. Sci. Polymer Chem. Ed. 1977; 15, 1081.
5. Hedberg, F. L.; Arnold, F. E. J. Appld. Polym. Sci. 1979; 24, 763.
6. Hergenrother, P. M. In "Chemistry and Properties of Crosslinked Polymers"; Zabana, S. S., Ed.; Academic: New York, 1977; p. 139.
7. Landis, A. L.; Bilow, N.; Boschan, R. H.; Lawrence, R. E.; Aponyi, T. J. Polym Prep. 1974, 15, 533.
8. Lau, K.S.Y.; Kellaghan, W. J.; Boschan, R. H.; Bilow, N. J. Polym. Sci. Polym. Chem. Ed. 1983, 21, 3009.
9. Reinhardt, B. A.; Loughran, G. A.; Soloski, E. J.; Arnold, F. E. In "New Monomers and Polymers"; Culbertson, B. M., Ed.; Plenum Publishing: New York, 1983; p.29.
10. Helminiak, T. E.; Jones, W. D. Org. Coat. and Plast. Prepr. 1983, 48, 221.
11. Harrison, J. T.; Sabourin, E. T.; Selwitz, C. M.; Feld, W. A.; Unroe, M. R.; Hedberg, F. L. Polym. Prepr. 1982, 23, 189.

RECEIVED January 17, 1985

4

Network Structure in Polymers from Bisphthalonitriles

JEFFREY A. HINKLEY[1]

Naval Research Laboratory, Washington, DC 20375

>Bisphthalonitrile monomers were cured neat, with nucleophilic and redox co-reactants, or in combination with a reactive diluent. Dynamic mechanical measurements on the resulting polymers from -150 to +300°C turn up several differences attributable to differences in network structure. Rheovibron results were supplemented with solvent extraction, differential scanning calorimetry (DSC), vapor pressure osmometry, and infrared spectroscopy to characterize the state of cure.

Network-forming (thermosetting) polymer systems have the advantage of being processable at an initial, low-viscosity stage, before they undergo a tranformation to a solid which resists creep at high temperatures. This processing advantage is obtained at the expense of added chemical complexity. The properties of the product may depend on the nature and extent of the curing reactions, so techniques which probe network structure beyond the gel point are essential.

In this work, bis-phthalonitrile networks (1, 2) were examined by dynamic mechanical and dielectric methods, supplemented with infrared measurements of state of cure, DSC, vapor pressure osmometry, and solvent extraction. For resins cured with 4,4'-methylene dianiline as co-reactant, a simple network model rationalizes the data.

Experimental

Ether-linked bisphthalonitriles were synthesized by Mr. T. R. Price of the Naval Research Lab. A number of monomers containing various aliphatic and aromatic "linking groups" between the phthalonitrile functions are available; three representative aromatic monomers were selected for this study

[1] Current address: NASA, Langley Research Center, Hampton, VA 23665

(Table I). A fourth monomer (structure IV in Table I) was designed with a single reactive phthalonitrile nucleus to serve as a "reactive diluent" in these network-forming systems.

Polymerization is usually carried out by heating in an open aluminum mold in an air-circulating oven. The crystalline monomer melts and, over a period of days, forms a dark green "B-staged" resin, which eventually solidifies into a rubber and finally a glass.

Except as otherwise noted, the resins studied were cured at 280°C for 7 days. In the experiments which used the monofunctional phthalonitrile, the aluminum foil mold was placed in a closed steel ointment can to prevent evaporative loss of the more volatile monophthalonitrile.

Bulk samples were sectioned with a diamond saw to provide samples for the Rheovibron DDV II Viscoelastometer operated at 11 Hz. In some cases a thin sheet cured between aluminum plates, was heated to the rubbery state, cut while hot, then returned to the oven to complete the cure.

Dielectric data came from a Tetrahedron dielectrometer operated at a 5°/min heating rate in air.

DSC scans were performed at 20°C/min after a rapid quench from above the sample glass transition temperature. A nitrogen atmosphere was used.

Number-average molecular weight was measured in 1,2-dichloroethane at 40°C using a Wescan vapor pressure osmometer.

Solvent extraction was done by placing small pieces of cured polymer in vials of chloroform which were shaken daily. After one week, the chloroform was poured off and replaced with fresh solvent. After another week, the extract was combined with the first and dried to yield the weight of sol.

For the measurements of extent of cure (3), the monomer mixture was melted onto a salt plate, then cured in a closed can under conditions identical to those used for the mixtures in the sol/gel experiments. The plate was removed from the oven periodically and the spectrum recorded on a Perkin-Elmer Model 267 spectrometer.

Dynamic Mechanical Properties

Figure 1 shows the 11 Hz modulus and loss tangent of monomer I cured for seven days at 280°C. (This was the standard cure cycle adopted during development of these resins). Three relaxation processes are evident as peaks in tan delta:

1) A major peak at 280°C matches exactly the glass transition temperature (Tg) determined by DSC. It corresponds with a drop of several orders of magnitude in Young's modulus,

Table I. Monomer Structures

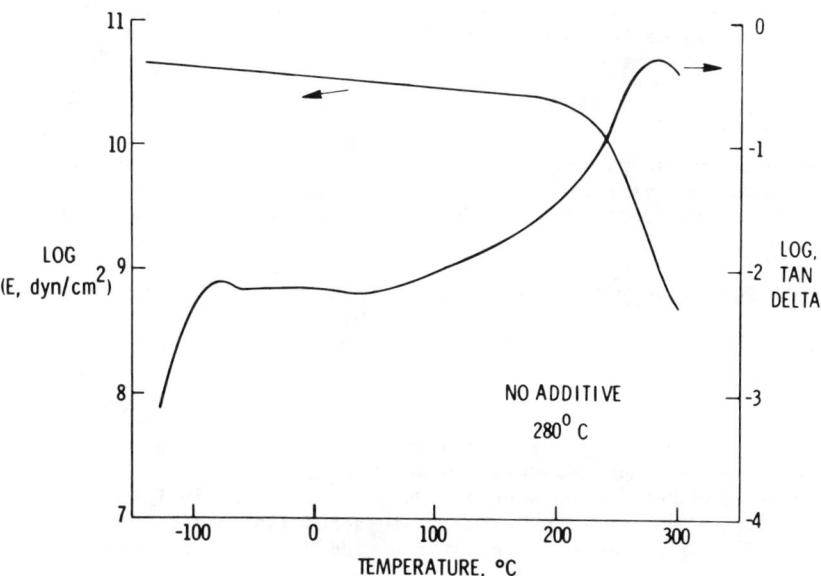

Figure 1. Temperature dependence of dynamic modulus and loss tangent (11 Hz) of monomer I cured for 7 days at 280°C.

and is assigned to the glass-rubber transition. The bisphthalonitrile resins, typical of thermosets, show glass temperatures, at most, equal to the cure temperature. Once the specimen has vitrified at the cure temperature, further reaction is effectively prevented. In practice this means that a post-cure at temperatures above the intended use temperature must be employed.

2) A small peak (tan δ ≈ .02) occurs near -80°C. By analogy to similar peaks in thermoplastics containing p-phenylene backbone structures (4, 5) this peak is assigned to relaxation of monomer-sized units. These motions are highly coupled through the bulk (5, 6) and cannot be assigned to a single process.

3) In the intermediate temperature range (-50 < t < 150°C) there is a broad region of damping associated with relaxations of larger structures. In samples where the cure is well advanced, a clear peak appears near +50°C. This is circumstantial evidence that the peak is associated with crosslink structures.

Although the intermediate peak and the low-temperature (-80°C) peak have similar magnitudes in a mechanical measurement, in a dielectric experiment, the intermediate peak is over twice as large (Figure 2). This seems to be due to moisture, since heating briefly to 300°C decreases the peak loss tangent substantially.

Monomer Structure

In order to establish the effect of varying monomer structure on dynamic mechanical results, three films were cured as thin sheets under identical conditions. No significant differences appear in the Rheovibron plots (Figure 3). Thus the mechanical properties (and by inference, such properties as strength and toughness) appear to be insensitive to monomer structure. The dynamic mechanical properties should be regarded as influenced primarily by the network connectivity and extent of cure.

Network Structure

A number of reactions have been proposed to explain the viscosity increase and gelation which occur when bisphthalonitriles are heated. Possible products include phthalocyanine, triazine rings, and isoindoline chains (7), depending on the reaction conditions and co-reactants used.

In the absence of co-reactants, it is supposed that the polymerization is promoted by traces of water or other nucleophiles, since very pure monomer does not gel even after extended heating. Conversely, gelation may be accelerated by addition of phenols such as bisphenol A. Dynamic mechanical analysis of cured resins confirms that they are practically identical whether or not the phenol is added.

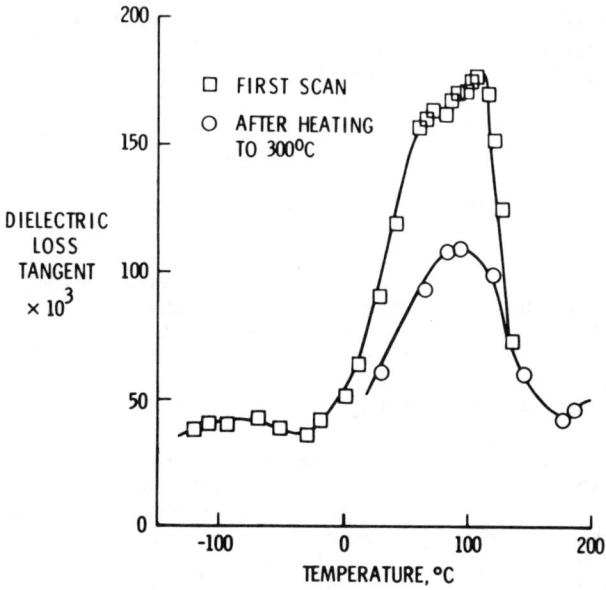

Figure 2. Secondary dielectric loss peaks (1 kHz) in cured I.

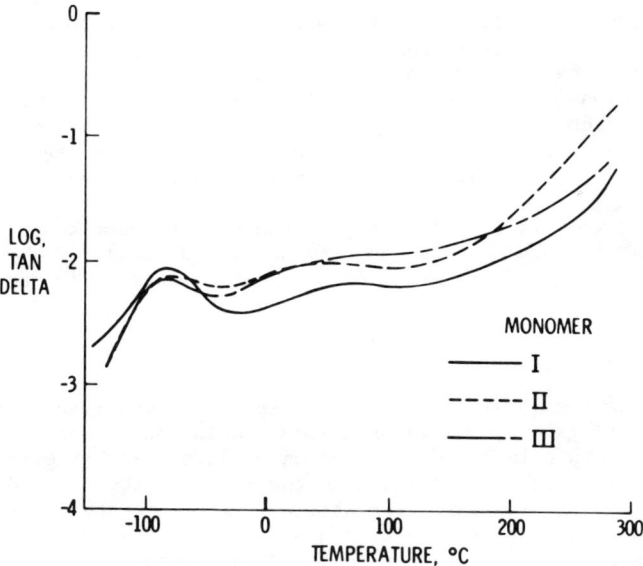

Figure 3. Mechanical damping of thin films cured at 280°C. Solid line, monomer I; dashed, monomer II; dotted, monomer III.

Another class of co-reactants leads to entirely different behavior, however. By providing a redox pathway to phthalocyanine (7), hydroquinone promotes a very different network structure, and this difference shows up clearly in Figure 4. This resin still shows the cryogenic damping peak, the +50°C peak which has been attributed to crosslinked structures is very prominent, but the T_g is hardly visible. The glassy-state modulus is very high and E shows no sign of approaching a "rubbery" value even at 300°C.

Although this change in crosslinking chemistry holds promise for increased use temperatures, a tougher product is desirable. Experiments were therefore designed to decrease the crosslink density by the addition of a monofunctional reactant (Structure IV in Table I). For these experiments, an aromatic diamine co-reactant was used to accelerate the cure (8).

As expected, addition of IV tended to increase the sol fraction in the cured specimens. The statistical theory of gelation (9) can be used to calculate the expected gel fraction as a function of crosslink functionality, extent of reaction, and proportion of difunctional units in the mixture. For these mixtures, an adequate fit to the extraction data was obtained by assuming that the primary reaction is isoindoline formation (Figure 5). Triazine or phthalocyanine formation alone cannot account for the data.

The "best fit" line in Figure 6 was obtained by assuming 85% reaction. This is consistent with infrared spectroscopic results showing 80-90% consumption of functional groups. Furthermore, the predicted number average degree of polymerization in the absence of crosslinks is $(1-p)^{-1}$, where p is the extent of reaction. This yields, for p = 0.85, x = 6.7, which agrees exactly with the observed number average molecular weight of 1.47×10^3 for cured model compound IV.

This result is significant. It means that under these conditions (11 mole percent added amine, 280°C cure) approximately 15% of the nitrile groups are inaccessible and hence unreactive even though the mixture has not vitrified.

Crosslink Density

If we accept the model proposed for these mixed monofunctional/ difunctional systems, we can draw some conclusions about the network structure in polymers based on I alone. For example, Fig. 7 shows how the Tg varied with the relative crosslink density in the mixed systems. The abcissa represents the probability that a monomer chosen at random is linked to the network at both ends. At moderate degrees of crosslinking, the expected relationship between T_g and crosslink density is linear, so the data were approximated by a straight line (10). From the extrapolation in Fig. 7, one concludes that a typical bisphthalonitrile cured to a T_g of 280°C has a relative crosslink density of 0.5, or about 70% reaction of nitrile groups.

Figure 4. Dynamic mechanical properties of monomer I cured with stoichiometric excess of hydroquinone.

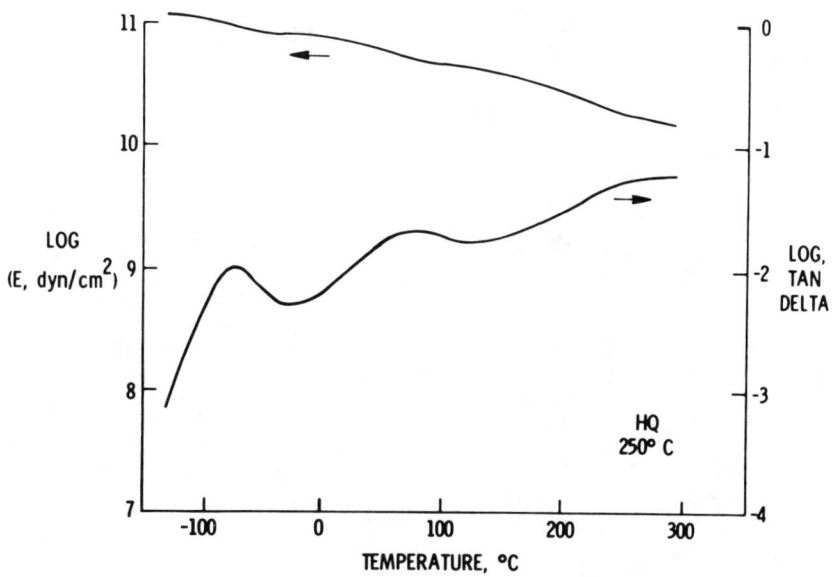

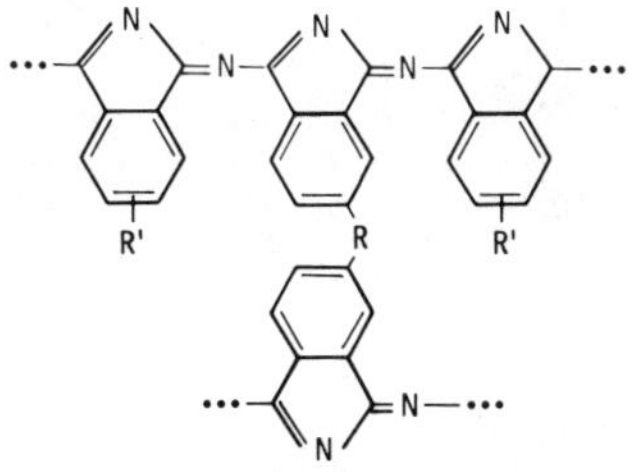

Figure 5. Hypothesized network structure in amine-cured bisphthalonitrile polymers.

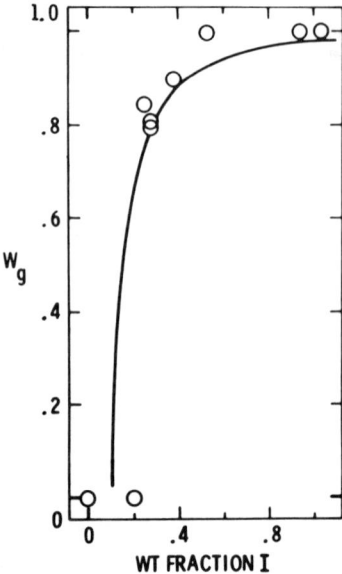

Figure 6. Weight fraction gel in cured mixtures of mono- (IV) and difunctional (I) phthalonitriles.

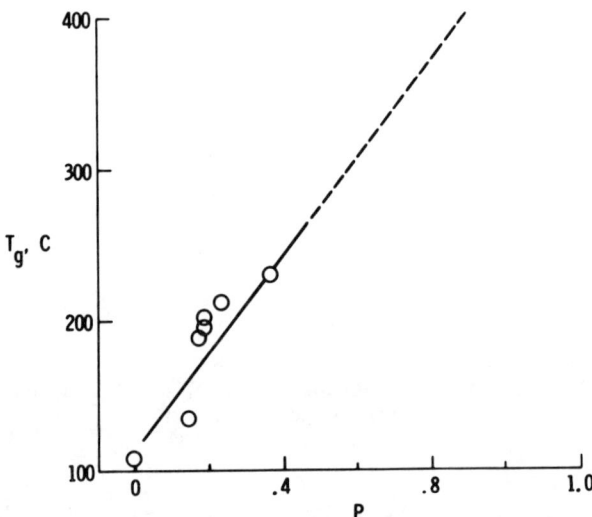

Figure 7. Glass transition temperatures of cured mono (IV) + difunctional (I) phthalonitrile mixtures. P is the probability that a randomly-chosen monomer is reacted at both ends, assuming 85% consumption of functional groups.

The assumption is that a dangling unreacted functionality on a bis-phthalonitrile molecule would affect T_g in the same way as does a molecule of the model compound IV.

A further consequence of this line of reasoning is that significantly higher T_gs (at least 350°C) might be obtained if the extent of reaction could be driven as high as 85% in the difunctional case.

The statistical theory of crosslinking used in the last section also gives the theoretical concentration of elastically-active chains, N, which in turn determines the rubbery modulus E = 3NRT (R is the gas constant and T is the absolute temperature). At 70% reaction one calculates E = 2 x 10^8 dyn/cm^2, in agreement with the apparent level in Figure 1.

Acknowledgments

Thanks are due to T. Price for supplying monomers, and to T. Keller and A. Snow for helpful suggestions.

Literature Cited

1. Keller, T. M.; Griffith, J. R. ACS Org. Coat. Plast. Chem. Preprints 40, 781 1979.
2. Ting, R. Y.; Keller, T. M.; Marullo, N. P.; Price, T. R.; Poranski, C. F. Jr. Polymer Preprints 22 (1), 50, 1981.
3. Marullo, N. P.; Snow, A. W., ACS Symposium Series No. 195, p. 325, 1982.
4. Allen, G.; McAinsh, J.; Jeffs, G. M.; Polymer 12, 85 1971.
5. Yee, A. F.; Smith, S. A., Macromolecules 14, 54 1981.
6. Hinkley, J. A.; J. Polymer Sci. - Polym. Lett. Ed. (in press).
7. Snow, A.; Griffith,, J.; Marullo, N.; Macromolecules July 1984 (in press).
8. Keller, T. M.; Price, T. R.; J. Macromol. Sci. - Chem., A18 (6), 931, 1982.
9. Miller, D. R.; Macosko, C. W.; Macromolecules 9 (2), 206, 1976.
10. Fox, T. G.; Loshaek, S.; J. Polym. Sci., 15, 371, 1955.

RECEIVED March 1, 1985

Intermolecular and Intramolecular Reactions of Substituted Norbornenyl Imides

CHAIM N. SUKENIK, VINAY MALHOTRA, and UDAY VARDE

Department of Chemistry, Case Western Reserve University, Cleveland, OH 44106

> The present study reports the synthesis, characterization and thermal reactions of phenyl and carbomethoxy substituted norbornenyl imides. These substrates were designed to model the reactive end-caps of the PMR-15 resin and allow an assessment of the effect that conjugating substituents would have on the high temperature cure of such systems. The effect of these substituents on both monomer isomerization and polymerization is reported and a possible use of the phenyl substituent as a probe of polymer structure is suggested.

The thermal polymerization of reactive polyimide oligomers is a critical part of a number of currently important polymers. Both the system in which we are interested, PMR-15, and others like it (LARC-13, HR-600), are useful high temperature resins. They also share the feature that, while the basic structure and chemistry of their imide portions is well defined, the mode of reaction and ultimately the structures that result from their thermally activated end-groups is not clear. Since an understanding of this thermal cure would be an important step towards the improvement of both the cure process and the properties of such systems, we have approached our study of PMR-15 with a focus only on this higher temperature thermal curing process. To this end, we have used small molecule model compounds with pre-formed imide moieties and have concentrated on the chemistry of the norbornenyl end-cap (1).

The specific compounds that we used are shown below. They can be grouped in the following way. The simplest models (Parent Endo: PN and Parent Exo:PX) have unsubstituted norbornenyl rings and differ from each other only in the stereochemistry of their ring fusion. These compounds have been elaborated by the incorporation of either a phenyl (φ) or a carbomethoxy (C) group in the bridgehead (B) or vinyl (V) positions. Thus, the notation φVN represents PN with a phenyl substituent at a vinyl position, while CBX represents PX with a carbomethoxy substituent at the bridgehead position. The isomerization and polymerization chemistry of these nine compounds (PN, PX, φBN, φVN, φVX, CBN, CBX, CVN, CVX) are the main concerns of our work.

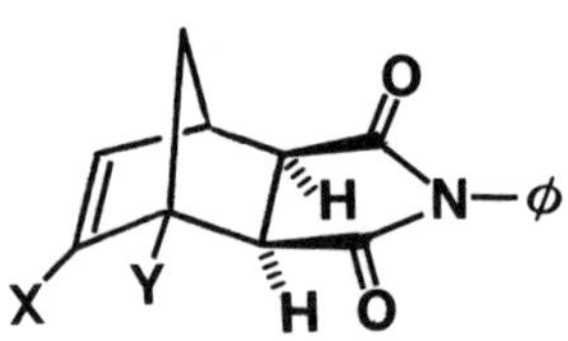

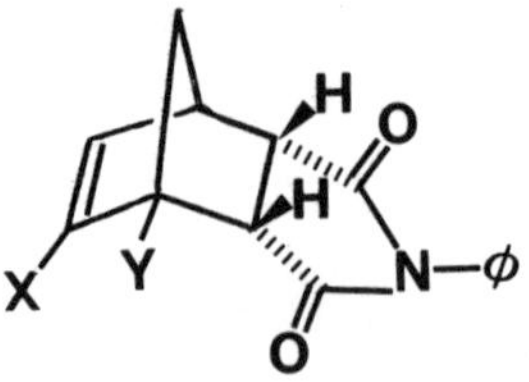

PN(X = H, Y = H)
CBN(X = H, Y = COOMe)
CVN(X = COOMe, Y = H)
φBN(X = H, Y = φ)
φVN(X = φ, Y = H)

PX(X = H, Y = H)
CBX(X = H, Y = COOMe)
CVX(X = COOMe, Y = H)
φVX(X = φ, Y = H)

Substrate Synthesis and Characterization

The first phase of our efforts was the unambiguous synthesis of each model substrate. PN and PX were already well characterized materials (1). While direct synthesis of the phenyl and carbomethoxy compounds from PN and/or PX was attempted, this approach was unsuccessful due to the sluggish reactivity of the norbornenyl double bonds in these molecules (2). A successful approach to CBN and φBN based on N-phenyl maleimide (NPMI) trapping of the respective thermodynamically favored 1-substituted cyclopentadienes is shown in Equation 1. Similarly, kinetic trapping of 2-phenyl cyclopentadiene, from the in situ dehydration of 3-hydroxy, 3-phenyl cyclopentene, gives a clean yield of φVN (Equation 2). The remaining phenyl isomer (φVX) and the three other carbomethoxy isomers (CBX, CVN, CVX) were all obtained by the thermal isomerization chemistry described in the next section of this paper. They were each isolated in pure form by liquid chromatography. We were unable to obtain any φBX or any of the 7-substituted isomers by any means.

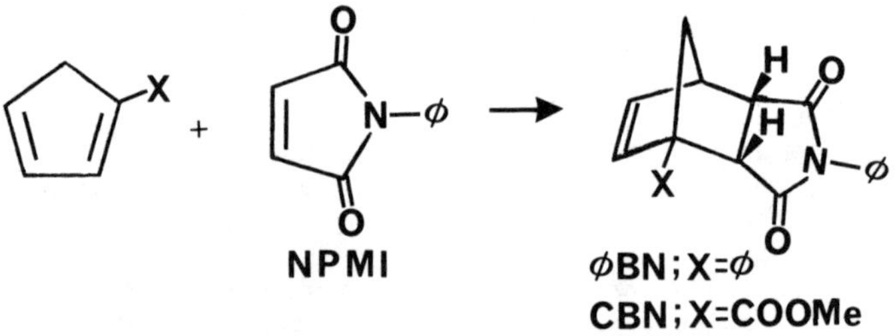

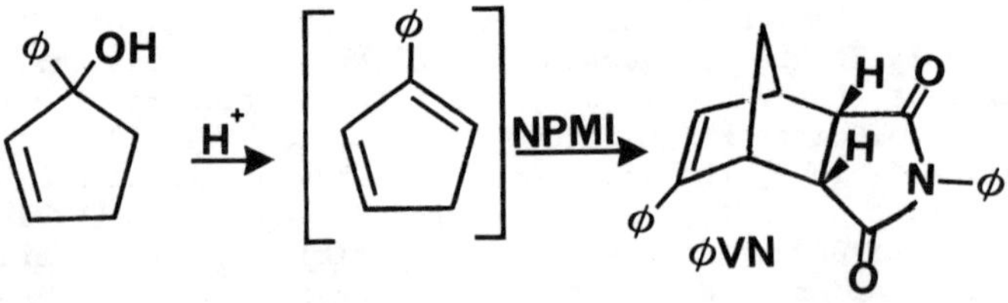

All compounds studied were sharp melting solids showing imide carbonyls between 1702 and 1712 cm^{-1}. The bridgehead substituted compounds were distinguished from their vinyl substituted isomers by a shorter λ_{max} in their UV spectra and each isomer had distinctive H^1 and C^{13} NMR spectra. These spectral data (fully assigned) as well as a delineation of patterns for distinguishing sets of isomers are reported in detail elsewhere (3). For present purposes, it will suffice to note that all isomers for a single substituent are readily separable (for both preparative and analytical purposes) and that assignments of both substituent regiochemistry and ring fusion stereochemistry have been made with a high degree of certainty.

Monomer Isomerization

One of the complexities in the direct study of the PMR cure is the superposition of monomer isomerization on the polymerization chemistry of interest. To ensure our ability to dissect these two kinds of processes, we first studied the thermal behavior of each of our model compounds in the absence of any polymer forming process: in dilute solution.

All solvents for these solution thermolysis reactions were freshly distilled and all reactions were done in sealed glass tubes heated in a thermostatted oven. Over a wide range of solvents (DMF, naphthalene, diphenylmethane, benzene, toluene, and decalin) there was no significant variation in either isomerization rate or product composition. Reactions were done at 125°C, 155°C and 195°C and the only limitation was that DMF could not be used as the solvent in reactions at 195°C; it led to substantial substrate destruction (polymer forming reactions of substrate with DMF?). Isomer compositions were ascertained both by HPLC and by H^1 NMR.

The results of these isomerization studies can be summarized as follows. At 125°C and 155°C both parent isomers (PN and PX) and the vinyl substituted compounds (CVN, CVX, φVN, and φVX) were all largely stable; even after 24 hrs at 155°C each of them showed <15% conversion to any other isomer. On the other hand, the bridgehead substitued compounds (CBN, CBX, and φBN) all readily isomerized at temperatures as low as 125°C. Specifically, after 24 hrs at 125°C CBN is 50% isomerized to CVN and φBN is 80% isomerized to φVN. At 155°C it only takes 3 hrs for φBN to be 95% isomerized.

At 195°C isomerization of the parent and vinyl substituted isomers is observed. This higher temperature thermal reaction is an exo/endo equilibration and for each set of compounds yields a bonafide equilibrium that can be achieved starting with any of its isomers. This equilibrium is dominated (>97%) by the vinyl substituted isomers for both the phenyl and carbomethoxy series. In all cases (including the parent system), although the endo compounds were produced exclusively by our Diels Alder synthetic routes, the exo isomer dominates the final equilibrium mixture (56% PX/44% PN; 62% φVX/37% φVN/1% φBN; 71% CVX/27% CVN/2% CBN/1% CBX).

All of the isomerization data shown above is consistent with the normal electrocyclic reaction chemistry expected for such substrates (4). That such fused norbornenyl systems undergo exo/endo isomerization via Diels Alder/retro Diels Alder reactions has been explicitly proven for simple cyclopentadiene-maleic anhydride adducts (5) and should apply here as well. Moreover, our monosubstituted substrates provide additional confirmation that this pathway is operative by their low temperature isomerization of BN to VN. Their selective formation of the kinetically preferred (endo) Diels Alder product rather than VX, the thermodynamically favored product is inconsistent with other possible radical mechanisms (6).

The other necessary reaction for a BN to VN isomerization is a well precedented 1,5 H shift to convert the linearly conjugated substituted cyclopentadiene (LCC) into the cross conjugated cyclopentadiene (CCC). The relative lability of BN relative to VN is thus a reflection of the stabilizing conjugation of the substituent in the vinyl isomers and the fact that the formation of LCC from BN is more favorable than the formation of CCC from the retro Diels Alder of VN. The relative energetics for all of these processes is represented in a combined reaction profile diagram shown in Figure 1.

The construction of this overall reaction profile is based on the following considerations. The lower energy (5-10 kcal/mole) of VN and VX versus BN is due to the conjugation of their substituent with the norbornenyl double bond. Linearly conjugated cyclopentadiene derivatives are lower in energy (1-2 kcal/mole) than their cross conjugated counterparts. VX is sterically less crowded than VN and is thus of lower energy ($\sim$1kcal/mole). The observed kinetic preference for endo Diels Alder adducts requires that TS_4 be higher in energy (2-4 kcal/mole) than TS_3. Similarly, since the selective generation of CCC for x = ϕ (Equation 2 above) allowed the selective synthesis of ϕVN, TS_3 (Diels Alder to form VN) must be somewhat lower than TS_2 (1,5 H shift to form LCC). And lastly, the selective Diels Alder synthesis of BN from LCC and NPMI (Equation 1 above) requires that TS_1 be lower in energy than TS_2.

A consequence of our observation that the bridgehead isomers are easily isomerized and thermodynamically disfavored is that they may be largely irrelevant to the actual PMR cure. Moreover, the approach of each set of monomers to a bonafide equilibrium under conditions that are milder than typical thermal curing conditions (see below) suggests that regardless of the isomeric composition present in the PMR resin in its low temperature imidization stage, the final high temperature cure probably occurs with a mixture that reflects substantial, if not complete, isomer equilibration. Lastly, the observation that both the phenyl and carbomethoxy substituted monomers undergo exo/endo equilibration at about the same rate as the PN/PX system, indicates that these vinyl substituents have little or no effect on the ease of the retro Diels Alder process. Their effect on polymerization rates will be detailed below, but it is important to note in advance that any effect these substituents do have on polymerization is not a reflection of a change in the propensity for retro Diels Alder chemistry. Despite the importance of Diels Alder reactions in our isomerization processes, its relevance to any polymer forming reactions must still be critically evaluated.

Polymer Formation

The polymerization of our model substrates was studied by heating neat solid samples of each monomer in sealed glass tubes. No pretreatment of the glass was needed to achieve reproducible results and the sealed

5. SUKENIK ET AL. *Substituted Norbornenyl Imides* 57

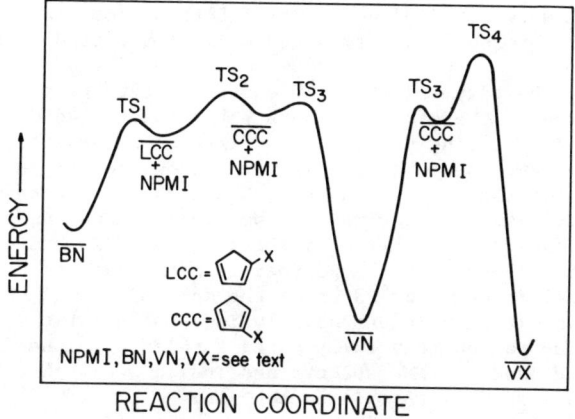

Figure 1. Combined reaction profile diagram for monomer isomerization.

tubes prevented any loss of volatiles. All cured samples (heated to 195°, 250°, or 330°C) were completely soluble in THF and in chloroform. This greatly facilitated analysis by both size exclusion chromatography (SEC) and by NMR.

Our first interest was to assess the polymer forming ability of each of our monomers. This was done by analyzing fully cured (250°C for 24 hrs or 330°C for 2 hrs) samples of each monomer. In all cases no residual monomeric substrate could be detected and SEC results showed that substantial oligomerization had occurred. For all monomers the bulk of the cured material was in the form of oligomers with n = 5-8 and the longest chains were n = 24 $\pm$ 3. The SEC profiles for all three kinds of polymer (unsubstituted, phenyl and carbomethoxy) were similar enough to justify a comparison of the polymer forming process for each kind of monomer. This comparison is outlined below. A more detailed analysis of both oligomer distribution and structure is presently underway and will be reported in full elsewhere (7).

Monitoring the polymerization of each substrate provided an informative picture of the effect of both substituents and of isomer distribution on the curing process. We first addressed the question of the relative rates of substrate isomerization and polymerization. We found that, for the parent monomer (PN and PX), the rate of isomerization greatly exceeds the rate of polymerization. Under conditions where PN and PX are fully equilibrated (195°C/15 hrs or 250°/1 hr) there is still less than 20% polymer formation in the neat sample. We conclude that for PN or PX the composition of the mixture undergoing polymerization is essentially independent of the starting isomer. The observation that fully cured samples of either PN or PX show identical H^1 and C^{13} NMR spectra and indistinguishable SEC analyses, is consistent with this contention.

While exo/endo isomerization of the substituted monomers had been found to be comparable to that of the parent monomer, polymerization of the substituted monomers was much faster. Under the conditions indicated above for <20% polymer formation in neat samples of PN or PX, both the phenyl and carbomethoxy monomers show >80% polymer formation. This greatly complicates any attempt to dissect isomerization from polymerization in these neat samples. Nevertheless, we were able to establish that, in the lower temperature (195°C) polymerization, the isomeric mixture undergoing polymerization was the same (for the carbomethoxy system) whether we started from CBN or from CVN (195°, 1 hr: 71% VN, 23% VX, 6% BN + BX). A similar result was obtained by comparing the polymerization of ϕVN and ϕBN.

However, in the higher temperature experiments (>250°C), both polymerization and isomerization were proceeding so rapidly that precise analyses of isomer ratios became difficult. Equilibration and polymerization were occurring at competitive rates. We do have some tentative evidence that in the higher temperature cures there may be some differences in the rate at which polymer forms depending on the starting isomer. This data is currently being reinvestigated. Even if such an effect is verified, it is small and in no way affects our major conclusion that substitution greatly enhances the rate of polymerization of norbornenyl imides. Moreover, since even samples of substituted isomeric imides cured at 330°C showed similar SEC profiles, the probability of major differences as a function of isomeric composition is small.

In terms of an interpretation of the enhancement of polymerization rate provided by our vinyl substituents, we can only speculate. It is clear that any kind of addition polymerization of the norbornenyl double bond will benefit from the electronic stabilization provided by a conjugating substituent. A simple radical addition process such as is known for both styrene and acrylate monomers may be a reasonable analogy to our system. Whether this effect alone is enough to account for our observations is not clear. A possible additional effect, at least in the case of the phenyl substituted monomers, is suggested below as part of our work on polymer structure.

Polymer Structure

Having established the effect of substitution on the rates of both monomer isomerization and polymerization, we addressed the question of polymer structure. Specifically, are norbornenyl imide units incorporated into the fully cured polymer with their norbornyl rings intact? If so, does the polymer also reflect the equilibrium ratio of exo and endo ring fused monomers? For our parent monomers, PN and PX, this question has been unanswerable. We have not found any direct probe that allows an unambiguous assessment of specific substructures within the cured polymer. We do, however, have some evidence bearing on this question for the phenyl substituted monomer. This evidence is attributable in part to our discovery of an unexpected side-reaction in the cure of the phenyl substituted monomer, and in part to the presence of a unique NMR diagnostic for phenyl substituted, endo norbornyl N-phenyl imides. Both of these results are detailed below.

In the course of our analysis of cured samples of phenyl substituted monomers, we found that many of these samples yielded as much as 15-20% of a monomeric material identified as the hydrogenated analog of ϕVX (HYVX). Even more interestingly, none of the corresponding HYVN was found in any sample. While we do not yet fully understand this unusual pair of observations, a number of points are clear. First of all, the formation of HYVX requires a donor of hydrogen. We have verified that this is not arising from adventitious contamination of our reaction vessels. It must therefore be true that some of our phenyl monomer molecules are acting as the hydrogen donors and are themselves being oxidized. A possible avenue by which this could occur is by increasing unsaturation and possible aromatization of some fraction of our norbornyl rings. This possibility is depicted in the scheme below. It provides not only a source of hydrogen for the production of HYVX, but also suggests a new way in which the phenyl substituent could enhance the generation of radical initiators and thus enhance the polymerization process as well.

The observed absence of any HYVN despite the formation of HYVX is interpretable in either of two ways. It is possible the ϕVN is polymerizing while the ϕVX is hydrogenating and thus no hydrogenated ϕVN is seen. Alternatively, all the chemistry of our monomers, both polymerization and any side reactions are only occurring on the exo isomer. Disappearance of endo isomer is only via isomerization to exo. This second interpretation is consistent with the observation of enhanced double bond reactivity of exo isomers reported for the

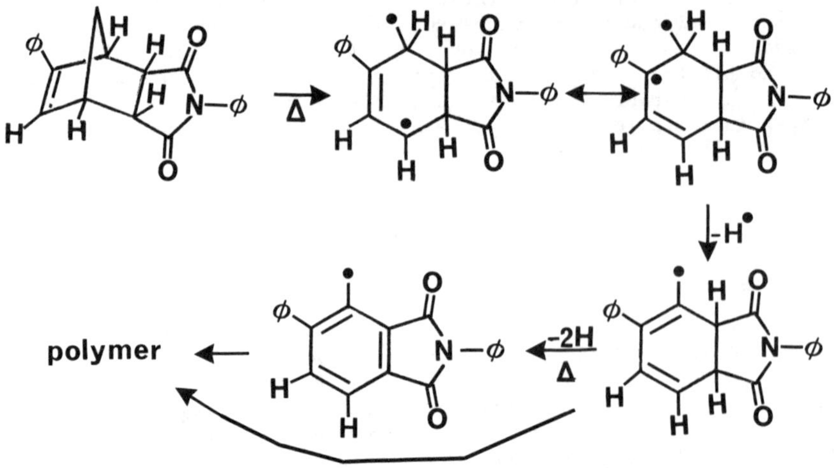

Scheme 1

reactions of a variety of electrophiles and dienes with cyclopentadiene-maleic anhydride adducts (8). It is also consistent with a totally separate line of reasoning which we have developed based on the NMR data described below.

To verify our structural assignment of HYVX we independently synthesized it, along with HYVN. For further comparisons, we also synthesized the epoxides of φVN and φVX, VNE and VXE respectively.

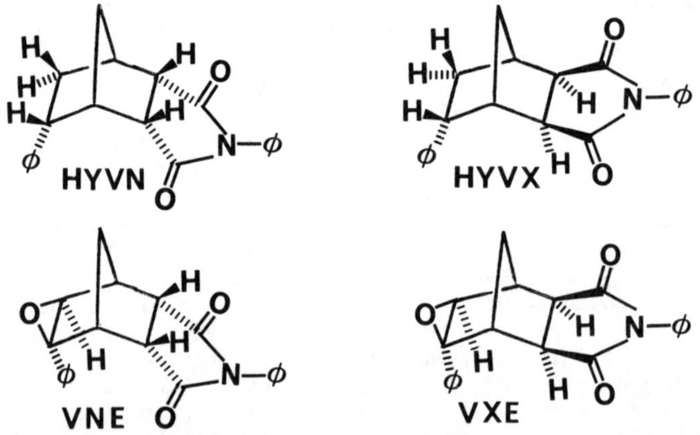

Our syntheses of these new materials showed that both hydrogenation and epoxidation occurred exclusively from the exo face of both φVN and φVX. This is consistent with extensive literature precedent for a variety of reactions on norbornenyl double bonds (9). H^1 NMR spectroscopy of these materials also revealed an interesting pattern. All phenyl substituted compounds with endo imide rings (φVN, HYVN, VNE) evidenced signals due to one of the ten aromatic protons (five on each phenyl ring) being shifted to unusually high field (6.4 ± .1δ). This is consistent with a field effect on a proton of one of the phenyl rings due to its position in the space over the center of the ring current of the other phenyl ring. While we cannot yet indicate which proton is shifted, this effect seems to be a consistent diagnostic for the presence of a phenyl substituted norbornyl ring fused to an endo N-phenyl imide.

On the basis of this NMR pattern we would suggest that, if any endo fused phenyl monomers are being incorporated into the phenyl substituted PMR polymer chain, we should see H^1 NMR signals between 6.6 and 6.2δ. Cured samples of φVN and φVX show identical H^1 NMR spectra and neither has any signals in this region. This may mean that only φVX polymerizes. Alternatively, neither φVN nor φVX is incorporated into the polymer with its norbornyl ring intact and the polymer does not contain any norbornyl sub-structures at all. This might at least partly result from the kind of radical process shown in Scheme 1 above. An approach to these structural and mechanistic questions is presently being developed in our laboratory (10).

Acknowledgments

Extensive collaboration with Professor W. M. Ritchey at Case Western Reserve University and with Dr. Tito Serafini and Dr. Richard Lauver at the NASA Lewis Research Center are gratefully acknowledged. Financial support of this work was provided by NASA (NAG 3-163).

Literature Cited

1. For earlier work based on this approach see: a) Wong, A. C.; Ritchey, W. M. Macromolecules 1981, 14, 825; b) Serafini, T. T.; Delvigs, P.; Lightsey, G. R. J. Appl. Polym. Sci. 1972, 16, 905.
2. a) Liotta, D.; Zima, G. Tetrahedron Lett. 1978, 4977; b) Gray, A. P.; Heitmeier, D. J. Org. Chem. 1969, 34, 3253; c) Kwart, H.; Miller, L. J. J. Am. Chem. Soc. 1961, 83, 4552.
3. Malhotra, V. Ph.D. Thesis, Case Western Reserve University, Cleveland, 1984.
4. Kwart, H.; King, K. Chem. Rev. 1968, 68, 415.
5. Granter, C.; Scheidegger, U.; Roberts, J. D. J. Am. Chem. Soc. 1965, 87, 2771.
6. a) Mironov, V. A.; Fadeeva, T. M.; Stepaniantz, A. U.; Akhrem, A. A. Tetrahedron Letters 1966, 5823; b) Craig, D. J. Am. Chem. Soc. 1951, 73, 4889.
7. Varde, U. Ph.D. Thesis, Case Western Reserve University, Cleveland, in preparation.
8. See reference 2a and 2c and a series of papers by Salakhov and coworkers; Org. React. (Tartu), 1979, 16, 54, 65, 75, 378, 386, 394, 402.
9. Brown, H. C.; Kawakami, J. H. J. Am. Chem. Soc. 1970, 92, 201.
10. It has been called to our attention by a referee that Young and Chang(11) have recently observed the endo-exo thermal isomerization in addition polymer prepregs of LARC-160 by diffuse reflectance FTIR. This technique may prove useful in the present work as well.
11. Young, P. R.; Chang, A. C. SAMPE Technical Conf. 1984, 16, 136.

RECEIVED February 20, 1985

Thermally Stable Polymers for Electronic Applications

D. J. DAWSON, W. W. FLEMING, J. R. LYERLA, and J. ECONOMY

IBM Research Laboratory, San Jose, CA 95193

>High-temperature polymers which can be obtained from processable reactive oligomers would offer many unique opportunities, particularly to the microelectronics industry. The ideal oligomer would have properties such as low dielectric constant and good planarization and would be both thermally and photochemically curable to heat stable polymers. Initial research efforts have focused on melt processable diacetylenic oligomers with the hope that such units would thermally cure to condensed polyaromatics. Poly(triethynylbenzene) (PTEB) was chosen as a model compound and was prepared by the oxidative coupling of 1,3,5-triethynylbenzene with phenylacetylene. The composition and molecular weight of PTEB has been varied by incorporating different levels of phenylacetylene as a capping agent. The planarization of the oligomer was outstanding while the thermal and oxidative stability of the cured polymer was consistent with what one might expect for a network of aromatic rings.
>
>In order to study the cure behavior of the PTEB system, ^{13}C NMR of uncured and cured PTEB in the solid state was performed using cross-polarization magic-angle spinning techniques. The results show the polymerization to be via aromatization. The extent of cure versus cure temperature was determined quantitatively. It was found that the material was almost completely cured after one hour at 215°C. As the cure goes to completion, the ability to react decreases due to the corresponding rapid increase in T_g. Chemical shifts of the resonances in the cured material are consistent with a highly crosslinked condensed aromatic network.

The development of functional oligomers that can be thermally cured to heat-stable polymers has been the subject of considerable study over the past 15 years. Poly(aromatic acetylenes) were initially considered to be good candidates for such uses because the ethynyl groups were thought to undergo a thermally induced cyclotrimerization to a fully aromatic cured polymer. However, in recent years it has become evident ([1]) that ethynyl terminated oligomers such as Thermid 600 did not thermally cyclotrimerize. Presumably cyclotrimerization of ethynyl groups would require the presence of transition metal catalysts; however, these catalysts would remain in the cured polymer and could act as possible catalytic sites for subsequent thermal decomposition. As a result, the use of ethynyl terminated oligomers as precursors to polymers with high temperature thermo-oxidative stability appears questionable.

0097-6156/85/0282-0063$06.00/0
© 1985 American Chemical Society

Recently, the microelectronics industry has developed a need for thermally stable polymers for dielectric insulating layers. In addition to good thermal stability ($\geq 400°C$), these cured polymer films must exhibit a low dielectric constant, low moisture uptake, a high glass transition temperature (T_g), excellent planarization, and possibly lithographic sensitivity which would allow these films to be used as resists. Only a few of these properties can be achieved with polyimides. Other candidates for high temperature polymers also have drawbacks. Inorganic, fluoro-, and phenolic polymers have inadequate thermal stability. Heterocyclic and rigid-rodlike polymers are insoluble in solvents suitable for spin-coating.

A research program was initiated at IBM to design a model polymer that would meet all these requirements. The low dielectric constant and low moisture uptake requirements would best be achieved by a hydrocarbon structure. Optimal planarization would require a low-molecular-weight oligomer, and the required T_g would necessitate a crosslinked network of aromatic units in the cured polymer. It appears that diacetylenic oligomers could meet these specifications and at the same time obviate the potential oxidative instability inherent in the products of thermal crosslinking of terminal ethynyl resins. In 1969, Wegner ([2]) reported that monomeric diacetylenic single crystals could be polymerized by ultraviolet radiation. This finding suggested that polymers containing diacetylene units would have potential as resists. Somewhat earlier, Hay ([3]) prepared high molecular weight diacetylene polymers by the oxidative coupling of m-diethynylbenzene. These polymers underwent an apparent exothermic decomposition upon heating to 180°C and showed a weight retention of over 90% after heating to 800°C under inert conditions. Alternatively, the decomposition observed at 180°C could have been the result of an uncontrolled exothermic curing process. By working with very thin films typically used for dielectric insulators it was expected that the exotherm could be controlled to afford a material of high thermal stability.

Our initial work on diacetylenic polymers was directed to the synthesis of low-molecular-weight poly(m-diethynylbenzene) (Scheme 1). A low-molecular-weight resin was selected to permit the melt flow behavior essential for planarization. The molecular weight was kept low by including phenylacetylene in the polymerization reaction. Being monofunctional, phenylacetylene acted as a capping agent, terminating the growing polymer branch with a phenylbutadiyne group. The resulting oligomer exhibited melt flow properties below the curing temperature but formed poor films because of its low solubility in spin-coating solvents and its tendency to crack prior to curing. Both of these unfavorable properties were attributed to the crystalline nature of the oligomer. We found that these characteristics were eliminated by replacement of the linear oligomer with a branched 1,3,5-triethynylbenzene unit (III) in the oxidative coupling reaction ([4-5]) (Scheme 2).

In this paper methods are described for preparing the 1,3,5-triethynylbenzene monomer and oligomers. Evidence is then presented to support the theory that these diacetylene oligomers tend to thermally cure by an aromatization reaction.

Since the cured polymer is highly crosslinked and therefore insoluble, most characterization techniques, including solution NMR, are not suitable for studying curing mechanisms in glassy polymers. However, one can obtain substantial information at the molecular level using ^{13}C solids NMR techniques. Briefly, the carbon resonance signal from a solid sample experiences a proton-carbon dipolar coupling on the order of 40 KHz ([6]), an anisotropic chemical shift distribution of ~1-2.5 Khz at carbon frequencies of 15 MHz ([7]), and very long carbon spin-lattice, T_1, relaxation times that can be as long as 10^3 seconds ([8]). The signal resulting from a normal pulse-acquisition experiment as with solution samples is very broad and featureless. However, the large dipolar decoupling can be eliminated by the use of dipolar decoupling, DD, ([9]); the chemical shift anisotropies can be reduced to their singular values by spinning the sample rapidly at an angle of 54.7° with respect to the magnetic field (the so-called magic angle spinning, MAS) ([10]); and the limitations of long carbon relaxation times can be avoided by using the technique of cross-polarization, CP ([11]). The resulting spectrum consists of peaks of reasonably narrow line width and chemical shifts which correspond to the isotropic values observed in solution. The solid state line widths and chemical shift differences observed relative to those observed in solution samples are a result of solid state effects due to structure, stereochemistry, and motional effects. The solid state carbon

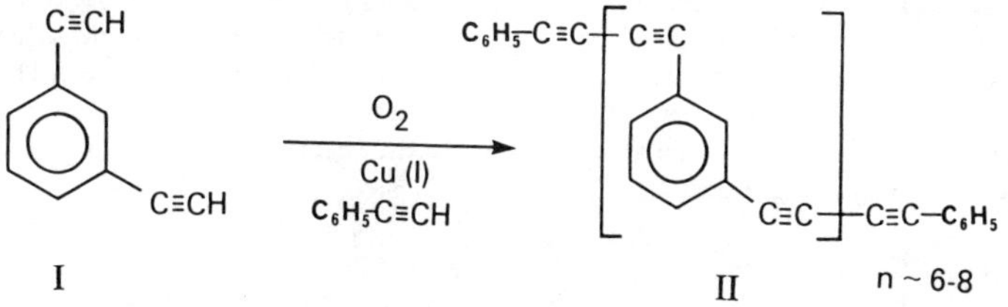

Scheme 1. Synthesis of poly(diethynylbenzene).

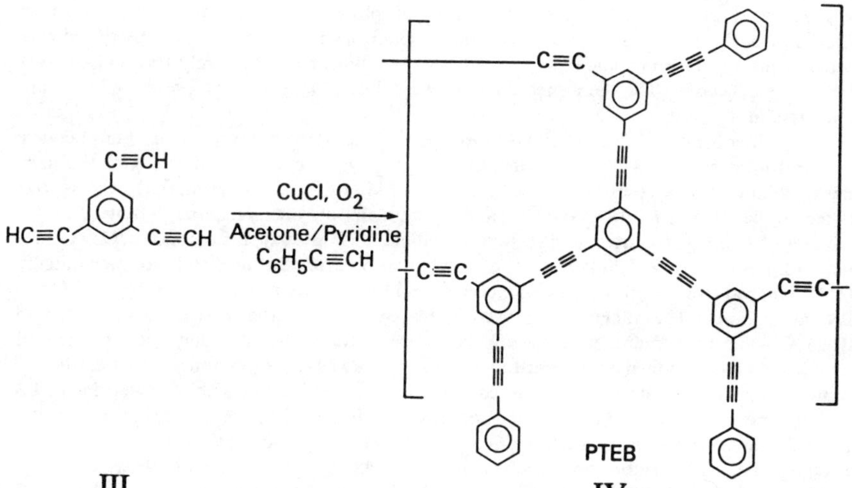

Scheme 2. Synthesis of poly(triethynylbenzene).

NMR experiment is generally termed the cross-polarization magic-angle spinning, CPMAS, experiment.

Solid state NMR has been used to study polymers of various classes over the past several years. In particular, the technique has been used to study curing reactions in epoxies (12), polyimides (1), and acetylenic terminated sulfones (13). The ability to observe the evolution of the carbons of the reacting species has been clearly shown to provide valuable information which has been difficult or impossible to obtain with other techniques. The use of ^{13}C solid state NMR techniques is essential for the understanding of curing reactions in high temperature polymers in order to be able to correlate the reaction chemistry with the structural and resulting physical properties.

Experimental

1,3,5-Triethynylbenzene (TEB, III). A 1-liter 3-necked flask equipped with heating mantle, overhead stirrer, thermocouple temperature sensor, condenser, and nitrogen/vacuum system was successively charged with 102.4 g (0.325 mol) of 1,3,5-tribromobenzene (VII); 45.11 g (0.536 mol, 0.55 equiv per Ar-Br) of 2-methyl-3-butyn-2-ol; 52.57 g (0.536 mol, 0.55 equiv) of 3-methyl-1-pentyn-3-ol; 1.83 g (6.98 mmol, 0.0072 equiv) of triphenylphosphine; 244 mg (1.22 mmol, 0.00125 equiv) of cupric acetate monohydrate; 108 mg (0.609 mmol, 0.000624 equiv) of palladium(II) chloride; and 400 mL (295 g, 2.92 mol, 3.0 equiv) of di-n-propylamine. The brown slurry was deoxygenated and then heated to reflux (113°C). After 3 h at reflux, TLC (silica gel, 1:1 ether/hexane) indicated that the reaction had gone to completion (R_f of the triol was 0.2, that of the nonexistent diols, 0.3-0.5). The dark-brown slurry was cooled to room temperature and vacuum filtered (C frit). The filter cake was washed carefully with two 160-mL portions of toluene. The combined filtrate was then filtered through a 160-g silica gel (40-140 mesh) column, which was subsequently rinsed with two 160-mL portions of toluene and then blown dry with nitrogen. Only the colored fractions were collected. The combined dark brown solution weighed 694 g and was used immediately in the next step.

A 2-liter 3-necked flask equipped with heating mantle, overhead stirrer, thermocouple temperature sensor, and Claisen distillation system was charged with 694 g of the dark brown solution prepared above. Potassium t-butoxide (8.0 g, 71.3 mmol, 0.073 equiv) was added to the stirring solution, which was then immediately deoxygenated, left at 10-in. Hg vacuum (20-in. Hg absolute), and heated to boiling. A skin temperature of 200°C was adequate to provide a slow distillation. The internal temperature during distillation was initially 91°C, dropped to 87°C over an 11-min period, and then slowly rose to 99°C after 1 h of total distillation time. The vacuum was then increased to lower the internal temperature to 80-85°C while the remainder of the solvents were distilled off. This operation was judged complete when the internal temperature rose to 95°C at 29-in. Hg vacuum. The mantle was immediately removed, and the molten reaction mixture was flooded with nitrogen. Next, 1.5 L of hexane was rapidly added to the distillation pot, followed by 8 mL of acetic acid in 100 mL of hexane. A column (5-cm × 36-cm) of 40-140 mesh silica gel was prepared in hot hexane. The crude reaction mixture was heated to 48°C and then quickly filtered through the column, followed by 1.5 L of hot hexane. The first 400 mL of eluant was discarded. TLC (silica gel, 10% ether/hexane) indicated that the TEB (R_f 0.75 with a small contaminant at R_f 0.55-0.7) was in the next 2.05 L of eluant. This fraction was concentrated to 430 mL and cooled overnight at 0°C. The crystallized TEB was isolated by filtration, washed with cold hexane, and dried under vacuum at ambient temperature for 5 h to yield 29.7 g of TEB as white needles, mp 101-102°C [lit. (16) mp 105-107°C]. The filtrate was concentrated to 100 mL, cooled, and filtered, and the crystals were washed and dried as above to give another 3.1 g of TEB, mp 100.5-101.5°C. The combined weight of 32.8 g represents a yield of 67%. ^1H NMR (CDCl$_3$) δ 3.10 (s, 3, C≡C-H), 7.55 (s, 3, Ar-H).

Poly(triethynylbenzene) (PTEB, IV), highly capped. A 5-liter, 4-necked flask equipped with overhead stirrer, internal thermocouple, addition funnel, condenser, and heating mantle was charged with 40.0 g (0.404 mol) of CuCl and 2.0 L of acetone. The agitated suspension was slowly heated to an internal temperature (T_i) of 30°C while the addition funnel was charged with a solution of 100 mL of acetone and 100 mL of pyridine; 25 mL of this solution was added to the flask. This dark green slurry was stirred for 10 min before a solution of 50.0 g (0.333 mol) of TEB, 340 g (3.33 mol, 10 equiv.) of phenylacetylene, and 25 mL of the acetone/pyridine mixture in 500 mL of acetone was added in one portion. The mantle was then removed. A small exotherm raised T_i to 32°C. After a further 10 min of stirring, oxygen bubbling through the reaction mixture was begun. The oxygen flow was continued throughout the rest of the reaction. The fritted-glass bubbler had a tendency to clog and was cleaned out twice. The remainder of the acetone/pyridine mixture was added over a period of 3.5 h, while T_i was kept below 45°C by ice-bath cooling as needed. After the acetone/pyridine addition was complete, a heating mantle was used to maintain T_i at 30-40°C for a further 2.5 h.

The reaction work-up was conducted under yellow light conditions in order to prevent UV-induced cross-linking. Isolation of the polymeric product began with the addition of the reaction mixture to a vigorously stirred solution of 330 mL of 12 N HCl in 10 L of methanol. The precipitated polymer was isolated by decantation and was washed with methanol (2 × 300 mL). The crude polymer was redissolved in 600 mL of chloroform, washed with 10% aq HCl (3 × 100 mL) and water (2 × 100 mL), dried (30 g $MgSO_4$), and filtered. The chloroform solution was slurried with 100 g of 60-200 mesh silica gel and filtered to give, after rinsing of the silica gel, 750 mL of a reddish-orange solution. Further filtration through a silica gel column had no apparent effect. The chloroform solution was vacuum stripped and the residue taken up in 200 mL of chloroform. This solution was precipitated from 3 L of methanol. Filtration of the solids, washing with two 300-mL portions of methanol, and vacuum drying (45-55°C) afforded 62.5 g of poly(triethynylbenzene) as a tan powder.

NMR Instrumentation. ^{13}C NMR spectra of uncured PTEB oligomer samples in solution were obtained using an IBM Instruments NR/80 spectrometer. Samples were dissolved in deuterated dichloroethane. Spectra were acquired quantitatively using heteronuclear scalar decoupling without nuclear Overhauser enhancement (NOE) while using a pulse width and pulse repetition rate that guaranteed full relaxation of carbon magnetization. The integrals of the resulting spectra could thus be used to determine the ethynyl to aromatic content for use in comparison with solid state NMR spectra. Heteronuclear coupled and j-modulated spin echo (14) experiments enabled the assignment of the peaks in the solution and, thus, the corresponding solid state spectra.

Solid state ^{13}C NMR experiments were carried out using a home-built spectrometer based upon a Varian Instruments DA-60 electromagnet, an IBM Instruments NR/80 NMR console, a Nicolet Instruments NIC-80 computer, a Diablo Model 44 disk drive, and home-built NMR probes. The console radio frequency circuitry was modified so that the proton and carbon irradiation frequencies were 60.03 and 15.01 MHz respectively. The probes used a magic-angle spinning apparatus which has been described previously (15). An external deuterium field frequency lock was used throughout the measurements. In order to adjust the angle setting to the magic angle accurately, the ^{79}Br quadrupolar resonance was observed and the spinning angle adjusted to optimize the resulting resonance.

Since the intensities of the resonances in the CPMAS experiment depend upon strengths of the dipolar couplings and molecular motions in the solid, it is not straightforward to obtain quantitative spectra. However, one may adjust the time during which magnetization is transferred from the proton to the carbon resonances, the CP time, and determine the time for which the solid state and solution integrals are equal. One can thereby obtain measurements of the aromatic to ethynyl ratio during the course of the cure.

Discussion

The original synthesis (16) used in our laboratories for the preparation of TEB is depicted in Scheme 3. 1,3,5-Triacetylbenzene (V) was treated with phosphorous pentachloride in carbon tetrachloride or heptane to give the tris(α-chlorovinyl) derivative **VI**, which, upon dehydrochlorination, afforded TEB in an overall yield of 34%.

Due to both the moderate yield and difficulties encountered in preparing large amounts of TEB by this procedure, a new route was developed based on the Hagihara (17) modification of the Stephens-Castro coupling (18). In this two-step procedure (Scheme 4), 1,3,5-tribromobenzene (**VII**) was coupled with a monoprotected acetylene using a Pd/Cu catalyst system. The protecting groups of the intermediate (**IX**) were then removed by a retro-Favorskii reaction (19).

This procedure included several novel modifications that enabled the reaction to be conducted rapidly, on a large scale, and in good yield. Use of the least expensive protected acetylene, 2-methyl-3-butyn-2-ol (**VIIIa**) (20), resulted in the highly polar and crystalline intermediate **IXa**, which had such poor solubility that high dilution was necessary both for the reactions and the intermediate work-up. Use of a 1:1 mixture of protected acetylenes **VIIIa** and **VIIIb**, as shown in Scheme 4, gave an intermediate **IX** that was a highly soluble four-component mixture. The original expensive catalyst, bis(triphenylphosphine)palladium dichloride, was replaced by the combination of $PdCl_2$ and extra triphenylphosphine with no loss of activity. When CuI was used as the copper co-catalyst, some of the palladium catalyst precipitated as palladium black, a form presumably ineffective in this reaction. Cupric acetate was found to function as well as cuprous iodide did as a co-catalyst and did not cause this palladium loss. Finally, di-n-propylamine was used as the reaction solvent instead of di- or triethylamine because of its higher boiling point and greater solvent power. Fortuitously, the hydrobromide salt of this amine, produced in large volume in the coupling reaction, formed a slurry which was much easier to stir than a slurry of triethylamine hydrobromide. The overall 67% yield of TEB represents an individual yield of 88% per ethynyl group.

Glaser oxidative coupling of TEB to give PTEB is conceptually simple but is difficult to control to give reproducible results. This difficulty is caused by the heterogeneity of the reaction mixture—solid CuCl, liquid solution of all reactants, and gaseous oxygen. Under these conditions, even parameters such as stirring rate become important. However, with starting materials and reagents of reasonable purity, the procedure reported in the experimental section is capable of good reproducibility. Control over both the molecular weight and degree of phenylacetylene capping of the PTEB can be achieved by adjustment of the reaction time and equivalents of phenylacetylene used.

As anticipated, PTEB exhibited no tendency to crystallize. The highly branched structure of this oligomer also rendered it quite soluble in a variety of solvents. PTEB could be spin-coated on silicon wafers, although an adhesion promoter proved to be necessary. The resultant films could be cured by heating to temperatures of 180°C or higher. The planarization properties of PTEB were excellent on a variety of surfaces. A surface roughness of less than 125Å was obtained as measured by reflected light interference microscopy. Thermogravimetric analysis (TGA) of cured PTEB (Figure 1) in helium indicated very high thermal stability up to 500°C and a weight loss of only 5% at 700°C. TGA in air revealed good oxidative resistance with a weight loss of 5% at 550°C.

Differential Scanning Calorimetry (DSC) proved to be a powerful tool for probing the nature of the PTEB curing process. Figure 2 depicts the effect of UV radiation on PTEB. Scan A is typical of PTEB and showed a small melting endotherm at 90°C, followed by a large sharp exotherm at 140° and a second, broader, exotherm centered at 200°C. The melt processing window of 40-50 C° resulted in excellent planarization behavior. Exposure to UV radiation caused sufficient reaction to eliminate the melting endotherm and greatly reduced the first sharp exothermic reaction (Scan B).

Figure 3 illustrates the effect of the degree of PTEB capping on the DSC. The highly uncapped polymer contained both ethynyl and diethynyl groups (Scan C). Both exothermic peaks occurred 20 C° lower than those of a fully-capped sample (Scan D). This data implies that the initial thermal curing reactions involve ethynyl groups, either with other ethynyl groups or with a diethynyl group. The shape of the DSC plots with the sharp initial peak at the leading edge of a broad one is suggestive of an artifact caused by the powerful cure exotherm. This question was addressed by an experiment in which the DSC sample was

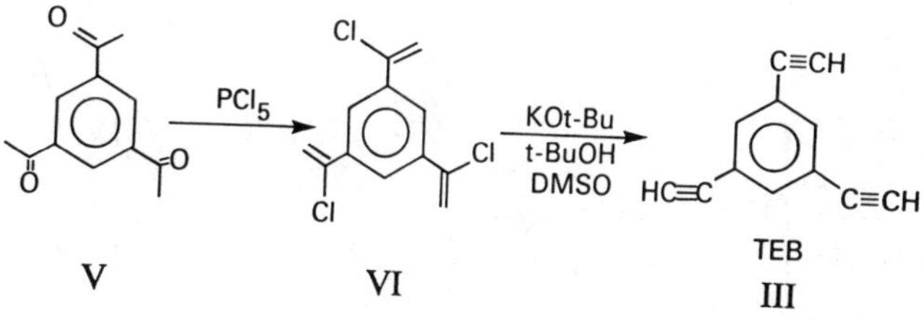

Scheme 3. Original synthesis of triethynylbenzene.

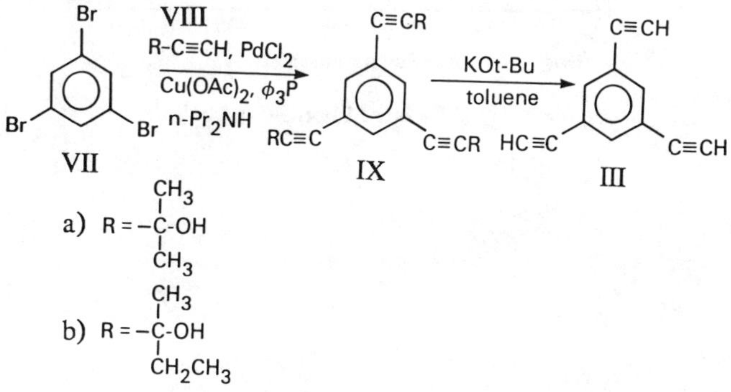

Scheme 4. Improved synthesis of triethynylbenzene.

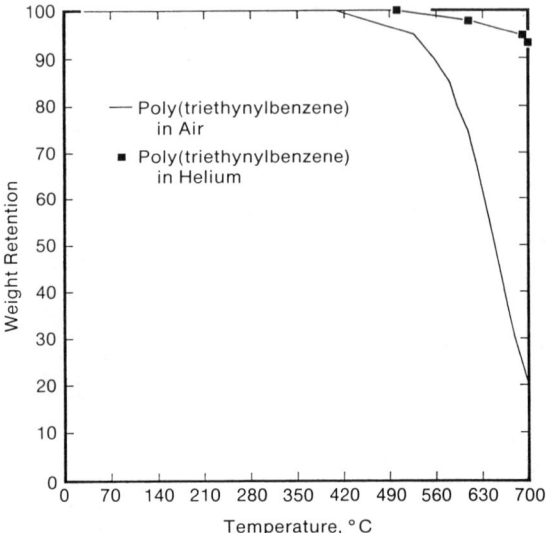

Figure 1. TGA of PTEB in air and helium.

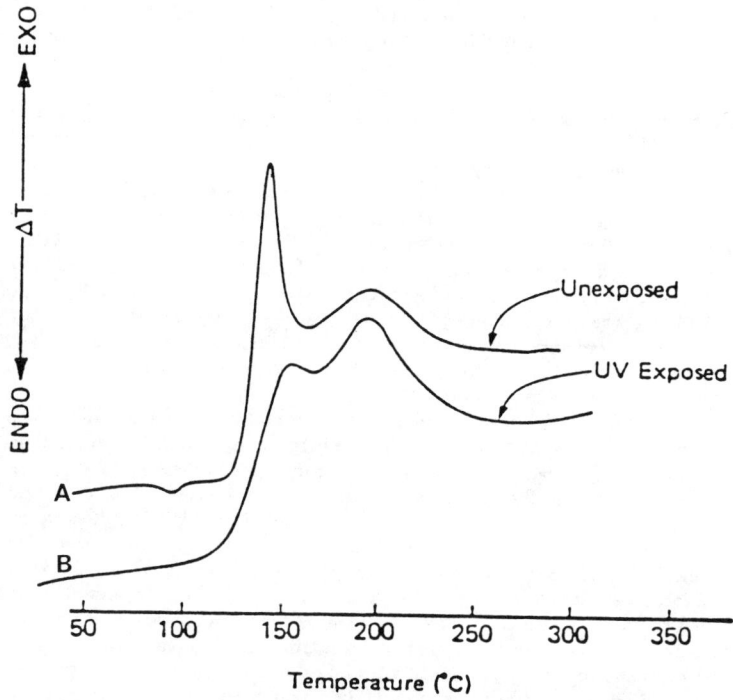

Figure 2. Effect of UV radiation of the DSC of PTEB.

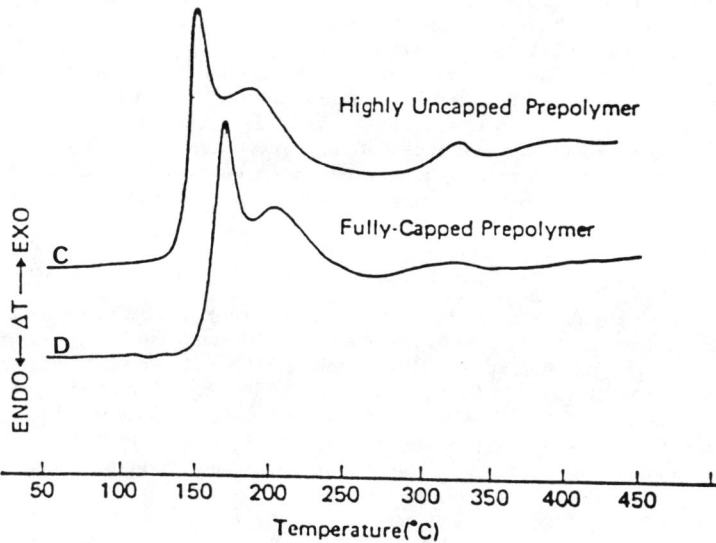

Figure 3. Effect of phenylacetylene capping on the DSC of PTEB.

quenched in liquid nitrogen as soon as the initial exotherm had started. If the first peak were an artifact, once it was blunted by this quenching treatment the second peak should increase in size. As the plots in Figure 4 indicate, this effect did not occur, thereby indicating that the two peaks were caused by two separate exothermic processes.

Infrared spectroscopy (Figure 5) was used to follow the thermal curing of a film of highly capped PTEB between two salt plates. The initial scan (RT) showed the ethynyl C-H stretch at 3300 cm^{-1} and the diethynyl C≡C stretch at 2220 cm^{-1}. After 2 hours at 150°C, the ethynyl absorption had decreased by 60%, whereas the diethynyl absorption exhibited only a 20% decrease. After an additional hour at 250°C, the ethynyl absorption was essentially gone whereas the diethynyl absorption retained 25% of its original strength. Another hour at 350°C caused the disappearance of both species. These findings offer further evidence that the initial reaction involved an ethynyl group. Quantitation using these IR absorptions is very difficult because the extinction coefficient of each group is affected by its substituents and the rigidity of its environment. Hence, the smaller initial decrease in the diethynyl absorption compared to that of the ethynyl absorption could represent a 1:1 reaction between these two groups to form some intermediate structure.

A high resolution ^{13}C NMR spectrum of the PTEB oligomer in 1,2 dichloroethane-d$_4$ is shown in Figure 6a along with structural assignments of the resonances. The spectrum was recorded quantitatively using broad-band scalar decoupling. The corresponding ^{13}C CPMAS NMR spectrum is shown in Figure 6b for the case where the cross polarization time was chosen so that the acetylenic to aromatic ratios of the solution and solid state spectra were equal. The wider line widths of the solid spectrum are primarily due to dispersion of the averaged chemical shift tensor resulting from stereochemical differences in the solid. However, one can still observe distinct resonances due to chemically different carbons. The CPMAS spectrum of a sample of PTEB cured at 215°C in air is shown in Figure 7. One can see that the acetylenic to aromatic ratio has dropped considerably. Correspondingly, the intensity of the broader aromatic region has increased with no other resonances being observed in the spectrum. By performing a CPMAS experiment in which decoupling is turned off for several microseconds prior to acquisition so that protonated carbons are allowed to preferentially relax, one can find an increase in the intensity of the nonprotonated carbons at around 132-135 and 140-145 ppm which is consistent with substituted carbons expected to be observed in aromatized cross-linked structures. These regions thus contain resonances typically associated with fused polycyclic aromatic hydrocarbons which suggests that a high degree of aromatization occurs in PTEB during curing.

The spectra of the samples cured at high temperature still show the presence of a small resonance in the acetylenic region which is probably unreacted diethynyl groups. The basis for this phenomena is clear from data on other crosslinked glasses. As the cure progresses the T_g of the system also increases and the ability of the remaining acetylenic groups to react drops. Eventually, the acetylenic groups remaining in the now highly crosslinked system are isolated from any sites with which they can react and will not undergo the cure reaction.

Conclusions

Several possible cure reactions for PTEB are shown in Scheme 5. Whether the reacting species are ethynyl-diethynyl or diethynyl-diethynyl, the same reactions are possible: [1,2]-addition, [1,4]-addition, and [2+4] cycloaddition. The [1,2]-addition process cannot be ruled out on spectral grounds, whereas the [1,4]-addition products (cumulenes or ene-ynes) are clearly absent from the ^{13}C-NMR spectra. Perhaps the most likely initial reaction is a [2+4] cycloaddition of an ethynyl group with a diethynyl group to give a benzene diradical (or benzyne) as shown in Scheme 5. This bimolecular aromatization process would explain both the spectral data and the highly exothermic nature of the initial cure reaction. These initial crosslinks would freeze the polymer structure until the temperature was raised above the new T_g, whereupon increased chain mobility would permit another diethynyl group to react with the intermediate structure, slowly building up polycyclic condensed aromatic sys-

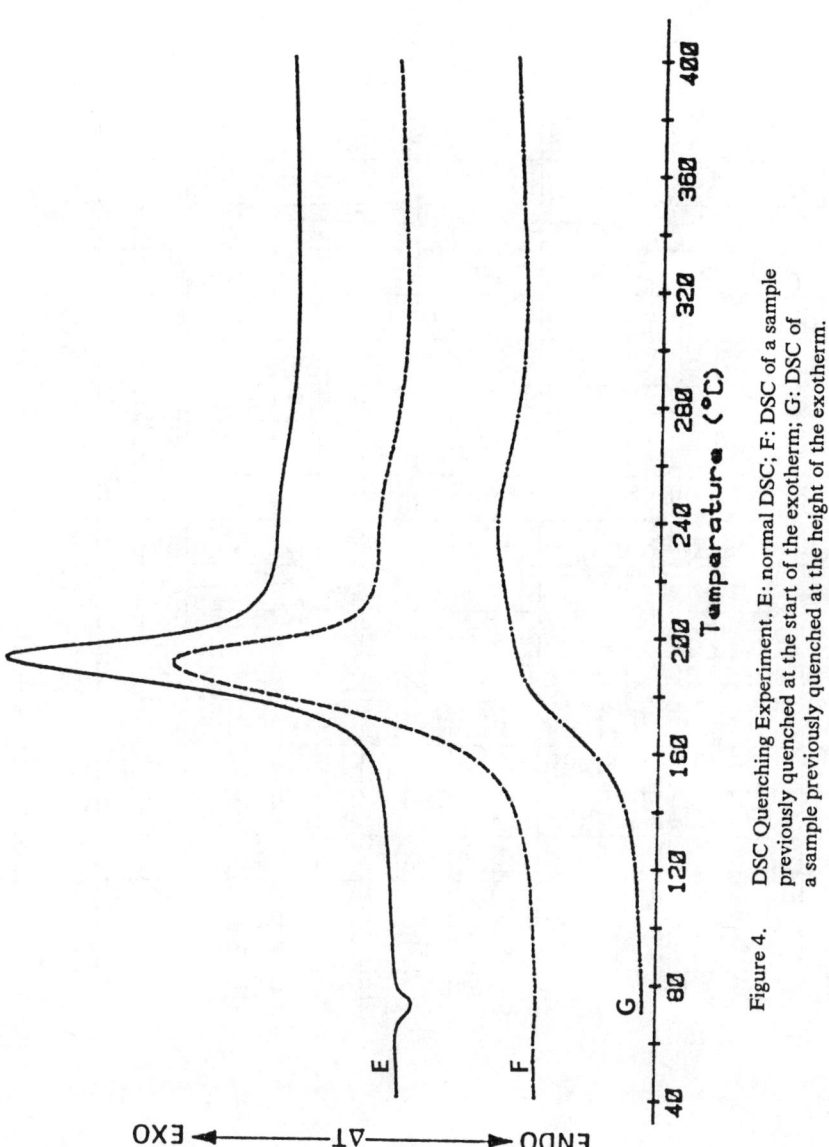

Figure 4. DSC Quenching Experiment. E: normal DSC; F: DSC of a sample previously quenched at the start of the exotherm; G: DSC of a sample previously quenched at the height of the exotherm.

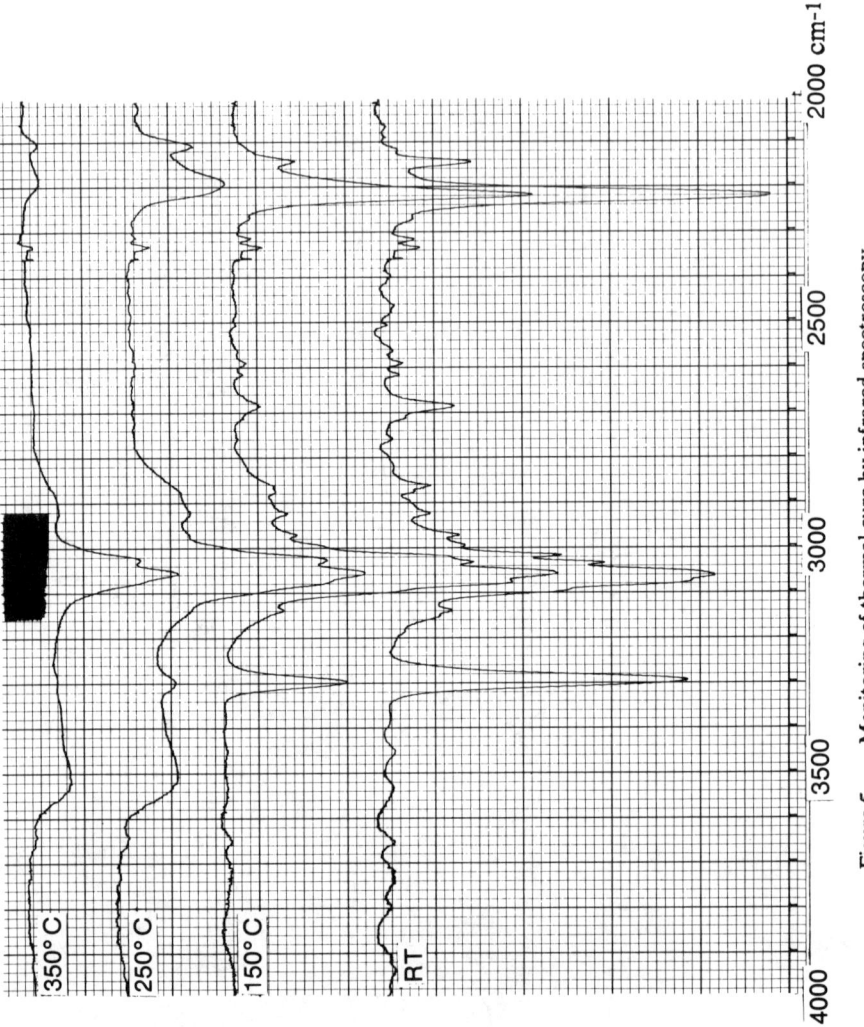

Figure 5. Monitoring of thermal cure by infrared spectroscopy.

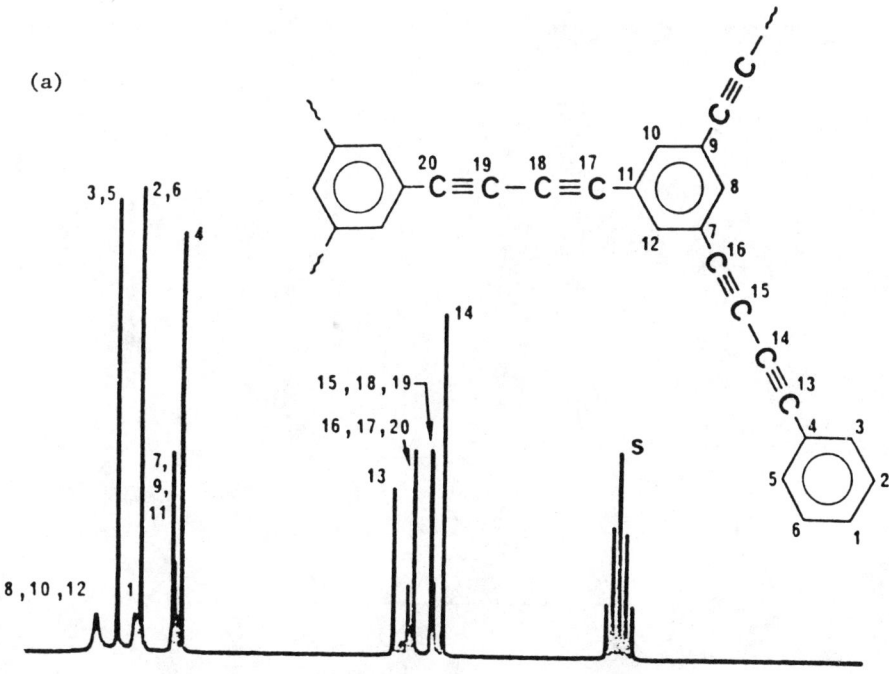
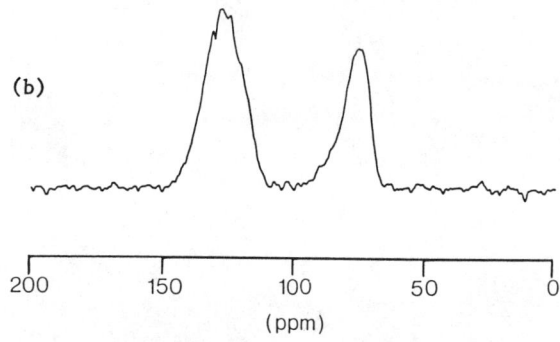

Figure 6. ^{13}C NMR of uncured PTEB. a) high resolution solution NMR in dichloromethane-d_4. The spectrum was run quantitatively with broad band heteronuclear decoupling. b) ^{13}C CPMAS NMR of uncured PTEB. Cross-polarization time was 4.5ms; the rotor speed was 2.3 kHz; the cross polarization field was 50 kHz. Chemical shifts are relative to TMS.

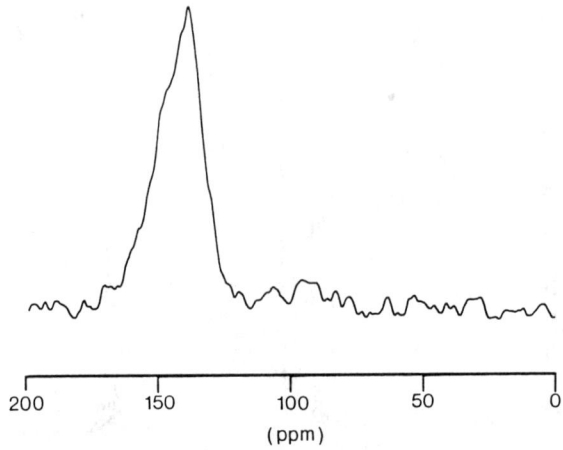

Figure 7. ^{13}C CPMAS NMR spectrum of PTEB cured in air at 215°C for 1 hour. Parameters are identical to those of the spectrum in Figure 6b.

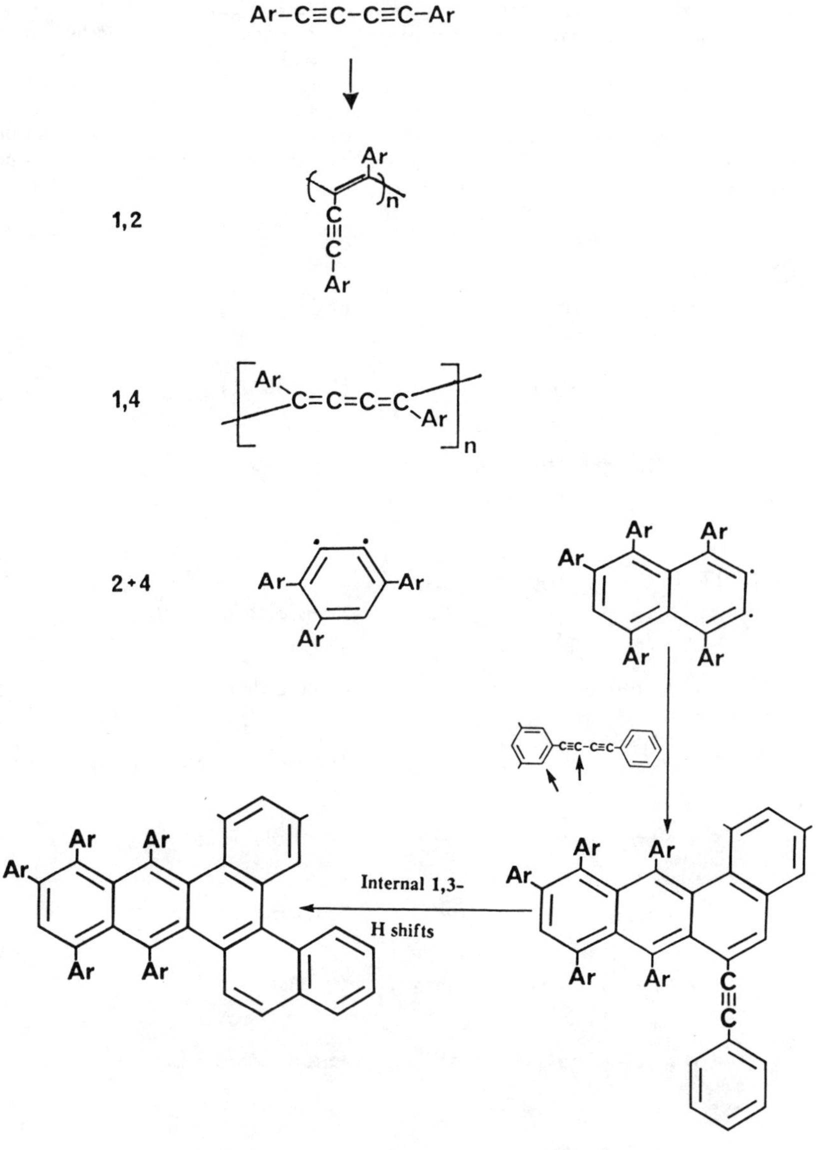

Scheme 5. Possible thermal curing products of PTEB.

tems. At higher temperatures, one might observe internal aromatization via phenyl or hydrogen shifts.

The properties of PTEB, both the oligomer and the cured film, are summarized in Figure 8. PTEB is soluble in most aromatic, ethereal, and halocarbon solvents. One micron coatings on aluminum substrates have reduced surface roughness from 3-4 µinches to <1 µinch. It photo-cures upon exposure to 280 to 330 nm radiation. The cured polymer has excellent thermal stability and a dielectric constant of only 2.7. PTEB's non-polar structure assures complete hydrophobicity, so it does not absorb and release water like polyimides do. As expected, the high crosslink density of the films causes the cured polymer to be brittle with a very low elongation at break. This single weakness is being addressed as we continue our research into this class of aromatic oligomers.

PTEB OLIGOMER

- Solubility   (Excellent)
- Planarization   (Excellent)
- Lithographic Sensitivity   (280-330 nm)

CURED PTEB

- Thermal Stability (He)   (>Polyimide)
- Adhesion   (Variable)
- High T_g (Follows Curing Temperature)
- Non-Porous   (Pinhole-free)
- Moisture Insensitive   (Pickup 10^{-3} X Polyimide)
- Low Dielectric Constant (2.7)
- Mechanical Strength   (Elongation <2%)

Figure 8.   Summary of physical properties of PTEB.

Acknowledgments

The authors express their appreciation of a number of co-workers who made significant contributions to the research reported above: P. J. Brock, M. A. Flandera, M. E. Oxsen, R. L. Siemens, and R. J. Twieg.

Literature Cited

1. Sefcik, M. D.; Stejskal, E. O.; McKay, R. A.; Schaefer, J. Macromolecules 1979, 12(3), 423.
2. Wegner, G. Z. Naturforsch. B 1969, 24(7), 824.
3. Hay, A. S. J. Pol. Sci. A-1 1969, 7, 1625.
4. Economy, J.; Flandera, M. A.; Liu, C.-Y. U.S. Patent 4 258 079, 1981.
5. Economy, J.; Flandera, M. A. U.S. Patent 4 273 906, 1983.
6. Waugh, J. S. Ann. N. Y. Acad. Sci. 1958, 70, 900.
7. Haeberlen, U. "Advances in Magnetic Resonance, Suppl. 1: High Resolution NMR in Solids. Selective Averaging"; Academic: New York, 1976; Chapter 3.
8. VanderHart, D. L. J. Magn. Reson. 1976, 24, 467.
9. Hartmann, S. R.; Hahn, E. L. Phys. Rev. 1962, 128, 2042.
10. Andrew, E. R. Progr. Nucl. Magn. Resonance. Spectrosc. 1971, 8, 1.
11. Pines, A.; Gibby, M. G.; Waugh, J. S. J. Chem. Phys. 1973, 59, 569.
12. Garroway, A. N.; Moniz, W.B.; Resing, H. A. ACS Symp. Ser. 1979, 103, 67.
13. Levy, R. L.; Lind, A. C.; Sandreczki, T. C. 15th Nat. SAMPE Tech. Conf. 1983, 21.
14. Le Cocq, C.; Lallemand, J.-Y. JCS Chem. Commun. 1981, 150.
15. Fyfe, C. A.; Mossbruger, H.; Yannoni, C. S. J. Magn. Reson. 1979, 36, 61.
16. Hübel, W.; Merenyl, R. Angew. Chem. Int. Ed. 1963, 2, 42.
17. Sonogashira, K.; Tohda, Y.; Hagihara, N. Tetrahedron Lett. 1975, 4467.
18. Castro, C. E.; Stephens, R. D. J. Org. Chem. 1963, 28, 2163, 3313.
19. Shchelkunov, A. V.; Muldakhmetov, Z. M.; Rakhimzhanova, N. A.; Favorskaya, T. A. Zh. Org. Khim. 1970, 6, 930.
20. Onopchenko, A.; Sabourin, E. T.; Selwitz, C. M. J. Org. Chem. 1979, 44, 1233.

RECEIVED March 5, 1985

7

Ethynyl End-Capped Polyimide Oligomers Containing Oxyethylene Linkages
Synthesis and Characterization

FRANK W. HARRIS and K. SRIDHAR

Institute of Polymer Science, The University of Akron, Akron, OH 44325

> Several ethynyl end-capped polyimide oligomers were prepared from 3,3',4,4'-benzophenonetetracarboxylic dianhydride, 1,2-bis(4-aminophenoxy)ethane (__1a__), bis[2-(4-aminophenoxy)]ethyl ether (__1b__), 1,2-bis[2-(4-aminophenoxy)ethoxy]ethane (__1c__), bis[2-(4-aminophenoxy)ethoxy]ethyl ether (__1d__), 1,1-bis(3-aminophenoxy)ethane (__2a__), bis[2-(3-aminophenoxy)ethyl]ether (__2b__) and 3-aminophenylacetylene. The oligomers prepared from the linear diamines __1a-d__ were semicrystalline and insoluble in all organic solvents. They also rapidly underwent crosslinking near 300°C. The DSC thermograms of the crosslinked samples showed a strong melting endotherm between 335 and 348°C. The oligomers prepared from the nonlinear diamines __2a__ and __b__ were amorphous and soluble in NMP, DMAC, __m__-cresol, and tetrachloroethane. Their Tg's ranged between 132 and 168°C. These oligomers underwent crosslinking at 300°C considerably slower than the crystalline systems. The Tg's of the crosslinked materials ranged between 147 and 245°C.

The overall goal of this continuing research effort has been the development of new polymeric systems for use as planarizing coatings in the microelectronics industry (__1-3__). In addition to low dielectric constants, these materials must display excellent thermal stability and high softening temperatures. They must also be quite soluble in organic solvents (20-40 wt %), in order to be applied by the preferred spin coating process. Our approach to such systems has involved the synthesis of soluble polyimide oligomers that are end-capped with functional groups that undergo thermally-initiated polymerizations. Thus, the materials can be applied from solution and then thermally crosslinked. Functional groups containing alkynyl moieties have been used so that the generated crosslinks are thermally stable. The systems have also been designed to undergo significant flow prior to the initiation of the cure process in order to aid the planarization process.

The major objective of this research was to investigate the use
of diamines containing oxyethylene linkages in the preparation of
oligomers of this type. Such diamines have been used in the preparation of soluble, high-molecular-weight polyimides that exhibited
glass transition temperatures (Tg's) as low as 140°C and good thermal
stability (4). The specific objectives of this work were: (a) to
prepare anhydride-terminated oligomers of 3,3',4,4'-benzophenonetetracarboxylic dianhydride (BTDA) and 1,2-bis(4-aminophenoxy)ethane
(1a), bis[2-(4-aminophenoxy)]ethyl ether (1b), 1,2-bis[2-(4-amino-
phenoxy)ethoxy]ethane (1c), and bis[2-(4-aminophenoxy)ethoxy]ethyl
ether (1d); (b) to end-cap the above oligomers with 3-aminophenyl-
acetylene (APA); (c) to characterize the oligomers with regard to
their solubilities, transition temperatures and effective curing temperature; and (d) to evaluate the thermal properties of the cured
resins.

Results and Discussion

Monomers. BTDA and APA were obtained from Gulf Chemicals. BTDA was
recrystallized from acetic anhydride prior to use, and APA was distilled under reduced pressure. Diamines 1a-d were prepared from
1-fluoro-4-nitrobenzene and the appropriate glycol according to the
known procedure (4).

As described in the following section, the initial oligomers prepared from 1a-d were semicrystalline and insoluble in organic solvents. Since it was postulated that these properties were associated
with the high degree of symmetry possessed by the diamines, new nonlinear monomers were sought that would afford soluble, amorphous
systems. Thus, 1,2-bis(3-aminophenoxy)ethane (2a) and bis[2-(3-amino-
phenoxy)ethyl]ether (2b) were prepared by treating the sodium salt of
m-aminophenol with 1,2-dibromoethane and 2-chloroethyl ether, respectively. The white diamines were purified by recrystallization from
ethanol.

$$H_2N-C_6H_4-OH \xrightarrow{NaOH, DMSO} H_2N-C_6H_4-ONa \xrightarrow{BrCH_2CH_2Br \text{ or } (ClCH_2CH_2)_2O}$$

$$H_2N-C_6H_4-O(CH_2CH_2O)_n-C_6H_4-NH_2$$

2a,b

Oligomers of BTDA and Linear Diamines. A series of ethynyl end-capped polyimide oligomers was synthesized by the following route.
First, anhydride-terminated, polyamic acid oligomers were prepared by
allowing excess BTDA to react with diamines 1a-d in DMAC at ambient
temperature. The molar ratios of BTDA to diamine employed were 2 to
1 and 4 to 3. Procedures were followed in which BTDA was added to

the diamine and vice versa. The terminal anhydride moieties were
then allowed to react with APA. The oligomers were dehydrated by
thermal and chemical methods. In the thermal method, after toluene
was added to the polyamic acid solutions, they were heated to near
100°C where the water of imidization was removed by azeotropic distillation. The polyamic acids were also dehydrated by treatment with

an equal-molar mixture of acetic anhydride and pyridine at ambient
temperature, and by stirring in refluxing acetic anhydride. The resulting polyimides, which precipitated as the reactions proceeded,
were insoluble in organic solvents at ambient temperature. They did
display very limited solubilities (1-2 wt %) in refluxing m-cresol
and NMP. Their infrared spectra contained strong absorption bands at
1780 and 720 cm^{-1}, characteristic of imides. The X-ray diffraction
patterns of 3b and c were characteristic of highly-crystalline solids.

The DSC thermograms of the oligomers showed barely perceptible
base line shifts between 140 and 175°C followed by broad exotherms
with maxima near 280°C (Table I, Figure 1). Oligomer 3a did not
display a transition prior to the onset of its exotherm. TGA thermograms of oligomers 3a and b obtained in air and nitrogen with heating
rates of 10°/min showed 5% weight losses near 410 and 470°C, respectively. Oligomers 3c and d displayed 5% weight losses near 365°C in
air and near 460°C in nitrogen (Figures 2 & 3). Although they did
darken slightly near 300°C, none of the oligomers showed any tendency
to flow when they were heated to 320°C on a Fisher-Johns melting
point apparatus.

In order to study their cure behavior, samples of each oligomer
were heated for 10, 20, and 30 min at 300°C under nitrogen. The DSC
thermograms of samples heated for 10 min did not contain curing exotherms, and were essentially identical to those of samples heated for
30 min (Figure 1). In fact, the thermogram of a sample of 3b that

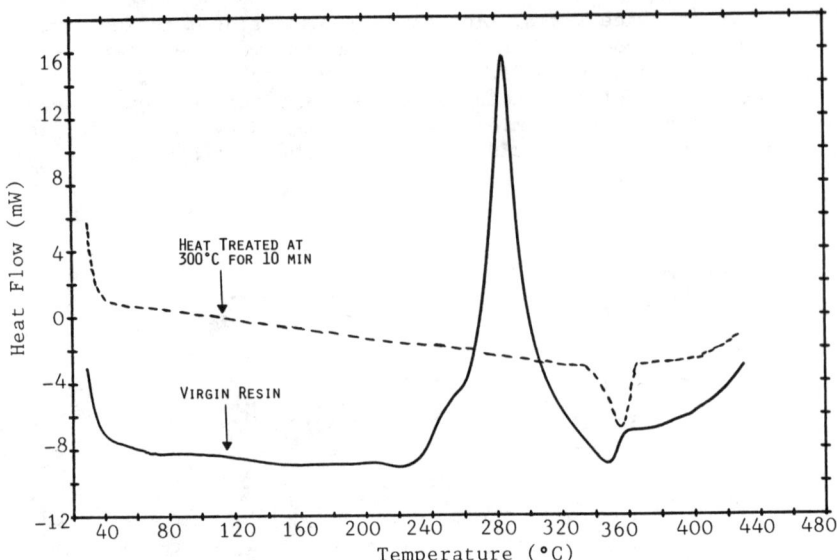

Figure 1. DSC thermograms of oligomer 3c obtained in nitrogen with a heating rate of 20°C/min.

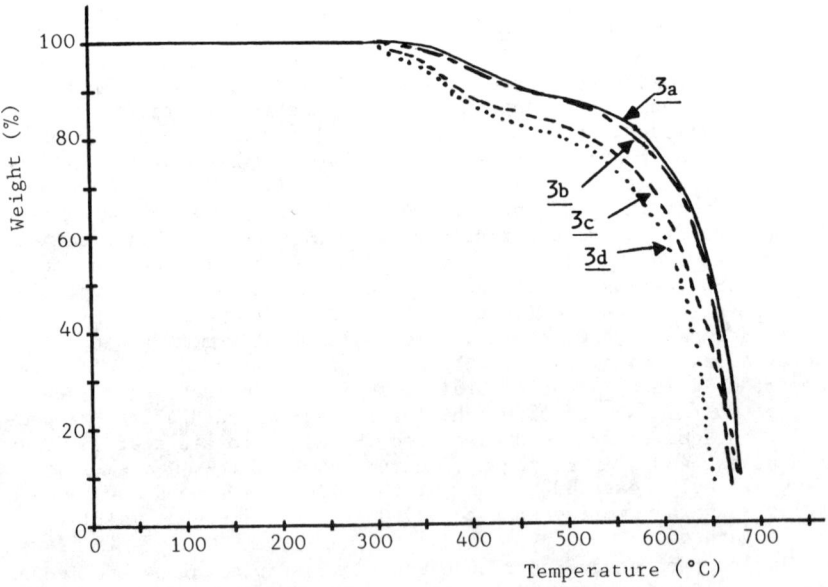

Figure 2. TGA thermograms of oligomers 3a-d obtained in air with a heating rate of 10°C/min.

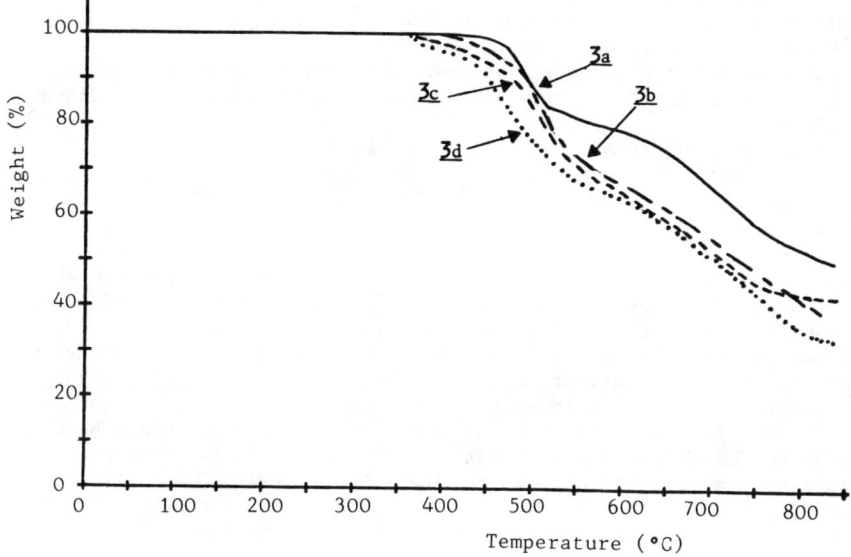

Figure 3. TGA thermograms of oligomers 3a-d obtained in nitrogen with a heating rate of 10°C/min.

Table I. Thermal Properties of Oligomers 3a-d [a]

Oligomer	T_g [b]	T_c (Onset) [c]	T_c (Max) [d]	TGA (Air) [e]	TGA (N_2)
3a	–	185	272	410	470
3b	172	235	277	410	465
3c1	167	240	280	365	460
3c2	170	235	308	360	410
3d	145	225	280	365	430

a. All of the oligomers except 3c2 were prepared from a 2:1 molar mixture of BTDA to diamine. Oligomer 3c2 was prepared from a 4:3 molar mixture. The polyamic acid intermediates were imidized by stirring in refluxing acetic anhydride.
b. Temperature at which a slight base line shift occurred in DSC thermogram (heating rate = 20°C/min). All temperatures reported are in °C.
c. Onset temperature of curing exotherm in DSC thermogram.
d. Maximum of curing exotherm in DSC thermogram.
e. Temperature at which a 5% weight loss occurred on TGA thermogram (heating rate = 10°C/min).

had been quickly heated (20°C/min) to 300°C in the DSC cell and then allowed to cool only contained a very weak exotherm. Thus, the ethynyl groups must have undergone unusually fast thermal polymerizations. This behavior is in marked contrast to that of previously prepared ethynyl-terminated oligomers, which have required heating cycles of several hours to complete the curing process (3). The thermograms also did not show any base line shifts that could be attributed to Tg's. However, with the exception of 3a, they did contain strong melting endotherms between 335 and 350°C (Table II).

Table II. Curing Study of Oligomers 3a-d

Oligomer	T_m [a]	T_m [b]	T_m [c]
3b	345	348	348
3c1	340	343	342
3c2	362	359	–
3d	337	335	335

a. Minimum of melting endotherm on DSC thermogram. Sample was heated for 10 min at 300°C under N_2 prior to determination.
b. Sample was heated for 20 min at 300°C under N_2 prior to determination.
c. Sample was heated for 30 min at 300°C under N_2 prior to determination.

Surprisingly, the X-ray diffraction pattern of a cured sample of 3c was considerably sharper than that of an uncured sample. All of these results suggest that the ethynyl groups were closely aligned in the oligomers' unit cells and underwent polymerization prior to the crystals melting. In the case of 3a, thermal crosslinking must have resulted in a material with a Tm higher than its decomposition temperature.

Oligomers of BTDA and Non-Linear Diamines. In an attempt to obtain
soluble, amorphous oligomers, BTDA was treated with various amounts
of diamines 2a and b. The reactions were carried out as described
for oligomers 3a-d with NMP as solvent. The molar ratios of BTDA to
diamine used were 2:1, 4:3, 8:7, 10:9, 14:13, 20:19 and 25:24
(Table III). After the addition of APA, the polyamic acids were

Table III. Thermal Properties of Oligomers 4a,b[a]

Oligomer	Molar Ratio of BTDA to Diamine	Tg	Tc (Onset)	Tc (Max)	Tg[b]	TGA (N_2)
4a	4:3	168	270	308	218	465
4b1	2:1	132	240	272	245	457
4b2	4:3	135	227	280	188	448
4b3	8:7	137	235	307	168	446
4b4	10:9	143	250	315	163	452
4b5	14:13	-	255	318	147	445
4b6	20:19	143	255	323	162	441
4b7	25:24	146	265	330	157	436

a. Polyamic acid intermediates were imidized by stirring in refluxing NMP for 2 h under N_2. For explanation of column headings see Table I.
b. Tg of cured sample. Sample was heated for 20 min at 325°C under N_2 prior to determination.

thermally imidized by heating the reaction mixture to reflux and then
slowly distilling off the evolved water with solvent. Fresh solvent
was continuously added so as to maintain a constant oligomer concentration of 20 to 25 wt %. If the distillation-addition cycle were
carried out for more than 3 h, the oligomeric products formed semi-
solid, gel-like masses upon cooling. However, if the heating was
stopped after 2 h, the products remained completely in solution. It
is likely that this difference in behavior was due to different degrees of imidization. Although infrared analysis indicated that the
imidization process was over 90% complete in both cases, no quantitative correlations were made. The NMP-soluble oligomers, which were
isolated by precipitation in ethanol, were also soluble in DMF, DMAC
and tetrachloroethane. The higher molecular weight materials (4b6,
4b7) could be cast into tough, flexible films from NMP solutions. An
X-ray diffraction pattern of oligomer 4b1 did not show any indication
of crystallinity in the sample.

The DCS thermograms of the oligomers contained base line shifts
between 132 and 146°C, followed by strong curing exotherms with
maxima between 258 and 330°C. As seen in Table III, the oligomers'
Tg's and curing maxima increased as the molecular weight increased.
Visual observations on the Fisher-Johns apparatus revealed that the
resins underwent considerable flow between 200 and 220°C. In fact,
yellow transparent films could be compression molded from the neat
resins at 220°C. The TGA thermograms obtained in nitrogen with heating rates of 10°C/min showed 5% weight losses between 436 and 465°C.
The systems' thermal stabilities decreased as their molecular weights
increased. This was most likely due to the decrease in the number of
crosslinks generated as the distance between crosslinking sites
increased.

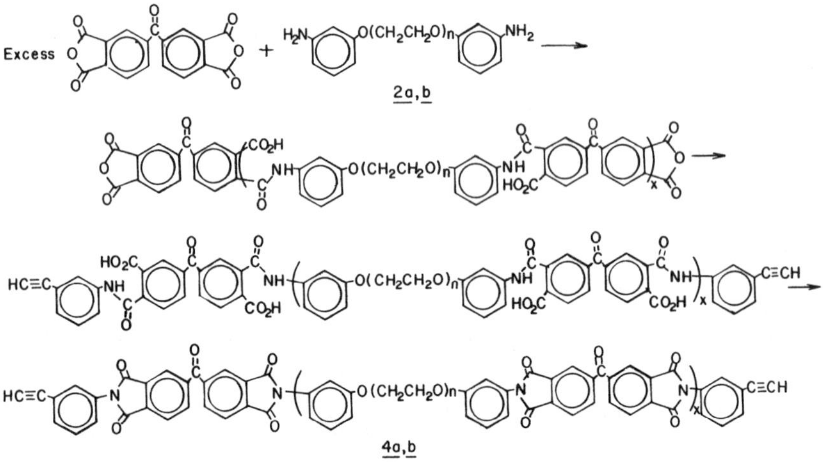

The amorphous oligomers underwent thermal crosslinking considerably slower than their crystalline counterparts. Although heating at 325°C for 20 min under nitrogen produced significant increases in their Tg's, their DSC thermograms still contained curing exotherms indicating that the curing process was not complete (Table III). The thermal treatment produced smaller increases in the oligomers' Tg's as their molecular weight increased. This can also be attributed to a decrease in crosslink density and to a decrease in the curing rate as the concentration of sites decreased.

Conclusions

Ethynyl end-capped polyimide oligomers can be prepared from BTDA and the linear diamines 1a-d that are highly crystalline and insoluble in organic solvents. The oligomers undergo rapid crosslinking in the crystalline state near 280°C. The crosslinked resins are also highly crystalline and undergo melting transitions near 340°C. Isomers of the oligomers can be prepared from BTDA and the non-linear diamines 2a,b that are totally amorphous and soluble in several organic solvents. These oligomers will also soften and flow at temperatures below those needed to affect their thermal cure. The dramatic differences in oligomer morphology must be associated with the catenation of the amine component in the repeat unit. All of the highly-crystalline materials are para catenated, while the amorphous systems contain meta catenation.

Experimental

IR spectra were obtained with a Perkin-Elmer Model 1330 grating spectrophotometer. DSC thermograms were obtained with a DuPont 900

Thermal Analyzer equipped with a differential calorimetric cell. TGA thermograms were obtained on a DuPont 1090 Thermal Analyzer. Elemental analyses were performed by Galbraith Laboratories, Knoxville, TN.

1,2-Bis(3-aminophenoxyethane)(2a). To a 250-mL, 3-necked flask equipped with a Dean-Stark trap, a condenser, a N_2 inlet, a thermometer and a magnetic stirring bar were added 25.0 g (0.23 mol) of m-aminophenol, 50 mL of DMSO and 50 mL of toluene. After the solution was purged with N_2, 9.16 g (0.23 mol) of 50.2% NaOH was added. The mixture was heated with stirring to 110°C. The water that evolved from the reaction mixture was removed by azeotropic distillation. After most of the water had been removed, an additional 20 mL of toluene was added. The remainder of the water was then removed by increasing the temperature to 120°C. The reaction mixture was cooled to 50°C, and 21.5 g (0.11 mol) of 1,2-dibromoethane was slowly added over a period of 0.5 h. The ensuing exothermic reaction, which began immediately upon the addition of 1,2-dibromoethane, was accompanied by considerable salt formation. After this reaction had subsided, the mixture was heated at 60°C for 2 h. The solution was then cooled to 50°C and filtered to remove NaBr. The filtrate was poured into ice cold water to precipitate 15.3 g (57%) of an off-white solid that was recrystallized from methanol to yield white crystals: mp 133-134°C; IR (KBr) 3380 and 3300 cm^{-1}(NH$_2$). Anal. calcd. for $C_{14}H_{16}N_2O_2$: C, 68.85; H, 6.55; N, 11.47. Found: C, 68.80; H, 6.89; N, 11.34.

2,2'-Bis(3-aminophenoxy)ethyl ether (2b). A solution of 25.0 g of the sodium salt of m-aminophenol in DMSO, which was prepared by the method described in the procedure for 3a, was heated to 110°C. 2-Chloroethyl ether (16.37, 0.11 mol) was then slowly added over a period of 15 min. After the reaction mixture was heated at 110-130°C for an additional 2 h, it was filtered to remove the NaCl. The filtrate was cooled and poured into ice water containing 1% Na_2SO_3 and 2% NaOH. The white gummy mass that separated was triturated with hexane and then with water to afford 22.17g (70%) of white powder. The crude product was recrystallized from ethanol containing decolorizing carbon to yield white crystals: mp 98-100°C; IR (KBr) 3410 and 3320 cm^{-1}(NH$_2$). Anal. calcd, for $C_{16}H_{20}N_2O_3$: C, 66.66; H, 6.94; N, 9.72. Found: C, 66.60; H, 6.89; N, 9.80.

General Procedure for the Preparation of Polyamic Acid Oligomers. The diamine (0.009 mol) was dissolved in 35 mL of DMAC contained in a 100-mL, 3-necked flask equipped with a thermometer, a magnetic stirring bar, a N_2 inlet, and a $CaCl_2$ drying tube. BTDA (0.018 mol) was then added in one portion. The temperature of the mixture quickly rose to 35 to 38°C and then slowly returned to ambient (26-28°C). After the mixture was stirred for 3 h under N_2, 0.018 mol of APA was added and the stirring continued for an additional 3 h. The polyamic acid solution was stored at 0°C.

General Inidization Procedures. Procedure 1. Acetic anhydride (25 mL) was heated to reflux in a 3-necked flask, fitted with a condenser, a N_2 inlet, and a dropping funnel. The polyamic acid solution (10 mL) was then added dropwise over a period of 0.5 h. After the addition was complete, the mixture was heated at reflux for an

additional 2 h and then poured into ethanol. The precipitate that
formed was collected by filtration and washed several times with
water to remove acetic anhydride. The product was dried overnight
under reduced pressure at 55°C. Procedure 2. To 20 mL of an equimolar mixture of acetic anhydride and pyridine contained in a 100-mL
flask fitted with a N_2 inlet tube was added 5 mL of the polyamic acid
solution. After the mixture was stirred at ambient temperature for
24 h, it was poured into 200 mL of water. The dark yellow oligomer
was collected by filtration, washed several times with water, and
then dried under reduced pressure. Procedure 3. The polyamic acid
solution was heated to 100°C in a 3-necked flask equipped with a condenser, a magnetic stirring bar, a thermometer, and a N_2. After 5 mL
of toluene was added, the toluene-water azeotrope, which immediately
began to reflux, was removed by distillation. The toluene addition-
distillation cycle was repeated 5 to 7 times over a period of 4 h.
The imidized product was isolated by precipitation in ethanol.

General Procedure for the Preparation of Oligomers 4. BTDA (0.25 mol)
was dissolved in NMP (20% solids) contained in a dry 100-mL, 3-necked
flask equipped with a thermometer, a N_2 inlet and a short-path distillation apparatus. After the appropriate amount of amine was added
in a single portion the solution was stirred at room temperature for
2 h. The appropriate amount of APA was added, and the reaction mixture was stirred for an additional 2 h. The temperature of the flask
was then increased until distillation commenced. The oligomer concentration was maintained at approximately 20% by continually replac-
ing the distillate with fresh NMP. The distillation-addition cycle
was carried out for 2 h. The mixture was then poured into ethanol.
The precipitate that formed was collected by filtration, washed several times with ethanol, and dried under reduced pressure.

Acknowledgments

The support of this research by IBM Research Laboratory, San Jose,
California, is gratefully acknowledged.

Literature Cited

1. Harris, F.W.; Pamidimukkala, A.; Gupta, R.K.; Das, S.; Wu, T.;
   Mock, G. ACS Polym. Div., Polym. Preprints 1983, 24(2), 324.
2. Harris, F.W.; Sridhar, K.; Das, S. ACS Polym. Div., Polym. Preprints 1984, 25(1), 110.
3. Harris, F.W.; Pamidimukkala, A.; Gupta, R.K.; Das, S.; Wu, T.;
   Mock, G. J. Macromol. Sci.-Chem. 1984, A21, 1117.
4. Feld, W.A.; Ramalingam, B.; Harris, F.W. J. Polym. Sci., Polym.
   Chem. Ed. 1983, 21, 319.

RECEIVED March 5, 1985

8

Polyaromatics with Terminal or Pendant Styrene Groups
A New Class of Thermally Reactive Oligomers

VIRGIL PERCEC and BRIAN C. AUMAN

Department of Macromolecular Science, Case Western Reserve University, Cleveland, OH 44106

> Synthetic methods for the preparation of α,ω-bis(vinyl-
> benzyl) polyaromatic oligomers, and polyaromatics with
> pendant substituted styrene-type vinyl groups are pre-
> sented. In the first case, the synthesis involves a
> quantitative etherification of an α,ω-bis(hydroxyphenyl)
> oligomer with chloromethylstyrene in the presence of
> tetrabutylammonium hydrogen sulfate as phase transfer
> catalyst. In the second case, the method entails a
> phase transfer catalyzed Wittig vinylation of polyaro-
> matics containing pendant phosphonium salt groups from
> precursor chloromethylated or bromomethylated polymers.
> Examples are presented for α,ω-bis(hydroxyphenyl) aro-
> matic polyether sulfone and poly(2,6-dimethyl-1,4-
> phenylene oxide). Both terminally and pendantly func-
> tionalized oligomers show very high thermal curing
> reactivity and can be used to tailor the properties of
> the final networks.

Functional polyaromatics containing either pendant or terminal groups which undergo thermally initiated crosslinking reactions without releasing volatile by-products, have received much attention recently. Presently, there are only a few reactive groups which can be used to provide crosslinking reactions without the evolution of volatile by-products. Nadimide (1), acetylenic (2), biphenylenic (3), and male-imide (4) are among the functional groups most frequently employed for the preparation of thermally crosslinkable polymers. All these functional groups give rise to networks through a very slow and complicated polymerization mechanism. The unelucidated mechanism of polymerization(except for maleimide functional groups) provides polymer networks whose microstructure is strongly dependent on the thermal history of the polymerization reaction.

The most successful class of thermally reactive oligomers consists of the functional polymers containing either pendant or terminal triple bonds, especially ethynyl groups. This area of research was recently reviewed (2), and the more recent developments in the field of α,ω-bis(ethynylphenyl) aromatic polyether sulfones were summarized in two recent papers (5,6).

0097-6156/85/0282-0091$06.00/0
© 1985 American Chemical Society

Although intensive research was done on the polymerization mechanism of the reactive oligomers containing ethynyl groups, it still seems unclear (5). The only clear information about the thermal polymerization of phenylacetylene, a model for the chain ends of these oligomers, is as follows. Thermal polymerization of phenylacetylene gives rise to a mixture of poly(phenylacetylene)oligomers, triphenylbenzene, and phenyl naphthalene derivatives (7). There is no available mechanism which can account for the formation of the last two series of products through a thermal polymerization reaction. In a recent series of publications, we have advanced the idea that triphenylbenzene derivatives are obtained through the intramolecular cyclization of a cis-polyphenylacetylene, while phenyl napthalene derivatives are obtained in the same manner from a trans-polyphenylacetylene (8-10). This mechanism would lead to crosslinked polymers whose microstructure would depend strongly on their thermal history.

In order to tailor the physical properties of the polyaromatic networks obtained by thermal curing, it is important to ascertain the relationship between structure and physical properties for both the starting oligomer and its resultant network. We therefore sought a reactive group whose mechanism of thermal initiation, kinetics, and thermodynamics of polymerization are known.

This is one of the reasons we decided to prepare oligomers containing styrene-type functional groups. Styrene's thermal initiation mechanism is fairly well understood, and the same is true for the kinetics and thermodynamics of its radical polymerization. In addition, thermal and radical polymerization of styrene is much faster than any of the other previous classes of reactive groups; and at the same time, the microstructure of the crosslinking points is known.

The literature on thermally reactive oligomers contains only one example of a polymer containing styrene chain ends, i.e., p,p'-divinylbenzyl end-capped phenylquinoxaline oligomers (11).

This paper reviews some of our work on general methods for the synthesis of polyaromatics containing either terminal or pendant styrene groups and their thermal polymerization. The examples provided in this paper refer to an aromatic polyether sulfone (PSU) and poly-(2,6-dimethyl-1,4-phenylene oxide) (PPO).

Experimental

Materials. A series of α,ω-bis(hydroxyphenyl)PSU oligomers with different molecular weights were synthesized and characterized as was previously reported (12). Two samples of PPO (one from Aldrich and one from General Electric Co.) were both purified by precipitation from chloroform solution into methanol. A commercial sample (Dow Chemical) of an isomeric mixture of chloromethylstyrenes (ClMS, 40% para, 60% meta) was used as received. 1-Chloromethoxy-4-chlorobutane (CMCB) was prepared according to a procedure developed by Olah et al. (14) and modified by Daly et al. (15), i.e., from paraformaldehyde, tetrahydrofuran and anhydrous HCl.

Synthesis of α,ω-bis(vinylbenzyl)PSU. The synthesis of α,ω-bis-(vinylbenzyl)PSU is outlined in Figure 1.

Procedure A. To a stirring solution of 6g (0.0047 moles-OH) α,ω-bis(hydroxyphenyl)PSU ($\overline{M}n$=2,550) in 30 ml CH_2Cl_2, 4 ml of a 50%

aqueous solution of NaOH were added at room temperature. The sodium salt of the α,ω-bis(hydroxyphenyl)PSU precipitated immediately. After the addition of 1.6g (0.0047 moles) tetrabutylammonium hydrogen sulfate (TBAH), the reaction mixture became homogeneous once more. The dropwise addition of 1.0 ml (0.007 moles) chloromethylstyrene (ClMS, 40%p, 60%m) created a dark blue colored reaction mixture. After 30 min. of stirring at room temperature, the reaction mixture turned to a light green and shortly thereafter to yellow. ^{1}H-NMR analysis proved complete reaction at this point. In order to avoid incomplete reaction, in all cases the reaction was continued for one more hour. The reaction mixture was then diluted with 20 ml CH_2Cl_2 and the organic phase was separated. It was washed with water and precipitated in methanol acidified with a few drops of dilute HCl.

Procedure B. The same procedure as above except that the reactants were mixed in the following order: CH_2Cl_2 solution of α,ω-bis-(hydroxyphenyl)PSU, TBAH, ClMS, and NaOH 50% water solution. In this way the reaction mixture was always homogeneous and the bisphenolate of the α,ω-bis(hydroxyphenyl)PSU was not in contact with CH_2Cl_2 in the absence of ClMS.

Procedure C. The same procedure as B except that chlorobenzene (ClBz) was used as solvent instead of CH_2Cl_2.

Synthesis of PSU and PPO containing pendant vinyl groups. The syntheses of both PPO and PSU containing pendant vinyl groups are outlined in Figures 2 and 3. Experimental details were published elsewhere ([13]).

Techniques. 200 MHz ^{1}H-NMR spectra were recorded on a Varian XL-200 spectrometer ($CDCl_3$ or CCl_4 solutions and TMS internal standard). A Digilab FTIR spectrometer was used to record IR spectra of polymer films on KBr plates or powders in KBr pellets. DSC curves and glass transition temperatures (Tg) were determined by a Perkin-Elmer Differential Scanning Calorimeter (Model DSC-2). Heating and cooling rates were 10°C/min. The Tg values were recorded during the second heating cycle, except as noted. GPC analyses were carried out with RI and UV detectors using μ-Styragel columns of 10^5, 10^4, 10^3Å and a calibration plot constructed with polystyrene standards.

Results and Discussion

Synthesis and Characterization of α,ω-bis(vinylbenzyl)PSU. In a series of previous papers we have demonstrated that Williamson phase transfer catalyzed etherification of α-(hydroxyphenyl) oligomers with monomers containing electrophilic groups or α,ω-bis(electrophilic) oligomers can be preformed quantitatively, providing convenient methods for the preparation of macromonomers ([16,17]) and ABA triblock copolymers ([18]), respectively. At the same time, the phase transfer tatalyzed polyetherification of α,ω-bis(hydroxyphenyl) oligomers with monomers or oligomers containing electrophilic groups in α,ω-positions led to a new avenue for the synthesis of regular copolymers and alternating block copolymers ([12,20,21]), respectively.

The major difference between our reaction conditions and the conventional phase transfer catalyzed Williamson etherification ([22]) is the use of stoichiometric amounts of phase transfer catalyst versus the nucleophilic chain ends in the former case. Under these where X=O for ClBz as solvent, X=O or $(OCH_2O\text{-}PSU)_m O$- for CH_2Cl_2

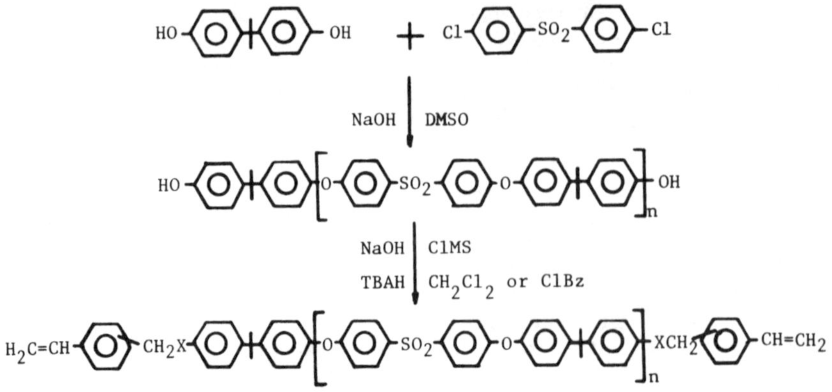

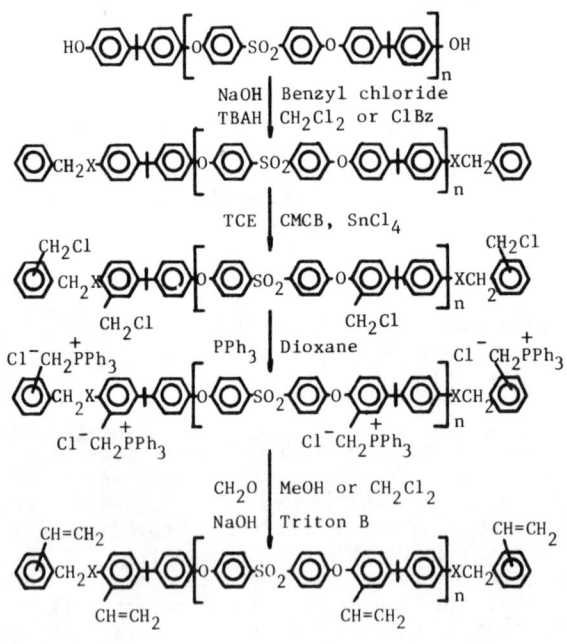

Figure 1. Synthetic routes used for the preparation of α,ω-bis-(vinylbenzyl)PSU.

where X=O for ClBz, X=O or $(OCH_2O\text{-}PSU)_m O$- for CH_2Cl_2

Figure 2. Synthetic route used for the preparation of PSU with pendant vinyl groups.

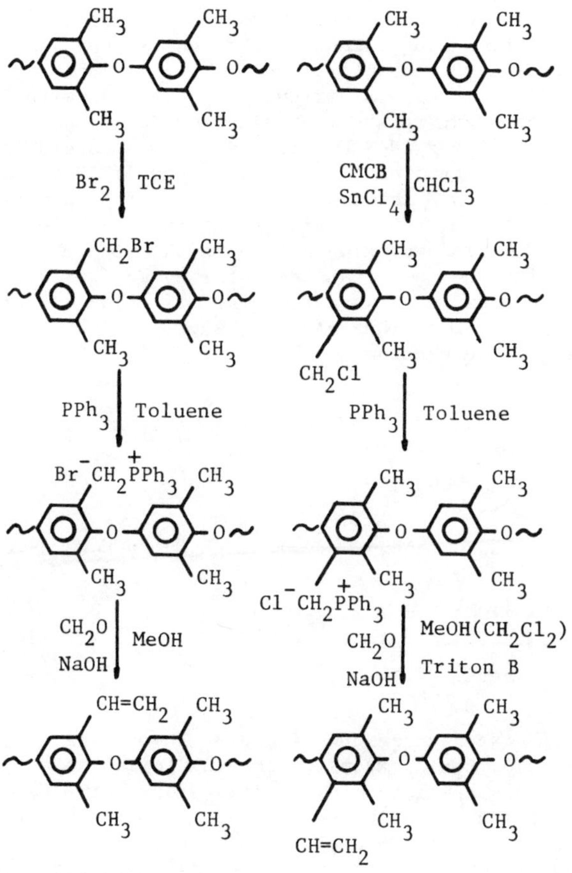

Figure 3. Synthetic routes used for the preparation of PPO with pendant vinyl groups.

reaction conditions, both the nucleophilic and electrophilic groups are present in the organic phase. The basic function of the phase transfer catalyst (the most common one used by us is tetrabutylammonium hydrogen sulfate, TBAH) is to transfer the α-onium or α,ω-bis-(onium) oligomers into the organic medium in the form of ion pairs. In non-polar aprotic solvents like chlorobenzene and CH_2Cl_2, these are virtually unsolvated and unshielded (except by their counterions) and are consequently very reactive. Therefore, the etherification takes place in solution; the reaction is very fast at room temperature; and the reaction course can be followed by the disappearance of the green color of the phenolate anion.

The etherification of α,ω-bis(hydroxyphenyl)PSU with ClMS was demonstrated to be quantitative (for all procedures: A, B and C), based on FTIR and 200 MHz ^{1}H-NMR analyses (6). A careful analysis of the 200 MHz ^{1}H-NMR spectra of α,ω-bis(vinylbenzyl)PSU prepared according to procedures A and B shows a signal assigned to aromatic formal protons ($-OCH_2O-$, δ=5.64 ppm). This is due to a chain extension reaction of the α,ω-bis(hydroxyphenyl)PSU with methylene chloride. The explanation for the fact that the obtained PSU contains only vinylbenzyl chain ends, even when the etherification is performed in methylene chloride as reaction solvent, is provided by the kinetic particularities outlined below:

$$...-PhO^- + CH_2Cl_2 \xrightarrow{k_1} ...-PhOCH_2Cl + Cl^-$$

$$...-PhO^- + ...-PhOCH_2Cl \xrightarrow{k_2} ...-PhOCH_2OPh-... + Cl^-$$

$$...-PhO^- + ClCH_2PhCH=CH_2 \xrightarrow{k_3} ...-PhOCH_2PhCH=CH_2 + Cl^-$$

$$...-PhOCH_2Cl + OH^- \xrightarrow{k_4} ...-PhO^- + CH_2O + HCl$$

where: $k_2 > k_3 > k_1$

Obviously, between procedures A and B, B would have to provide the lowest degree of chain extension through formal bonds. This is because in B the bisphenolate of α,ω-bis(hydroxyphenyl)PSU is not in contact with methylene chloride in the absence of ClMS (ClMS is a stronger electrophile than CH_2Cl_2). Additional evidence was obtained from the GPC analysis of α,ω-bis(vinylbenzyl)PSU prepared according to procedures A, B, and C (6) and confirms the validity of this assumption. Consequently, the method to be used for the preparation of α,ω-bis(vinylbenzyl)PSU with well-known molecular weight is that presented by the procedure C, using chlorobenzene or any other solvent which is "inert" under phase transfer catalyzed reaction conditions. At the same time, the other procedures provide us with a simple avenue for the preparation of α,ω-bis(vinylbenzyl)PSU with different molecular weights starting from only one sample of α,ω-bis(hydroxyphenyl)PSU.

Differential Scanning Calorimetry Analysis of the Thermal Curing of α,ω-bis(vinylbenzyl)PSU. Typical DSC curves for α,ω-bis(hydroxyphenyl)-PSU and α,ω-bis(vinylbenzyl)PSU are presented in Figure 4. The

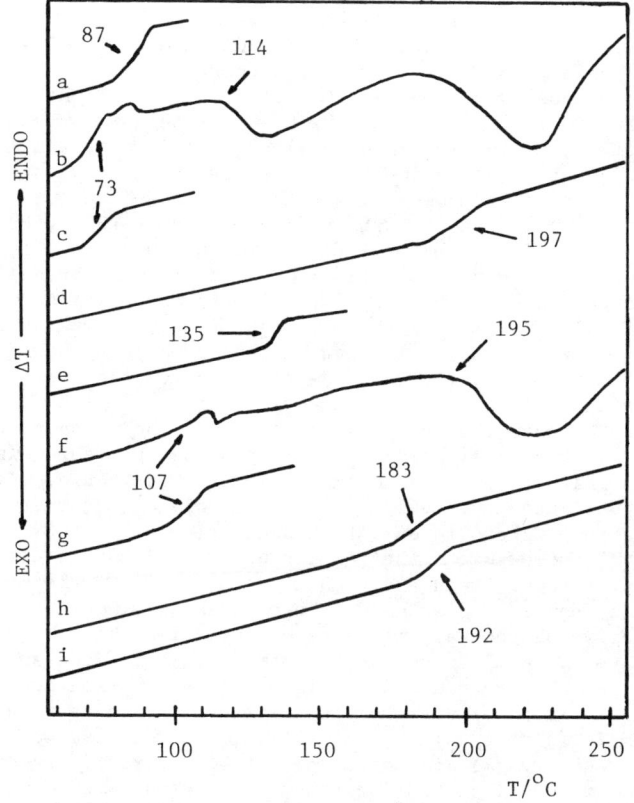

Figure 4. DSC curves of: a) α,ω-bis(hydroxyphenyl)PSU-5 (second heating run); b) α,ω-bis(vinylbenzyl)PSU-5B (first heating run); c) α,ω-bis(vinylbenzyl)PSU-5B (second heating run, first heating run up to 97°C); d) α,ω-bis(vinylbenzyl)PSU-5B (second heating run, first heating run as in (b), and then held for one hour at 257°C; e) α,ω-bis(hydroxyphenyl)PSU-1 (second heating run); (f) α,ω-bis(vinylbenzyl)PSU-1C (first heating run); g) α,ω-bis-(vinylbenzyl)PSU-1C (second heating run, first heating run up to 147°C); h) α,ω-bis(vinylbenzyl)PSU-1C (second heating run, first heating run as shown in (f) and then held 0.08 hour at 257°C); i) α,ω-bis(vinylbenzyl)PSU-1C (third heating run, second heating run as shown in (h) and then held for 0.17 hour at 257°C).

quantitative data of the thermal characterization of α,ω-bis-(hydroxyphenyl)PSU and α,ω-bis(vinylbenzyl)PSU prepared through the procedures A, B and C is summarized in Table I.

The first important observation to notice is that the Tg values of all the α,ω-bis(vinylbenzyl)PSU's (procedures A, B, C) are much lower than those of the initial α,ω-(hydroxyphenyl)PSU (Figure 4 (a), (b), (c) for PSU-5B; (e), (f), (g) for PSU-1C). This data is presented in Table I as Tg [α,ω-bis(hydroxyphenyl)PSU] and T_g^i [initial Tg of the α,ω-bis(vinylbenzyl)PSU]. The difference T_g^i- Tg is strongly dependent on the synthetic procedure used for the preparation of the α,ω-bis(vinylbenzyl)PSU. It decreases in the order C>B>A, i.e., in the order of increasing chain extension during the functionalization of the α,ω-bis(hydroxyphenyl)PSU. This decrease in the Tg values after functionalization is due to the lack of hydrogen bonding which, in the case of the α,ω-bis(hydroxyphenyl)PSU-5, provides an increase in the Tg value as high as 31°C (Table I, PSU-5, PSU-5C).

All the α,ω-bis(vinylbenzyl)PSU's present one or two exothermal peaks on their DSC curves (Figure 4). Two exothermal processes were observed especially in the case of the lowest molecular weight α,ω-bis(vinylbenzyl)PSU. The exothermal peak from low temperature is due to the polymerization of m-vinylbenzyl groups while the other is due to p-vinylbenzyl groups (23). An explanation for the presence of only one exotherm in the case of the oligomers with high molecular weight could be that the Tg's of the oligomers, being above the temperature of the first polymerization process, prevent this from occurring. This process does occur above Tg, but then overlaps the second polymerization process, thus giving only one exotherm. Research is in progress to answer this question.

Table I also presents the temperature at which this exothermal process starts (as T_S) and at which it ends (as T_E). The difference $T_S - T_g^i$ can be considered as being the processing window of the thermally reactive oligomers, and is also listed in Table I.

It is very important to notice that after the first scan and quenching, the DSC trace no longer shows the exothermal process. At the same time, it shows a tremendous increase in the Tg value of the polymer sample (Figure 4 (d), (h), (i)). This very fast curing reaction was not observed in the other classes of thermally reactive oligomers. After the first heating scan, all the analyzed samples were held at 257°C as long as it was necessary in order to obtain a constant value of the Tg. Usually, after the first heating scan, although the exothermal process was no longer observable by DSC, one could still observe an increase in the Tg value (Figure 4 (h), (i)). The final Tg value T_g^f, is presented in Table I, together with the required curing time at 257°C and the difference between the initial and final Tg values, i.e., $T_g^f - T_g^i$.

The Tg value of a conventional, high molecular weight, commercial aromatic poly(ether sulfone) is 180°C. From Table I one can see that by thermally curing an α,ω-bis(vinylbenzyl)PSU not only is its Tg value increased by as much as 150°C (sample PSU-5C), but also a value is almost always obtained which is higher than that of the high molecular weight commercial PSU. The highest Tg value obtained was 205°C (sample PSU-5C), which is 25°C higher than that of high molecular weight PSU. The expected dependence of the initial molecular weight of the α,ω-bis(vinylbenzyl)PSU on its increase in Tg by curing

Table I. Thermal Characterization (DSC) of α,ω-Bis(hydroxyphenyl)PSU and α,ω-Bis(vinylbenzyl)PSU

α,ω-bis(hydroxyphenyl)PSU			α,ω-bis(vinylbenzyl)PSU*							
No.	$\overline{M_n}$ (NMR)	T_g (°C)	Type	Exothermal Process			Processing Window		$T_g^f - T_g^i$ (°C)	Reaction Time to T_g^f at 257°C (hr)
				T_g^i (°C)	T_S (°C)	T_E (°C)	$T_S - T_g^i$ (°C)	T_g^f (°C)		
PSU-5	1,210	87	A	77	117	257	40	172	95	0.5
			B	73	114	"	41	197	124	1.0
			C	56	139	"	83	205	150	2.5
PSU-1	2,550	135	A	125	166	"	41	184	59	0.42
			B	115	196	"	71	191	76	0.25
			C	107	195	"	88	192	85	0.25
PSU-2	3,070	138	A	137	174	"	34	180	43	0.25
			B	132	193	"	61	185	53	0.5
			C	117	166	"	49	189	72	0.33
PSU-4	3,890	146	A	136	173	"	37	170	34	0.08
			B	138	196	"	58	178	40	1.67 **
			C	126	176	"	50	180	54	0.33

* T_g = T_g of α,ω-bis(hydroxyphenyl)PSU; T_g^i = initial T_g of α,ω-bis(vinylbenzyl)PSU; T_g^f = final T_g of the thermally cured α,ω-bis(vinylbenzyl)PSU, i.e., constant value which does not increase by subsequent curing at 257°C; T_S = temperature at which the exothermal process starts; T_E = temperature at which the exothermal process ends. ** Does not reflect time necessary to complete reaction, only reflects length of time held at 257°C after the first heating run.

is clearly shown by the data presented in Table I. The higher the molecular weight of the initial α,ω-bis(hydroxyphenyl)PSU, the lower is its final Tg (T_g^f). For the same initial α,ω-bis(hydroxyphenyl)PSU the Tg of the cured α,ω-bis(vinylbenzyl)PSU depends on the synthetic procedure A, B, or C. It decreases in the same order as the increase in the degree of chain extension during preparation. This can be even better observed from Figure 5, which demonstrates the dependence between the Tg value of α,ω-bis(hydroxyphenyl)PSU (Tg), α,ω-bis(vinylbenzyl)PSU (T_g^i), and thermally cured α,ω-bis(vinylbenzyl)PSU (T_g^f). This clearly demonstrates that it is possible to tailor the T_g^i of the reactive oligomer, and the T_g^f of the cured oligomer through the $\overline{M}n$ value of the α,ω-bis(hydroxyphenyl)PSU.

Synthesis of PSU and PPO Containing Pendant Vinyl Groups. The synthetic routes used for the preparation of PSU and PPO containing pendant vinyl groups are outlined in Figures 2 and 3. PSU containing pendant vinyl groups was prepared by chloromethylation of an α,ω-bis-(benzylether)PSU, followed by conversion of the chloromethyl groups into phosphonium salt groups, and subsequent phase transfer catalyzed Wittig vinylation of the phosphonium salt residues (Figure 2). Two different procedures were used for the functionalization of PPO. The first one involved the radical bromination of the PPO methyl groups, followed by phosphonylation and then Wittig reaction. The second one required the chloromethylation of PPO, followed by phosphonylation and the Wittig reaction (Figure 3). Details of these synthetic methods are presented elsewhere (13). No hydrolysis of the phosphonium groups by OH⁻ could be observed during any of the phase transfer catalyzed Wittig reactions.

Attempts to transform both PSU and PPO containing high concentrations of chloromethyl or bromomethyl groups into the corresponding polymers containing pendant vinyl groups led to crosslinked polymers. A combination of structural analysis by 200 MHz ^{1}H-NMR spectroscopy, coupled with FTIR spectroscopy and GPC, demonstrated that this is due to the fact that steric hindrances, as well as the difference in solubility between the chloro(bromo)methylated and phosphonylated polymers did not allow complete phosphonylation of the electrophilic groups. When these groups are present during the phase transfer catalyzed Wittig reaction, displacement of -Br or -Cl from -CH$_2$Br or -CH$_2$Cl with OH⁻ occurs, and the resulting benzyl alcohol groups are etherified with remaining -CH$_2$Br or -CH$_2$Cl groups by a phase transfer catalyzed etherification. In this way, branched and insoluble polymers are produced. It was shown by these experiments, that the upper limit of pendant vinyl groups attached to polymers which are easily prepared is approximately 1.2 per repeat unit in the case of PSU and 0.6 per repeat unit in the case of PPO.

Differential Scanning Colorimetry Analysis of the Thermal Curing of PSU and PPO Containing Pendant Vinyl Groups. The thermal behavior of all these thermally reactive oligomers is summarized in Table II. In this table, T_S, T_g^i and T_g^f have the same meaning as in Table I. Typical DSC traces for these reactive polymers are presented in Figure 6. This figure presents the DSC curves of α,ω-bis(hydroxyphenyl)PSU-2, α,ω-bis(benzylether)PSU-2 and α,ω-bis(benzylether)PSU-2 containing 0.57 vinyl groups per bisphenol-A unit. As was previously

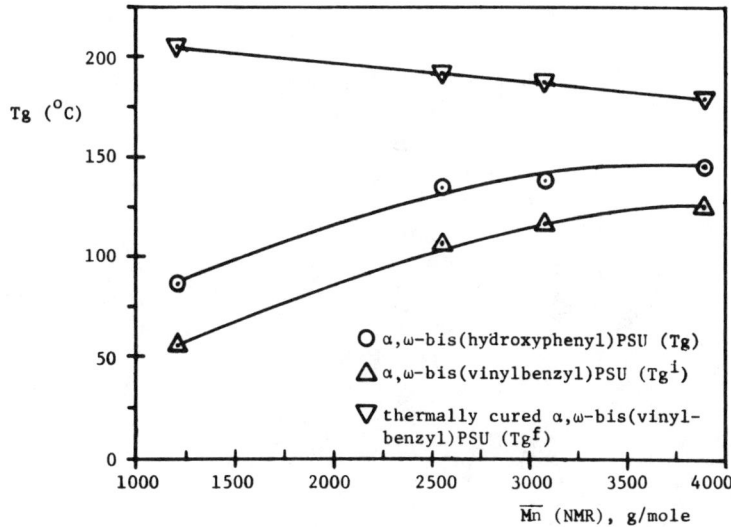

Figure 5. Dependence between the Tg and $\overline{M}_n$ of a α,ω-bis(hydroxyphenyl)PSU, and the Tg of α,ω-bis(vinylbenzyl)PSU (Tg^i) and thermally cured α,ω-bis(vinylbenzyl)PSU (Tg^f).

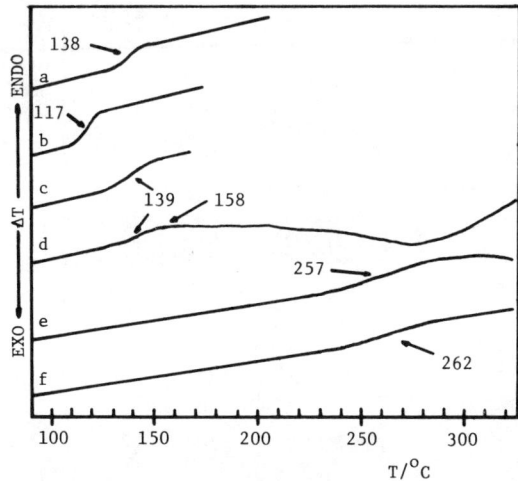

Figure 6. DSC curves of: a) α,ω-bis(hydroxyphenyl)PSU-2 (second heating run); b) α,ω-bis(benzylether)PSU-2 (second heating run); c) PSU-2 with pendant vinyl groups (second heating run, first heating run up to 165°C); d) PSU-2 with pendant vinyl groups (first heating run); 3) PSU-2 with pendant vinyl groups (second heating run, first heating run as in (d) then held for 0.02 hour at 327°C); f) PSU-2 with pendant vinyl groups (third heating run, second heating run as in (e), then held for 0.08 hour at 327°C).

Table II. Thermal Characterization by DSC of PSU and PPO with Pendant Vinyl Groups

	PSU and PPO Starting Materials					Functionalized PSU and PPO							Reaction Time to T_g^f at 327°C
Sample	$\overline{M_n}$ (GPC)	$\overline{M_w}$	$\overline{M_w}/\overline{M_n}$	$\overline{M_n}$ (NMR)	T_g (°C)	Type (a)	Degree of Functionalization (b)	T_g^i (°C)	T_s (°C)	$T_s-T_g^i$ (°C)	T_g^f	$T_g^f-T_g^i$	
PSU-3	3,270	5,570	1.7	3,410	148	VC	1.20	164	202	38	282 (c)	118	--
PSU-2	2,820	4,410	1.6	3,050	138	VC	0.57	139	158	19	262	123	0.1 hr.
PSU-2Bz	--	--	--	3,230	117								
PPO-1	21,000	53,000	2.5	--	217	VC	0.35	228	275	47	272 (e)	44	--
PPO-2	19,000	49,000	2.6	--	209	VC	0.09	215	273	58	244	29	0.2 hr.
						VC	0.60	241	263	22	None (d)	--	--
						VB	0.53	185	225	40	None (d)	--	--

a. VC = vinyl from chloromethylation, VB = vinyl from bromination.
b. Mole fraction of chloromethyl or bromomethyl groups per repeat unit.
c. After only one heating to 327°C and subsequent quenching at 320°/min.
d. No T_g within range tested (27 to 327°C).
e. T_g after four runs to 327°C, but not held at 327°C.

shown, the Tg values of α,ω-bis(hydroxyphenyl)PSU are higher than those of the corresponding benzylated polymer. This is the reason for the decrease in the PSU's Tg value from 138°C to 117°C after benzylation (Figure 6a, b). The further modification of the PSU backbone through the introduction of pendant double bonds, and/or through branching which results from the chloromethylation reaction (15) increases the Tg back to the region of the starting α,ω-bis(hydroxyphenyl)PSU (Figure 6a, c). It is interesting to observe that although the concentration of the double bonds is half that of the sample PSU-3 (which has 1.2 vinyl groups per bisphenol-A unit), the increase in Tg after the first heating scan is the same, i.e., 118°C (Table II). Consequently, it seems important to conclude that a high concentration of pendant vinyl groups does not provide a much higher increase in the T_g^f of the cured polymer, as it does in the case of the α,ω-bis(vinylbenzyl)oligomers. It is also interesting to observe that the cured PPO's containing 0.53 and 0.60 vinyl groups per structural unit show no Tg in the range 27 to 327°C.

The major difference between PPO with pendant vinyl groups obtained from bromomethylation and from chloromethylation is that, in the former case, the Tg of the uncured polymer is always lower than the starting material. In the latter case, the Tg of the uncured polymer is always higher than that of the starting material (Table II).

In conclusion, phase transfer catalyzed Williamson etherification and Wittig vinylation provided convenient methods for the synthesis of polyaromatics with terminal or pendant styrene-type vinyl groups. Both these polyaromatics appear to be a very promising class of thermally reactive oligomers which can be used to tailor the physical properties of the thermally obtained networks. Research is in progress in order to further elucidate the thermal polymerization mechanism and to exploit the thermodynamic reversibility of this curing reaction.

Acknowledgments

Financial support from the National Science Foundation (Grant DMR 82-13895) and B. F. Goodrich is gratefully acknowledged.

Literature Cited

1. Lubowitz, H. R. Polymer, Prepr. 1971, 12(1), 329.
2. Hergenrother, P. M. J. Macromol. Sci.-Rev. Macromol. Chem. 1980, C19, 1.
3. Vancraeynest, W.; Stille, J. K. Macromolecules 1980, 13, 1361.
4. Kwiatkowski, G. T.; Robeson, L. M.; Brode, G. L.; Bedwin, A. W. J. Polym. Sci., Polym. Chem. Ed. 1975, 13, 961.
5. Hergenrother, P. M. J. Polym. Sci., Polym. Chem. Ed. 1982, 20, 3131.
6. Percec, V.; Auman, B. C., Makromol. Chem. 1984, 185, 1867.
7. Jarre, W.; Bieniek, D.; Korte, F. Naturwissenschaften 1975, 62(8), 391.
8. Percec, V.; Rinaldi, P. L. Polym. Bull. 1983, 9, 548.
9. Percec, V.; Rinaldi, P. L. Polym. Bull. 1983, 9, 582.
10. Percec, V. Polym. Bull. 1983, 10, 1.
11. Wentworth, S. E.; Macaione, D. P. J. Polym. Sci., Polym. Chem. Ed., 1976, 14, 1301.

12. Percec, V.; Auman, B. C. Makromol. Chem. 1984, 185, 617.
13. Percec, V.; Auman, B. C. Makromol. Chem. 1984, 185, 2319.
14. Olah, G. A.; Beal, D. A.; Yu, S. H.; Olah, J. A. Syntheses 1984, 560.
15. Daly, W. H.; Chotiwana, S.; Neilson, R. Polym. Prepr. 1979, 20(1) 835. Daly, W. H.; Wu, S. J. Polym. Prepr. 1982, 23(2), 89. Daly, W. H. Makromol. Chem., Suppl. 1979, 2, 3.
16. Percec, V.; Rinaldi, P. L.; Auman, B. C. Polym. Bull. 1983, 10, 215.
17. Percec, V.; Rinaldi, P.L.; Auman, B. C. Polym. Bull. 1983, 10, 397.
18. Percec, V.; Nava, H., Makromol. Chem. Rapid Commun. 1984, 5, 319.
19. Percec, V.; Auman, B. C. Polym. Bull. 1983, 10, 385.
20. Percec, V.; Auman, B. C. Polym. Bull. 1983, 10, 391.
21. Percec, V.; Nava, H.; Auman, B. C. Polymer J. (Japan) 1984, 16, 681.
22. Freedman, H. H.; Dubois, R. A. Tetrahedron Lett. 1975, 38, 3251.
23. Percec, V.; Nava, H.; Auman, B. C., unpublished data.

RECEIVED February 4, 1985

9
N-Cyanourea-Terminated Resins

S. C. LIN

Washington Research Center, W. R. Grace & Company, Columbia, MD 21044

>A unique chemistry employing N-cyanourea-terminated reactive oligomers was developed to prepare three different types of polymeric materials. The reactive oligomers were obtained by allowing an isocyanate-terminated polycaprolactone oligomer to react with cyanamide in bulk or in aqueous alkaline solution at room temperature. Upon standing at room temperature, the di-N-cyanourea oligomers underwent homopolymerization to afford linear polymers. The oligomers formed thermosets upon heating above 100°C. When mixed with liquid epoxy resins, the oligomers slowly polymerized to yield linear polymers that were plasticized by the epoxy. These mixtures formed cross-linked thermosets upon heating. The linear polymers synthesized from the oligomers also served as curing agents for epoxy resins. The curing and polymerization mechanisms are discussed in the paper. The oligomers could be used for coatings, adhesives, and other applications needing thermosetting materials.

A thermosetting resin converts to an infusible, cross-linked plastic which is insoluble in any solvent after curing. Because of this irreversible cross-linking reaction, excellent physical properties such as heat resistance, creep resistance, mechanical strength, etc., are obtainable through design of the polymer structure.

Thermosetting resins have been used extensively in industry for applications such as structural adhesives, composites, RIM, coatings and sealants which need to resist severe service conditions. Thermosetting resins are generally composed of low molecular weight oligomers which allow fabrication convenience.

The study relates to new classes of thermosetting oligomers based on N-cyanourea-terminated resins, which are useful as a monomer, a thermosetting resin, and a cross-linking agent for epoxy resins.

Reactive oligomers such as epoxy resins, isocyanate-terminated compounds, and urethane-acrylates are extremely useful in the adhesive, coating, reaction injection molding (RIM), sealant and

composite industries. The application of reactive oligomers depends
on their chemical structures, which affect the rheology, physical
characteristics and curing speed of the resin mixture. Successful
preparation of a particular reactive oligomer which meets all the
desired specifications requires an understanding of the structure-
property relationships of the polymeric system, as well as the
control of the chemistry involved.

This paper deals with a particular chemistry based on N-cyanourea-terminated oligomers, which have been converted into high
molecular weight polymers, fast curing thermosets,(1) and curing
agents for epoxy resins.(2) The preparation of N-cyanourea-
terminated oligomers is simple, and the properties of materials can
be easily adjusted by structural change. Therefore, this chemistry
is very versatile for many industrial applications.

The synthesis of the oligomers involved the known reaction of
isocyanates and cyanamide (NH_2CN). For example, N-cyano-N'-phenyl
urea has been synthesized from phenyl isocyanate and an aqueous
alkaline solution of cyanamide in high yield.(3) Recently, similar
reactions were used to prepare various di-N-cyanourea compounds from
diisocyanates.(1) These monomers were also synthesized directly by
reacting diisocyanates with cyanamide at melt temperatures.

The difunctional N-cyanourea compounds were found to polymerize
into different polymeric materials at different temperatures. At
room temperature, a linear polymer was obtained either from the
polymerization of a di-N-cyanourea monomer or directly from the
mixture containing a diisocyanate and cyanamide. At elevated
temperature (>100°C), the di-N-cyanourea monomer, or the mixture of a
diisocyanate and cyanamide, cross-linked to a rigid foam or flexible
material, depending on the structure of the monomer.

Cyanamide and its aromatic derivative, such as 4,4'-methylene
bis(phenyl cyanamide), were reported to cure an epoxy resin at
elevated temperatures.(4) It is also well known that the dimer of
cyanamide (dicyandiamide) is the most important epoxy curing agent in
one-package epoxy compounding.(5) Unfortunately, this dimer
precipitates from the dispersion causing uneven mixing upon standing.

The goal of this research was to develop derivatives containing
-NHCN groups, retaining latency toward epoxides. The derivative
should have flexibility for chemical structure change and give
homogeneous one-package, thermosetting epoxy resin formulations.
N-cyanourea compounds and their polymers or oligomers were one of the
choice materials for study.

Experimental

Preparation of N-Cyanourea-Terminated Oligomers. To 0.2 moles of
toluene diisocyanate was added dropwise 0.1 mole of a polycaprolactonediol oligomer (Union Carbide, PCP 200), having a molecular
weight of 530 g/mole, over a period of 4 hours. After stirring at
room temperature overnight, the isocyanate-terminated resin was
warmed to 50°C, charged with 0.2 moles of cyanamide, and finally,
cooled to room temperature as soon as a homogeneous liquid had been
obtained.

Preparation of One-Package, Thermosetting Epoxy Resin. Varying
amounts of Epon 828 resin were added to di-N-cyanourea resin to form

thermosetting compounds having different equivalent ratios of epoxide to N-cyanourea. The IR spectra were taken right after mixing and after aging for 5 and 13 days to study the stability.

Curing of One-Package, Thermosetting Epoxy Compounds. The thermosetting epoxy resin mixture was cured at 160°C for 2 hours.

Adhesive Applications. Three substrates: steel, aluminum and fiber reinforced polyester mold compound (SMC) were used in study. In general, the thermosetting epoxy resin was applied between two pieces of substrates having 1/2 in^2 overlap for metals and 1 in^2 overlap for SMC. After curing, the lap shear strength of the samples was obtained on an Instron at a gear rate of 0.2 in/min. The results are summarized in Figure 1.

Coating Applications. The compositions and the performance coatings are described in Table I. The coatings are prepared by the same procedure as described in the oligomer preparation.

Table I. Application Examples of N-Cyanourea-Terminated Oligomers

Example	1	2	3	4
Diol (mole)	PCP-200 (a)	PPG-725 (b)	PCP-210	PCP-240
Diisocyanate (mole)	2 x MDI	2 x TDI	2 x TDI	2 x TDI
Cyanamide (mole)	2	2	2	2
Lap Shear for SMC (psi)	1070 (d)	-	-	-
Reverse Impact (in-lb) (c)	-	>160	>160	>160
MEK Rub (cycle) (c)	-	40	>100	>100
Adhesion (Tap Test) (c)	-	4	5	5
Pencil Hardness	-	3H	3H	2B

(a) Caprolactone diol oligomer
    TDI = Toluene Diisocyanate
    MDI = Di(p-isocyanato phenyl) methane
(b) Polypropylene glycol
(c) Coating application
(d) Substrate failure

Results and Discussion

Although N-cyanourea-terminated compounds could be synthesized directly from the reaction between isocyanates and cyanamide, to further modify the properties of an oligomer, the N-cyanourea-terminated compound having the desired characteristics were obtained by first reacting glycols with two molecules of a diisocyanate, and then reacting the diisocyanate oligomer with two molecules of cyanamide. The preparation of N-cyanourea oligomers is summarized by the following reactions:

$$\text{HO-R-OH} + 2\text{OCN-R'-NCO} \longrightarrow \text{OCN-R'-NH}\overset{\text{O}}{\underset{\|}{\text{C}}}\text{-O-R-O-}\overset{\text{O}}{\underset{\|}{\text{C}}}\text{-NH-R'-NCO} \qquad (1)$$

$$[\text{I}]$$

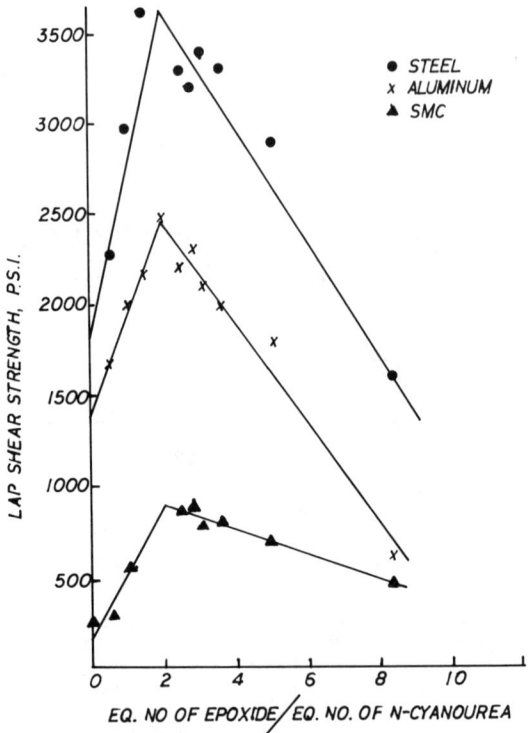

Figure 1. Lap shear strength as a function of equivalent ratio of epoxide to N-cyanourea.

$$[\text{I}] + 2\text{NH}_2\text{CN} \longrightarrow \underset{[\text{II}]}{\text{NCNHCNH-R'-NHC-O-R-O-C-NH-R'-NHC-NHCN}} \quad (2)$$

with carbonyl groups shown above each C.

Reactive Oligomers as Epoxy Curing Agents. An oligomer useful as a curing agent for epoxy resins was synthesized from toluene diisocyanate (TDI), a caprolactone diol (PCP) (MW = 530 g/mole) and cyanamide. The oligomer was mixed with Epon 828, a diglycidyl ether of bisphenol A, at various compositions. These mixtures, after applying as structural adhesives and curing at 160°C for 2 hours, provide excellent lap shear strength to steel, aluminum and SMC (a fiberglass-reinforced polyester molding compound), as shown in Figure 1. Maximum performance can be observed in the composition where the equivalent ratio of epoxy to N-cyanourea is 2. This implies that two epoxide groups react with one N-cyanourea group, giving the best adhesion to the substrates.

The reaction between the NCNH- of cyanamide and epoxide is reported to give a 1,3-oxazolidine, (4)

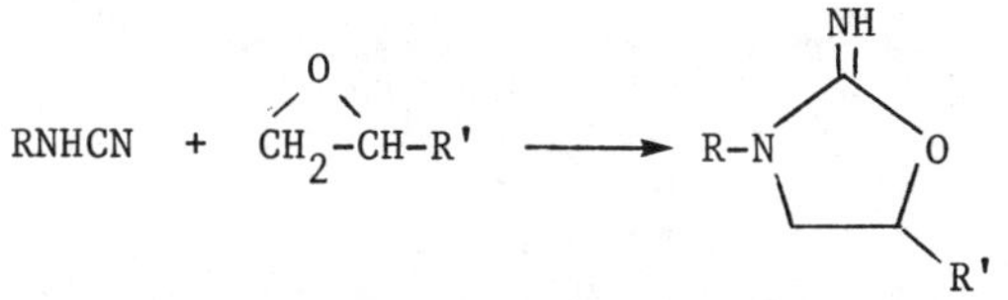

$$\text{RNHCN} + \underset{\text{CH}_2\text{-CH-R'}}{\overset{O}{\triangle}} \longrightarrow \text{R-N} \underset{R'}{\overset{NH}{\bigcirc}} O \quad (3)$$

which further reacts with diglycidyl ether of bisphenol A to form a cross-linked resin. This mechanism seems to agree with the conclusion that the maximum adhesion performance occurs at the equivalent ratio of epoxide to N-cyanourea equal to 2. However, this mechanism could not be used to explain facts obtained in the following studies.

To understand the reaction mechanism and to study the stability of the thermosetting system, the mixture of Epon 828 and N-cyanourea oligomer, synthesized from TDI, polycaprolactonediol and cyanamide, was stored at room temperature. It was found that the viscosity of the mixture increased with storage time and finally became a glassy solid after two months. Figure 2 shows the IR spectra of this mixture taken at different storage periods. A strong absorption of -C≡N group at 2270 cm^{-1} can be observed in Figure 2a. The N-cyanourea absorption in the resin mixture gradually weakens upon aging at room temperature, as shown in Figures 2b and 2c. In Figure 2, the epoxide absorption at 915 cm^{-1} shows no detectable change in intensity. This suggests that epoxide and NCNH- do not undergo any appreciable reaction upon standing at room temperature.

Recently, the difunctional N-cyanourea compound was discovered to polymerize into a high molecular weight polymer at room temperature. A tentative polymerization mechanism was proposed in the report, (1) based on IR and NMR studies. This polymerization is

$$\text{OCN-R-NCO} + \text{H}_2\text{NCN} \longrightarrow \text{NCNHCNH-R-NHCNHCN} \quad (4)$$

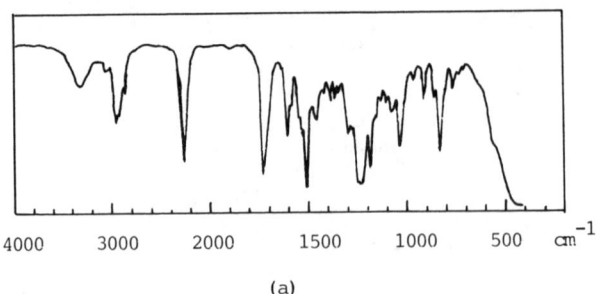

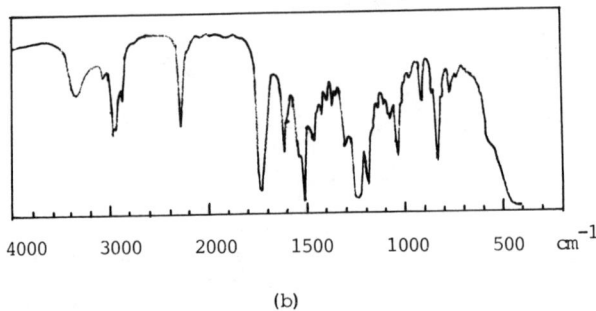

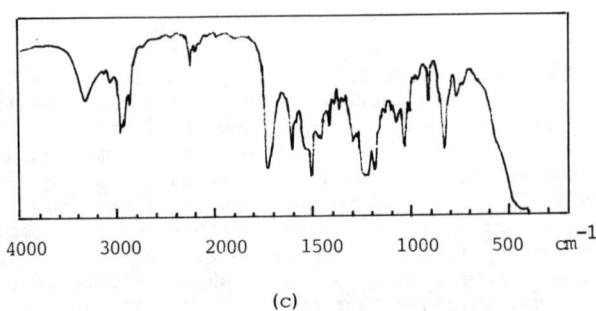

Figure 2. IR spectra of Epon 828-di(N-cyanourea) mixture at different aging periods, (a) 0 days; (b) 5 days; (c) 13 days.

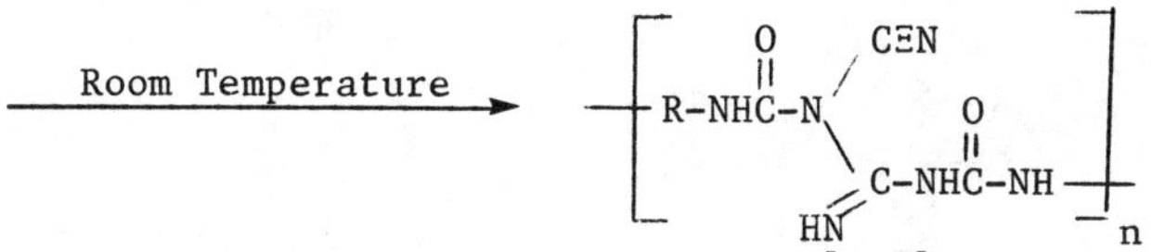

$$\xrightarrow{\text{Room Temperature}} \left[-R-NHC(O)-N(C\equiv N)-C(=NH)-O-C(O)-NHC-NH- \right]_n \quad (5)$$

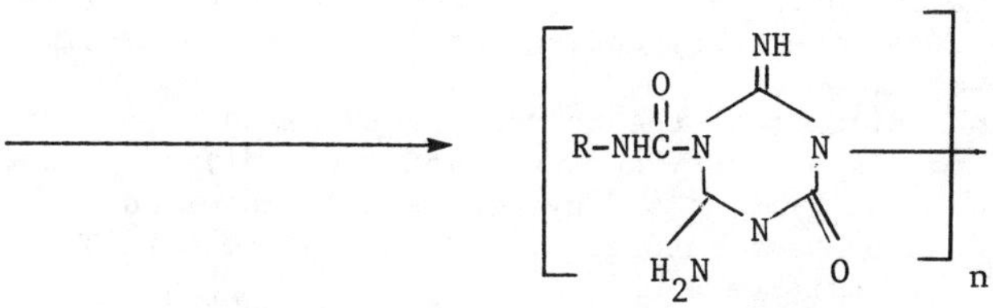

[III]

$$\longrightarrow \left[-R-NHC(O)-N\underset{H_2N}{\overset{NH}{\diagdown}}\text{triazine}\diagdown O- \right]_n \quad (6)$$

[IV]

well suited for illustrating the viscosity increase and the disappearance of –C≡N absorption in the IR spectra of Epon 828-di-N-cyanourea mixture. (Figure 2). Upon standing, the di-N-cyanourea oligomer slowly polymerized into a high molecular weight linear polymer which was plasticized by the liquid epoxy resin.

Upon heating, the linear polymer synthesized from a di-N-cyanourea compound was concluded to undergo a thermal degradation of the urea linkage, as shown in Equation 7.

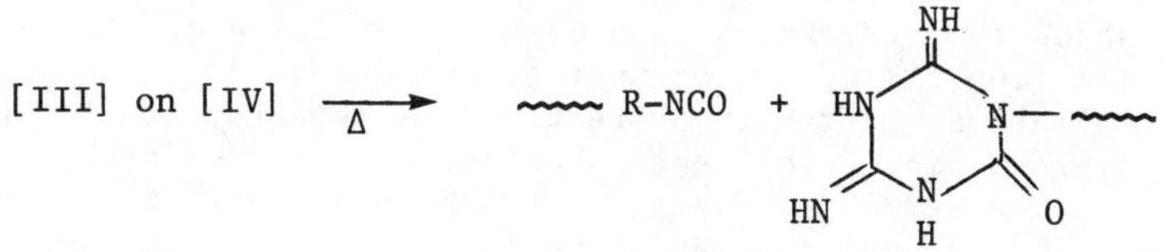

$$[III] \text{ on } [IV] \xrightarrow{\Delta} \sim\sim R-NCO + \text{HN-triazine(NH)} \sim\sim \quad (7)$$

[V]

or

$$\sim\sim R-NCO + \text{H}_2\text{N-triazine(NH}_2) \sim\sim$$

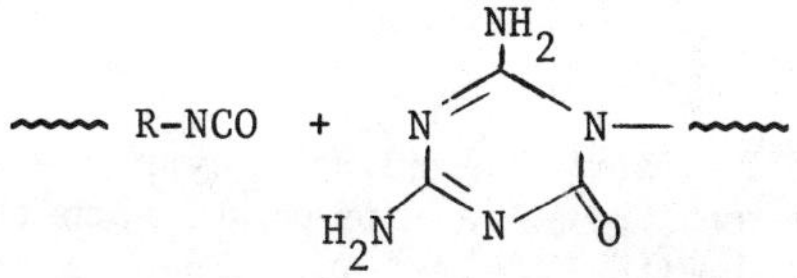

[VI]

Equations (5), (6), and (7) explain the curing of the aged Epon 828-di-N-cyanourea mixture. Figure 3 shows the epoxide absorption of the mixture disappearance upon heating at 165°C for 2 hours. At elevated temperatures the epoxide group should react with –NHCN, Equation 3, as well as function groups

$$H_2N-\underset{N-}{\overset{N=}{\left\langle\rule{0pt}{14pt}\right.}} \quad \text{and} \quad \underset{|}{HN}\overset{NH}{\diagdown\mkern-6mu\diagup}$$

in chemical structures II, IV, V and VI, to form a thermoset. The isocyanate groups shown in Equation (7) should react rapidly with -OH groups generated from the epoxy curing reactions, giving a highly cross-linked material.

In order to confirm that the polymer synthesized from a di-N-cyanourea compound was an epoxy curing agent, the polymer prepared for TDI, PCP (MW = 530 g/mole) and cyanamide was dissolved in acetone with Epon 828, cast into film, and then heated at 165°C for 2 hours. Similar results were obtained.

N-Cyanourea Oligomers as Coating Materials. Previously, the difunctional N-cyanourea-terminated oligomers were used to prepare thermoset plastics at a temperature higher than 100°C.(1) It was reported that the oligomers, upon heating, underwent the same major polymerization as equations (5) and (6) in linear polymer preparation. Some side reactions, based on IR, NMR and mass spectroscopic studies were proposed to illustrate the formation of cross-links, as shown in Figure 4.

Based on these results, application research was conducted to evaluate the feasibility of this chemistry in coatings utilization. Table I shows the structures of N-cyanourea-terminated oligomers synthesized from diols, diisocyanates, and cyanamide, and the use of these oligomers as corrosion protection coatings for steel substrate. The oligomers exhibit excellent adhesion, solvent resistance, flexibility and hardness, depending on the structure of the polymer backbone. In addition to coating applications, the oligomers also provide a substrate failure when they are used as SMC adhesives.

Conclusion

Novel chemistry regarding N-cyanourea-terminated reactive oligomers can be summarized as Scheme 1, when the previous publication is combined with this report. The oligomer can be prepared from a diisocyanate and cyanamide either in bulk or in an alkaline aqueous solution. The oligomer undergoes homopolymerization (Route a) to form high molecular weight, linear polymer at room temperature and cross-linking reaction (Route b) to generate a thermoset upon heating. The mixture of epoxy resin and the oligomer quickly forms a network plasticized by epoxy resin upon curing, and then, a totally cross-linked thermoset after final cure (Route e). The linear polymer derived from the oligomer also serves as cross-linking agent for epoxy resin (Route d).

The oligomers are, therefore, applicable to coatings, adhesives, and other uses needing thermosetting materials.

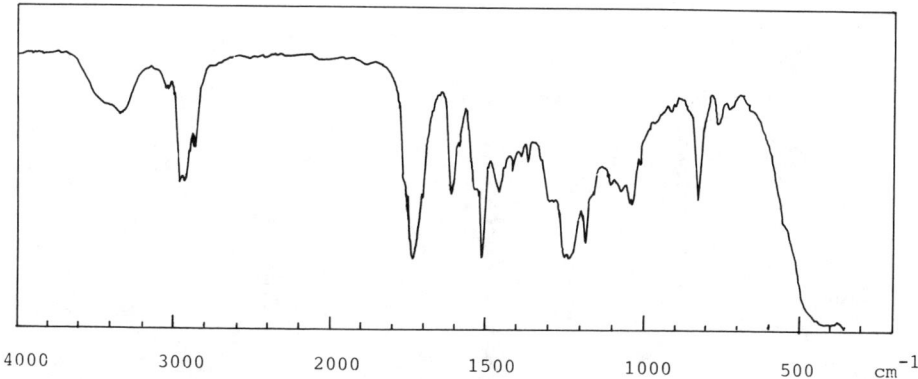

Figure 3. IR spectrum of Epon 828-di(N-cyanourea) mixture after being aged for 13 days, and then cured at 165°C for 2 hours.

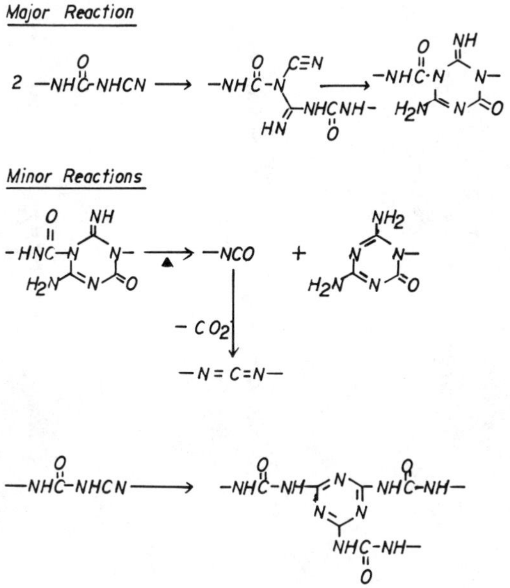

Figure 4. Possible cross-linking reactions.

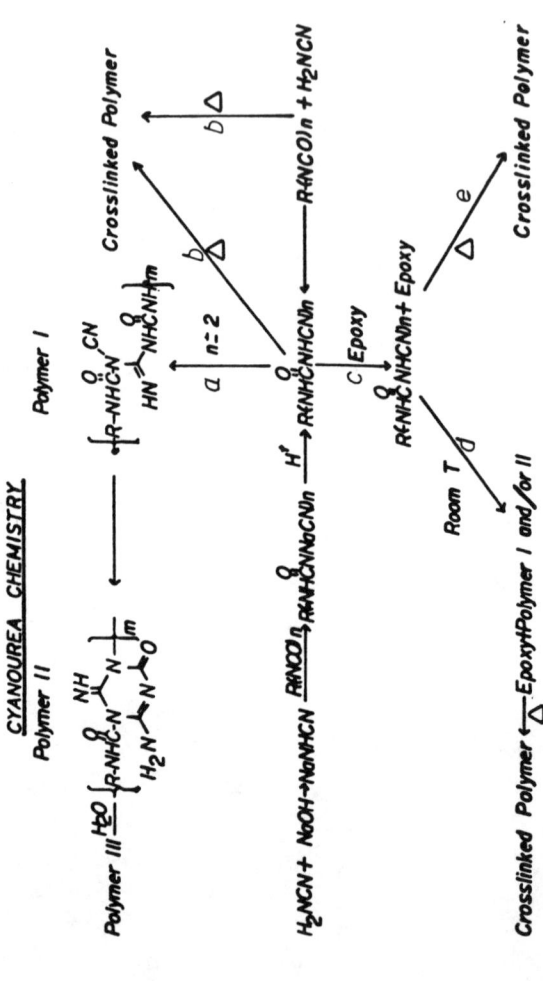

Scheme 1. Cyanourea chemistry.

Literature Cited

1. Lin, S. C., "New Polymers Prepared from N-Cyanourea Compounds," in "New Monomers and Polymers," Polymer Science and Technology, 1984, 25, 103.
2. Lin, S. C., "Novel Cross-linking System - N-Cyanourea and Its Adducts as Latent Curing Agent for Epoxy Resins," U.S. Patent 4 379 728, 1983.
3. Kurzer, F. and Powell, J. R., "Organic Syntheses," 1963, Coll. Vol. IV, p. 213.
4. Costsiff, E. H.; Dee, H. B.; Diprima, J. F.; and Seltzer, R.; ACS Polymer Preprints, 1981, 22 (2), 111.
5. Skeist, I., "Handbook of Adhesives," Van Nostrand Reinhold Company, New York, 1977, 2nd Ed.

RECEIVED March 12, 1985

10
Chain Extendable Oligomers for High-Solids Coatings

J. W. HOLUBKA

Research Staff, Ford Motor Company, Dearborn, MI 48121

>High solids coating systems usually employ low molecular weight multifunctional adducts or polymers in combination with multifunctional crosslinking agents. The networks that result upon curing differ substantially in structure and morphology from networks generated from conventional coating formulations that incorporate relatively high molecular weight polymers and crosslinkers. This paper describes three reactive oligourethane oligomers that incorporate chemical functionality that allows chain extension to occur during cure. The chain extension occurs preferably as a reaction independent of crosslinking and generates, in situ, during cure, a conventional high molecular weight polymer. One reactive oligomer incorporates blocked isocyanate-polyol chemistry that undergoes chain extension, during cure, to form hydroxy functional polyurethanes that are subsequently crosslinked, in situ, with conventional aminoplast crosslinkers. The remaining two reactive oligomers use blocked isocyanate-epoxy and t-butyl carbamate-epoxy chemistry for chain extension and subsequent self-crosslinking through formation of the corresponding oxazolidone and epoxy-amine linkages respectively.

High solids coating systems usually employ low molecular weight multifunctional adducts or copolymers in combination with multifunctional crosslinking agents. Typically, these formulations yield networks that differ greatly from those formed in conventional coating systems that incorporate relatively high molecular weight hydroxy functional polymers. In this paper, the synthesis and chemistry of three reactive urethane oligomers that can undergo chain extension during cure are reported. These oligomers referred to as urethane modified epoxy-diol oligomer, Type I urea-urethane oligomer and Type II urea-urethane oligomer, are also described in high solid coating formulations.

Experimental

Epoxy-diol Adduct. The epoxy-diol adducts were prepared using standard techniques (1) by heating a mixture of two moles of diol with one mole of epoxy resin at 130-140°C for 8 hours using 1% N,N-dimethylethanolamine as catalyst. Epoxy-diol adducts prepared in this manner showed the absence of epoxy absorption in the infrared spectrum.

Half-blocked Diisocyanates. The half-blocked diisocyanates were prepared using conventional methods (2) by adding dropwise, over a period of one hour, 1 mole of alcohol to 1 mole of diisocyanate and 100 mg dibutyl tin dilaurate in methyl amyl ketone under an inert atmosphere. After the addition of the alcohol, the reaction was heated at 60-80°C for 2 hours. For half-blocked diisocyanates prepared from tertiary alcohols, the heating period was replaced with room temperature stirring for 24-36 hours to prevent undesirable side reactions.

Chain Extendable Urethane Modified Epoxy Oligomer. The chain extendable urethane modified oligomers were prepared by combining equimolar amounts of epoxy-diol adduct and half-blocked diisocyanates, and heating the resulting mixture at 80°C for 4-6 hours until the isocyanate infrared band disappeared.

Type I Urea-Urethane Oligomers. To 30g (0.5 mole) ethylene diamine and 10g dimethoxy ethane in a three-neck flask equipped with an overhead stirrer was added over a one-hour period, 185g (0.5 mole) of butanol half-blocked diisocyanate. After the addition, the reaction was stirred overnight at room temperature.

Type II Urea-Urethane Oligomers. To 11.6g (0.1 mole) of melted 1,6-hexanediamine was added 76.9g of t-butanol half-blocked isophorone diisocyanate. During about a 30-minute addition, the temperature rose to 50-60°C. After an additional one hour of heating at 50-60°C, the oligomer solution was thinned with 40g of 2-hexoxyethanol.

Primer Formulations. Coatings were formulated using standard techniques. Mill bases were prepared by dispersing the oligomer solution with pigments (silica, carbon black, titanium dioxide and barium sulfate in a 1:1:1:10 ratio). The viscosity of the formulation was reduced to spray viscosity by addition of solvent.

Corrosion Test Procedures. Test coating samples were applied to cold rolled, unpolished bare steel panels and baked. Salt spray tests were conducted according to ASTM method B117 using a Singleton corrosion test cabinet operated at 35°C. Condensing humidity tests (ASTM D2246 and D2247) were conducted using a Cleveland Humidity cabinet (Q Panel Company). The cathodic polarization test was used to evaluate the resistance of the coatings to corrosion generated hydroxide. A detailed description of this experiment has been given elsewhere (3).

Results and Discussion

Urethane Modified Epoxy-diol Oligomers. Urethane modified epoxy-diol oligomers are easily prepared by reacting the free isocyanate functional group of a half-blocked diisocyanate with the epoxy-diol adduct. Although the oligomer formed in this reaction is undoubtedly a mixture, since the isocyanate could react with any of the hydroxy groups on the epoxy-diol adduct, it is expected that the predominate reaction would ocur at the most accessible location o the epoxy-diol adduct, namely the terminal hydroxy functionality. The reactive oligomer formed in this reaction would thus be end-capped with a blocked-isocyanate and a hydroxy group. During the cure, deblocking of the blocked-isocyanate generates a free isocyanate that subsequently reacts with hydroxy functionality to chain extend the oligomer and form a higher molecular polyurethane polymer having pendant hydroxy functionality. Infrared spectroscopy was used to study the mechanism of chain extension for a toluene diisocyanate modified hyantoin epoxy/2-ethyl-1,3-hexanediol oligomer. Spectra were obtained for samples baked 30 minutes at temperatures of 50-180°C. The mechanism of chain extension was studied by observing the OH and NH absorptions at 3460 cm-1 and 3340 cm-1 respectively. Since one hydroxy moiety is consumed for each chain extension while one urethane NH is formed, the ratio of OH/NH infrared absorptions will decrease as chain extension occurs. The results of this study are shown in Table I and Figure 1.

Figures 1a and 1b show the OH and NH infrared bands of the oligomer as a function of temperature in uncatalyzed and catalyzed formulations. The uncatalyzed urethane modified epoxy oligomer shows only small changes in the OH/NH absorbance ratio at temperatures below 165°C; only about a 60% conversion of the blocked isocyanate was observed. In contrast, sample of the oligomer catalyzed with 0.5% dibutyl tin dilaurate shows nearly complete chain extension at temperatures as low as 130°C.

Two types of ure-urethane oligomers were prepared. The alcohol used to block the diisocyanate distinguishes the two types of ureaurethane oligomers. Type I oligomers were prepared using a diisocyanate half-blocked with primary or secondary alcohols; Type II oligomers were prepared using diisocyanates half-blocked with tertiary alcohols. Well cured films (having good solvent resistance) are obtained when urea-urethane oligomers are combined with epoxy resins and baked without external crosslinking agents (e.g. melamine crosslinker). Current evidence indicates that the cure chemistry is complex and differs with changes in urea-urethane structure.

Type I Oligomers. Model studies indicate that the cure mechanism of Type I oligomers involve an initial chain extension resulting from the reaction of an oxirane functionality of the epoxy resin with isocyanate functionality (generated as a result of thermal deblocking of the urethane moiety in the urea-urethane oligomer) to form a polyoxazolidone. The polyoxazolidone formed as a result of oligomer chain extension has been found to crosslink through the reaction of N-H functionality on the backbone of the forming polymer with excess deblocked urea-urethane oligomer to form biuret crosslinks. Figures 2a and 2b show the infrared spectra of baked and unbaked Type I

Table I. Infrared Study of the Chain Extension Reaction

Experiment	Percent Catalyst	Temperature °C	A_{OH}/A_{NH}
A	0	50	0.651
B	0	130	0.589
C	0	150	0.536
D	0	165	0.519
E	0	180	0.385
F	0.5	50	0.702
G	0.5	130	0.438
H	0.5	180	0.440

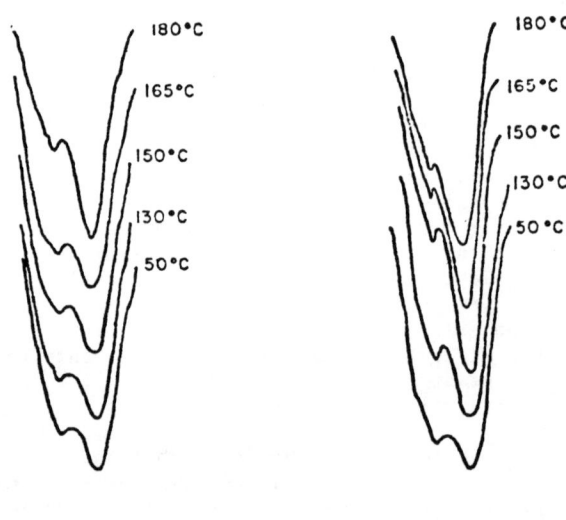

Fig. 1A Fig. 1B

Figure 1. Infrared spectra for a toluene diisocyanate modified hydantoin epoxy/2-ethyl-1,3-hexanediol oligomer showing the OH and NH bands of the oligomer as a function of temperature in uncatalyzed and catalyzed formulations.

urea-urethane/epoxy formulations respectively. The principal differences in the spectra involve a significant broadening and shifting of the carbonyl absorption to higher frequency (infrared absorptions at 1830-1840 cm-1, consistent with the formation of oxazolidone and biuret moieties). Also consistent with oxazolidone formation is a disappearance of the epoxy absorption at 910 cm-1.

Type II Oligomers. The cure mechanism of Type II oligomers relies on the thermal decomposition of tertiary carbamates (reaction products of tertiary alcohols with isocyanates) that affords carbon dioxide, alkene and free amine. In coating formulations Type II oligomers are also formulated with epoxy resins. Gas chromatographic evidence indicates that amine is liberated from the tertiary carbamate during the cure and crosslinks the coating through an epoxy amine reaction. The cure chemistry of a Type II urea-urethane/epoxy formulation was also studied by infrared spectroscopy. The results (Figure 3 and Table II) show decreases in the intensity of the urethane absorption at 1700 cm-1 and the disappearance in the epoxy band at 910 cm-1 as the temperature of the cure reaction is increased from 50-180°C. These results are consistent with the gas chromatographic data. Also noted in the infrared spectra is an increase in the urea absorption at 1630 cm-1 as the temperature is increased. In this latter case, the free amine reacts with isocyanate (liberated by normal deblocking of a blocked isocyanate) to generate urea linkages).

Table II. Infrared Absorption Data on the Cure of (t-Butyl Carbamate) Terminated Urea-Urethane Coating[a]

Temperature	Absorption Ratio[b]				
	A1700/ A1365	A1630/ A1365	A910/ A1365	A910/ A830	A1630/ A830
50	.95	1.6	.45	.68	1.43
150	.91	2.0	.38	.63	1.52
180	.83	2.6	0	0	3.26

[a] Coating composition consists of 80 weight percent 1,6-hexane diamine t-Butyl blocked isophorone diisocyanate based urea urethane resin.

[b] Absorbance ratios used peaks at 1365 cm^{-1} and 830 cm^{-1} as internal standards.

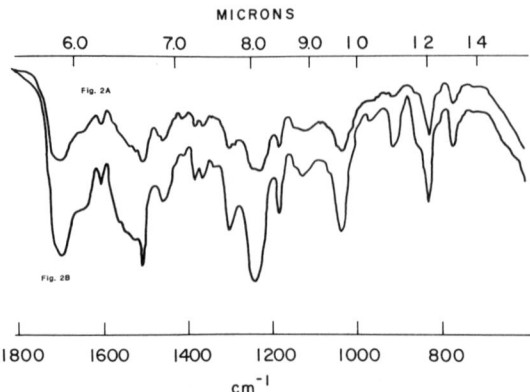

Figure 2. Infrared spectra of baked and unbaked Type I urea-urethane/epoxy formulations respectively.

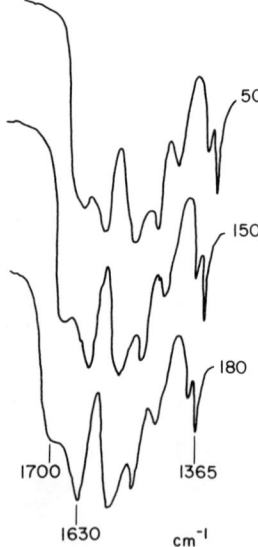

Figure 3. Infrared spectra for Type II urea-urethane/epoxy formulations, respectively, showing decreases in the intensity of the urethane absorption as the temperature of the cure reaction is increased from 50–180°C.

Conclusion

An alternate approach to high solids coating formulations has been presented. The synthesis of three reactive oligourethanes incorporating chemical functionality that allows chain extension to occur during cure in a reaction independent of crosslinking has been described. The proposed chain extension reaction occuring to generate, in situ, during cure, a conventional high molecular weight polymer that is subsequently crosslinked with usual crosslinking agents.

One reactive oligomer incorporates blocked isocyanate-polyol chemistry that undergoes chain extension, during cure, to form hydroxy functional polyurethanes that are subsequently crosslinked, in situ, with conventional aminoplast crosslinkers. The remaining two reactive oligomers use blocked isocyanage-epoxy and t-butyl carbamate-epoxy chemistry for chain extension and subsequent self-crosslinking through formation of the corresponding oxazolidone and epoxy-amine linkages respectively.

Literature Cited

1. Lee, E. "Epoxy Resins," American Chemical Society, Washington D.C., 1970.
2. Wicks, Z. W. Progress in Organic Coatings, 3, pp. 73-99 (1975).
3. Wiggle, R. R.; Smith, A. G.; and Petrocelli, J. V. Paint Tech., 40, 174 (1968).

RECEIVED February 19, 1985

New Telechelic Polymers and Sequential Copolymers by Polyfunctional Initiator-Transfer Agents (Inifers)
End Reactive Polyisobutylenes by Semicontinuous Polymerization

RUDOLF FAUST, AGOTA FEHERVARI, and JOSEPH P. KENNEDY

Institute of Polymer Science, The University of Akron, Akron, OH 44325

A novel semicontinuous cationic polymerization technique has been developed for the synthesis of linear and three-arm star *tert.*-chloro-telechelic polyisobutylenes by the use of inifers. In this technique a mixed monomer/inifer feed is added at a sufficiently low constant rate to a well stirred dilute BCl_3 charge. Stationary conditions are maintained throughout the experiment, and well defined telechelic products of identical (symmetrical) end-group structures, essentially theoretical end functionalities, and close to theoretical molecular weight dispersities are obtained. The polymerization kinetics are discussed and the $\overline{DP}_n$ equation is derived. The number average degree of polymerization is given by the ratio of monomer/inifer addition rate, $\overline{DP}_n = A_M/A_I$. The advantages of the semicontinuous technique are demonstrated by preparing telechelic, liquid isobutylene prepolymers at complete monomer conversion. The results obtained by the use of the semicontinuous technique are compared with those obtained in control experiments carried out by conventional batch polymerization which yield products with broad or multimodal molecular weight distributions and less uniform end groups at complete monomer conversion.

The inifer technique is a most convenient route for the preparation of well-defined end-reactive polyisobutylenes (PIB)(1). These materials may be linear or three-arm star telechelics carrying exactly 2.0 or 3.0 end-functions, respectively. The inifer technique yields *tert.*-chlorine-telechelic product, for example,

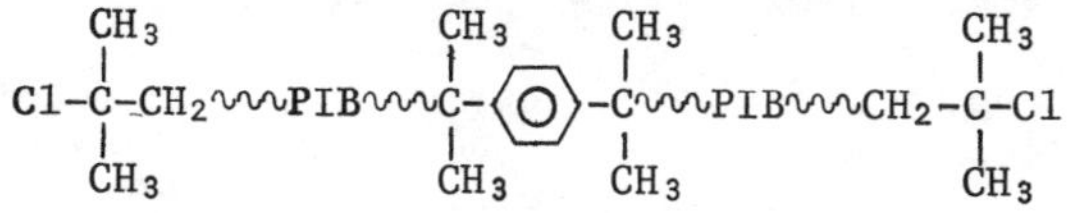

$$Cl-\underset{\underset{CH_3}{|}}{\overset{\overset{CH_3}{|}}{C}}-CH_2 \sim\sim PIB\sim\sim \underset{\underset{CH_3}{|}}{\overset{\overset{CH_3}{|}}{C}}-\!\!\left\langle\bigcirc\right\rangle\!\!-\underset{\underset{CH_3}{|}}{\overset{\overset{CH_3}{|}}{C}}\sim\sim PIB\sim\sim CH_2-\underset{\underset{CH_3}{|}}{\overset{\overset{CH_3}{|}}{C}}-Cl$$

0097–6156/85/0282–0125$06.00/0
© 1985 American Chemical Society

which can be quantitatively converted to olefin-, (2) alcohol-, (3) etc. ended prepolymers. The latters are valuable intermediates for the preparation of a large variety of new materials, i.e., PIB-based polyurethanes (4), epoxy resins (5).

A thorough investigation of the products obtained by the inifer technique has shown that sometimes low molecular weight materials ($\overline{M}_n$ = from ∿500 to ∿6000) may carry "unfired" or "once-fired" end-groups:

$$PIB\sim\sim C(CH_3)_2-C_6H_4-C(CH_3)_2-Cl$$

unfired end group

$$PIB\sim\sim C(CH_3)_2-C_6H_4-C(CH_3)_2-CH_2-C(CH_3)_2-Cl$$

once fired endgroup

The presence of these end-groups leads to a broadening of the molecular weight distributions; the theoretical molecular weight dispersity ratios $\overline{M}_w/\overline{M}_n$ were shown to be 1.5 for linear telechelics (6) and 1.33 for three-arm star telechelics (see later) prepared at low monomer conversions.

While the number average end-group functionality $\overline{F}_n$ of prepolymers carrying unfired or once-fired end groups is still 2.0 (for linear) or 3.0 (for three-arm star), the reactivity of the endgroups may be somewhat different from the ∿PIB∿∿CH$_2$-C(CH$_3$)$_2$Cl terminus. It has recently been shown that linear oligoisobutylenes may contain a significant amount of once-fired end-groups (7).

Another problem that occasionally arises and that is mostly due to insufficient reactor control (unsatisfactory stirring, insufficient temperature control, too high conversions) is broader than expected or multimodal molecular weight distributions.

The purpose of this paper is to outline semicontinuous reaction conditions under which perfectly symmetrical end-reactive materials can be obtained with close to theoretical molecular weight dispersities, i.e., $\overline{M}_w/\overline{M}_n$=1.5 for linear and 1.33 for three-arm star products, even at complete monomer conversions.

Experimental

The synthesis of p-di(2-chloro-2-propyl)benzene (dicumylchloride, binifer) and 1,3,5,tri(2-chloro-2-propyl)benzene (tricumylchloride, trinifer) has been described (1), (8). Isobutylene and methyl chloride were dried by passing the gases through columns packed with BaO and molecular sieves (4Å), condensing them under a dry nitrogen atmosphere. n-Hexane was refluxed with fuming sulfuric acid, washed with distilled water until neutral, dried over molecular sieves, refluxed, and subsequently distilled from calcium hydride under nitrogen.

Semicontinuous polymerization experiments were carried out in a stainless steel enclosure (dry box) under a dry nitrogen atmosphere in three neck flasks (11) equipped with overhead stirrer and an inlet for the continuous introduction of precooled inifer/isobutylene/solvent feeds to stirred, dilute BCl$_3$/solvent charges. Experiments with binifer were performed at -80°C by the use of CH$_3$Cl/n-hexane solvent mixtures (80/20 v/v), with trinifer at -40°C using CH$_3$Cl solvent. The input rate was controlled by applying constant nitrogen pressure on the reservoir that contained the feed. Solvent is needed in the feed to dissolve the inifers that are sparingly soluble in isobutylene. Polymerizations were terminated by quenching with prechilled methanol.

Conventional polymerizations were carried out by rapidly adding the BCl_3 coinitiator to stirred inifer/isobutylene/solvent charges. The composition of the final charges was identical to the final composition of the corresponding semicontinuous runs. Heat evolution could often be observed in conventional batch polymerization upon BCl_3 addition.

The products were dissolved in n-hexane, washed with dilute aqueous HCl, distilled water, dried over anhydrous $MgSO_4$, filtered, and recovered by removing the solvent (rotovap). Dehydrochlorination of the tert.-chlorine endgroups has been described (2).

Molecular weights were determined using a Waters high-pressure GPC instrument (Model 6000 A pump, a series of five μ-Styragel columns (10^6, 10^5, 10^4, 10^3, 500 Å), Differential Refractometer 2401 and UV Absorbance Detector Model (440) and a calibration curve made by well fractionated polyisobutylene standards.

While the calibration curve is strictly valid only for the linear PIBs, it was assumed that it provides sufficiently accurate $\overline{M}_n$ information for three-arm star products as well. This assumption was corroborated by determining the $\overline{M}_n$s of select liquid three-arm PIBs by GPC and VPO: The $\overline{M}_n$s were within experimental error. Evidently PIB calibration curves prepared with linear polymers can also be used for $\overline{M}_n$ determination of three-arm products also in the low molecular weight (<10,000) range. ^{1}H NMR spectra were taken by a Varian T-60 Spectrometer using concentrated (∿20% by weight) carbon tetrachloride solutions and TMS standard.

Results and Discussion

Kinetic Considerations. The following elementary reactions describe the polymerization of isobutylene by the inifer/BCl_3 system:

Ion generation

$$I + BCl_3 \underset{k_{-i}}{\overset{k_i}{\rightleftharpoons}} I^{\oplus} BCl_4^{\ominus} \tag{1}$$

Cationation

$$I^{\oplus} + M \xrightarrow{k_c} M_n^{\oplus} \tag{2}$$

Propagation

$$M_n^{\oplus} + M \xrightarrow{k_p} M_{n+1}^{\oplus} \tag{3}$$

Chain transfer to inifer

$$M_n^{\oplus} + I \xrightarrow{k_{tr,I}} I^{\oplus} + M_n \tag{4}$$

Termination

$$M_n^{\oplus}BCl_4^{\ominus} \xrightarrow{k_t} M_n + BCl_3 \qquad (5)$$

where I, M and M_n are inifer, monomer and polymer, $I^{\oplus}$ and $M_n^{\oplus}$ are inifer and polymer cations, and k_p, $k_{tr,I}$ and k_t are the corresponding rate constants. Chain transfer to monomer was shown to be absent up to -20°C in this system (6)(9).

The probability that the growing cation adds another monomer is

$$p = \frac{k_p[M]}{k_p[M]+k_{tr,I}[I]+k_t} \qquad (6)$$

The dispersion ratio for polymers obtained by bifunctional inifers (binifers) is (6):

$$\overline{M}_w/\overline{M}_n = 1 + \frac{2p}{(1+p)^2} \qquad (7)$$

and, by using a calculation procedure similar to that developed in ref. 6, those prepared by trifunctional inifers (trinifers) is:

$$\overline{M}_w/\overline{M}_n = (1+11 p + 11 p^2 + p^3)(1+p)/(1+4p+p^2)^2 \qquad (8)$$

Thus when $k_{tr,I}[I]+k_t \ll k_p[M]$, the polymer formed at each time element will exhibit $\overline{M}_w/\overline{M}_n = 1.5$ in polymerization by binifers and $\overline{M}_w/\overline{M}_n = 1.33$ in polymerization by trinifers. The number average degree of polymerization in inifer polymerization systems is:

$$\overline{DP}_n^{-1} = \frac{1-p}{1+p} = \frac{k_{tr,I}[I] + k_t}{k_p[M]} \qquad (9)$$

The $\overline{DP}_n^{-1}$ will remain constant only if [M] and [I] remain constant. Obviously, however, both [M] and [I] decrease with time during conventional batch polymerizations. Consequently $\overline{DP}_n$ could change with time, which in turn would broaden the molecular weight distribution ($\overline{M}_w/\overline{M}_n$ > theoretical value).

According to Equation 9 polymers with close to theoretical molecular weight distributions could be prepared even at very high conversions provided [M] and [I] remain constant throughout the polymerization. This condition can be fulfilled by continuously adding a mixed monomer/inifer feed at a sufficiently low constant rate to a coinitiator charge, making certain that the rate of monomer/inifer addition and that of monomer/inifer consumption are equal over the course of the polymerization.

Kinetics of the Idealized Semicontinuous Polymerization Technique.
Based on the kinetic scheme outlined by Equations 1 through 5 the following set of differential equations describe the changes in the concentrations during polymerization:

$$\frac{d[M]}{dt} = A_M - k_c[I^{\oplus}][M] - k_p[M_n^{\oplus}][M] \qquad (10)$$

$$\frac{d[I]}{dt} = A_I - k_i[I][BCl_3] - k_{tr,I}[M_n^{\oplus}][I] + k_{-i}[I^{\oplus}] \qquad (11)$$

$$\frac{d[M_n^{\oplus}]}{dt} = k_c[I^{\oplus}][M] - k_{tr,I}[M_n^{\oplus}][I] - k_t[M_n^{\oplus}] \qquad (12)$$

$$\frac{d[I^{\oplus}]}{dt} = k_{tr,I}[M_n^{\oplus}][I] - k_c[I^{\oplus}][M] + k_i[I][BCl_3] - k_{-i}[I^{\oplus}] \qquad (13)$$

$$\frac{d[M_n]}{dt} = k_{tr,I}[M_n^{\oplus}][I] + k_t[M_n^{\oplus}] \qquad (14)$$

where A_M and A_I are the addition rates of monomer and inifer, respectively.

In semicontinuous polymerization under stationary conditions the concentration of monomer and inifer are constant:

$$\frac{d[M]}{dt} = 0 \quad \text{and} \quad \frac{d[I]}{dt} = 0 \qquad (15)$$

and Equations 10 and 11 can be written as:

$$A_M = k_p[M_n^{\oplus}][M] + k_c[I^{\oplus}][M] \qquad (16)$$

$$A_I = k_i[I][BCl_3] - k_{-k}[I^{\oplus}] + k_{tr,I}[M_n^{\oplus}][I] \qquad (17)$$

From Equations 15 through 17 it also follows that under stationary conditions the concentration of cations are constant in the charge:

$$\frac{d[M_n^{\oplus}]}{dt} = 0 \quad \text{and} \quad \frac{d[I^{\oplus}]}{dt} = 0 \qquad (18)$$

The sum of Equations 11 through 13 gives:

$$A_I = K_{tr,I}[M_n^{\oplus}][I] + k_t[M_n^{\oplus}] \qquad (19)$$

Comparing Equations 14 and 19 the rates of polymer formation and inifer addition are equal under stationary conditions:

$$\frac{d[M_n]}{dt} = A_I \qquad (20)$$

Based on Equations 2 through 5 the number average degree of polymerization is:

$$\overline{DP}_n = \frac{k_p[M_n^{\oplus}][M] + k_c[I^{\oplus}][M]}{k_{tr,I}[M_n^{\oplus}][I] + k_t[M_n^{\oplus}]} \qquad (21)$$

By substituting Equations 16 and 19:

$$\overline{DP}_n = \frac{A_M}{A_I} \qquad (22)$$

Since the feed contains both the monomer and the inifer dissolved in a common solvent:

$$\overline{DP}_n = (\frac{[M]}{[I]} \text{ feed} \qquad (23)$$

This simple $\overline{DP}_n$ equation is generally valid for mono- or multifunctional inifers; however with multifunctional inifers $k_{tr,I}$ and k_t in Equation 21 correspond to one inifer/polymer molecule, i.e., the parameters are not normalized to one functional group.

The stationary concentration of monomer can be calculated by the following considerations. Comparison of Equations 13, 18 and 19 shows that the rate of cationation is equal to the rate of inifer addition:

$$k_c[I^\oplus][M] = A_I \qquad (24)$$

Substituting Equation 24 into Equation 21 and using Equation 22 gives:

$$\frac{A_M}{A_I} = \frac{k_p[M]}{k_{tr,I}[I] + k_t} + 1 \qquad (25)$$

From Equation 25 [I] can be expressed in terms of [M] and the addition rates A_M and A_I. Similarly, the polymer cation concentration can be expressed using Equation 19, and $[I^\oplus]$ can be expressed from Equation 24 in terms of the same variables [M], A_M and A_I.

A substitution of these stationary concentrations into Equation 17 gives:

$$k_i[BCl_3]\frac{k_p}{k_{tr,I}}\frac{A_I}{A_M-A_I}[M]^2 - k_i[BCl_3]\frac{k_t}{k_{tr,I}}[M] - \frac{k_t}{k_p}(A_M-A_I)\frac{k_{-i}}{k_c}A_I = 0 \qquad (26)$$

By introducing the abbreviations:

$$B = \frac{k_t}{k_p}(\frac{A_M}{A_I} - 1) \quad \text{and} \quad C = \frac{k_{tr,I}}{k_p}\frac{(A_M-A_I)}{k_i[BCl_3]} \qquad (27)$$

Equation 26 becomes:

$$[M]^2 - B[M] - (B + \frac{k_{-i}}{k_c})C = 0 \qquad (28)$$

so that

$$[M] = \frac{1}{2}B + \{\frac{1}{4}B^2 + (B + \frac{k_{-i}}{k_c})C\}^{1/2} \qquad (29)$$

The above expression contains the three rate constants involved in initiation, k_i (in C), k_c and k_{-i}, whose values are not known. (Recent attempts failed to determine the value of k_i due to experimental difficulties [10]). The value of [M] depends on both the relative and the absolute addition rates A_M and A_I, which means that at a given [M]/[I] in the feed, the [M] would increase with increasing addition rate. It is important therefore that the addition rate be sufficiently low so as to achieve stationary conditions rapidly, i.e., the time to reach stationary conditions should be negligible compared to the polymerization time. Unless this precaution is taken monomer/inifer may accumulate in the system which may lead to short or in worse cases even to an absence of stationary periods which in turn would yield products resembling those obtained in conventional batch polymerizations.

Experimental Results. A series of parallel experiments have been carried out using identical overall concentrations: one run was carried out by using the conventional batch technique (adding BCl_3 to monomer/inifer charges), and a corresponding semicontinuous experiment (adding monomer/inifer to BCl_3 charges). The ultimate reagent concentrations were identical in both sets of experiments. Both binifer and trinifer have been examined and various monomer/inifer ratios were used to prepare various molecular weight products. To prevent indanyl end-group formation experiments with binifer were carried out at -80°C by the use of methyl chloride/$\underline{n}$-hexane (80/20 v/v) mixed solvents (2). Experiments with trinifer were performed using pure methyl chloride at -40°C, since intramolecular ring formation leading to indanyl end-groups is impossible in trinifer systems (8). Essentially constant [M] and [I] could be maintained by using a minimum of solvent in the feed (needed to dissolve the inifers), otherwise due to the continuous dilution of the charge [M] and [I] would have continuously decreased. Results are compiled in Table I.

Evidently, polymers with molecular weight dispersities close to the theoretical values can be obtained even at 100% conversions by the semicontinuous technique, whereas much larger dispersities have been obtained in conventional runs. Inspection of GPC traces of products obtained in conventional batch binifer experiments showed broad distributions with a long tail toward the low molecular weight range. The GPC traces of products obtained in conventional batch trinifer experiments were broad and multimodal (see Figure 1) conceivably due to insufficient reactor control. In contrast, effective reactor control can be maintained throughout the semicontinuous experiments, for example, due to the slow and continuous feeding of the reagents sudden heat evolution can be easily avoided.

Efforts have been made to analyze the end-groups, particularly for unfired and once-fired end-groups, of linear and three-arm star polyisobutylenes obtained in binifer and trinifer experiments, respectively. Confirming our earlier observations and those of other researchers (7) products obtained in the presence of binifer do not exhibit unfired end-groups, however, may contain once-fired endgroups. The absence of unfired end groups is not surprising in view of the initiation activating effect of the p substituent ($Cl(CH_3)_2C-$) in the binifer or the p polymer residue ($PIB-(CH_3)_2C-$) in the "halfreacted" species.

The products obtained with trinifer by the conventional batch method contain both unfired and once-fired end-groups, as indicated by the representative ^1H NMR spectrum shown in Figure 2. To facilitate ^1H NMR spectroscopic analysis the tert.-chloro ended primary products have been quantitatively dehydrochlorinated in THF solution by $tBuO^\ominus$ to the corresponding exo olefins (2). In contrast, the products obtained by the semicontinuous technique with trinifer are virtually free of unfired or once-fired structures, as indicated by the ^1H NMR spectrum shown in Figure 3.

The molecular weight distribution of the corresponding polymer samples is also in accord with these observations. Products

Table I. Comparison of Semicontinuous and Conventional Polymerization Techniques for the Preparation of Telechelic Polyisobutylenes by Binifer and Trinifer

	BCl_3 in initial charge $M \times 10$	IB^*	Inifer in feed M	Feed rate ml/min	time min	$\bar{M}_n$	$\bar{M}_w/\bar{M}_n$
			Semicontinuous Binifer Runs:	$-80°C$,	$CH_3Cl/\underline{n}$-hexane	solvent mixture (80/20 v/v)	
1a	1.9	3.7	2.1×10^{-1}	11	6	980	2.0
2a	1.6	5.0	1.5×10^{-1}	8	92	2000	1.8
3a	4.7	3.3	6.5×10^{-1}	10	6	2900	1.8
			Semicontinuous Trinifer Runs:	$-40°C$,	CH_3Cl		
4a	4.7	6.4	1.7×10^{-1}	15	2.5	2500	1.4
5a	4.7	4.0	6.8×10^{-2}	21	2.1	3900	1.5
6a	4.7	6.0	5.2×10^{-2}	7	8	6700	1.3
7a	4.7	6.0	4.1×10^{-2}	11	5.1	8200	1.3
	$BCl_3\underline{M} \times 10$	$IB^*\underline{M}$	Binifer $\underline{M}$			$\bar{M}_n$	$\bar{M}_w/\bar{M}_n$
			Conventional Binifer Runs:	$-80°C$,	$CH_3Cl/\underline{n}$-hexane	solvent mixture (80/20 v/v)	
1b	1.6	0.8	4.5×10^{-2}			900	3.6
3b	3.1	1.3	2.6×10^{-2}			3200	3.0
			Conventional Trinifer Runs:	$-40°C$,	CH_3Cl solvent		
4b	3.3	1.7	4.6×10^{-1}			**	
5b	3.2	1.2	2.0×10^{-1}			**	
6b	2.9	2.1	1.8×10^{-1}			**	

* Monomer conversion was 100%   ** Multimodal molecular weight distribution, see Figure 1

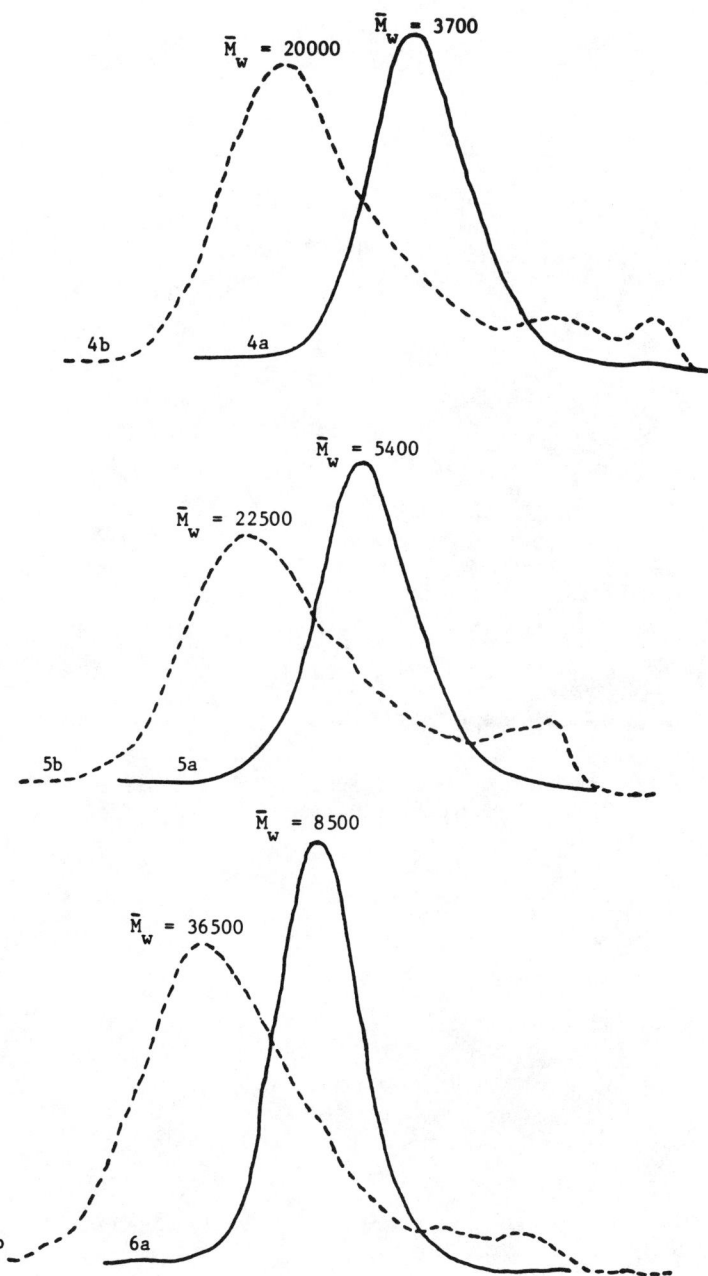

Figure 1. GPC Traces of Polymers by Batch and Semicontinuous Technique

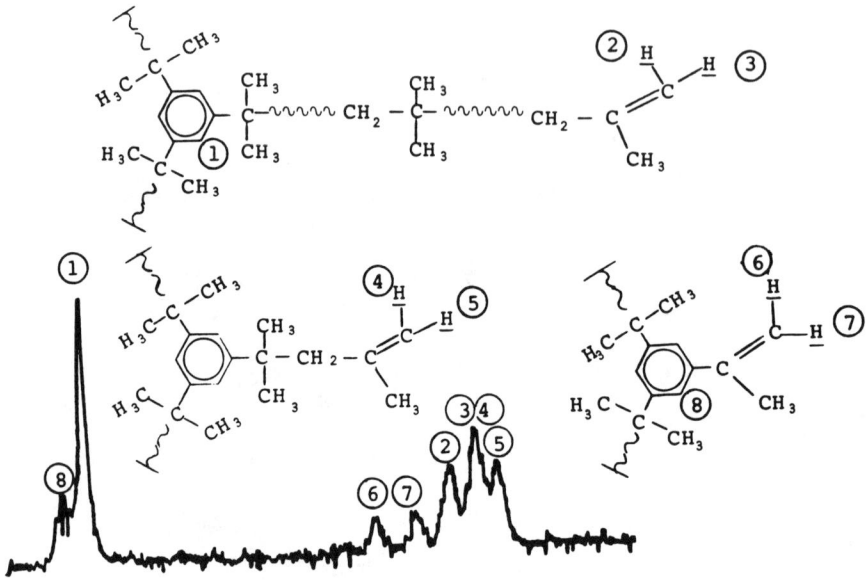

Figure 2. ^{1}H NMR Spectrum of Product 4b of Table I Obtained after Dehydrochlorination

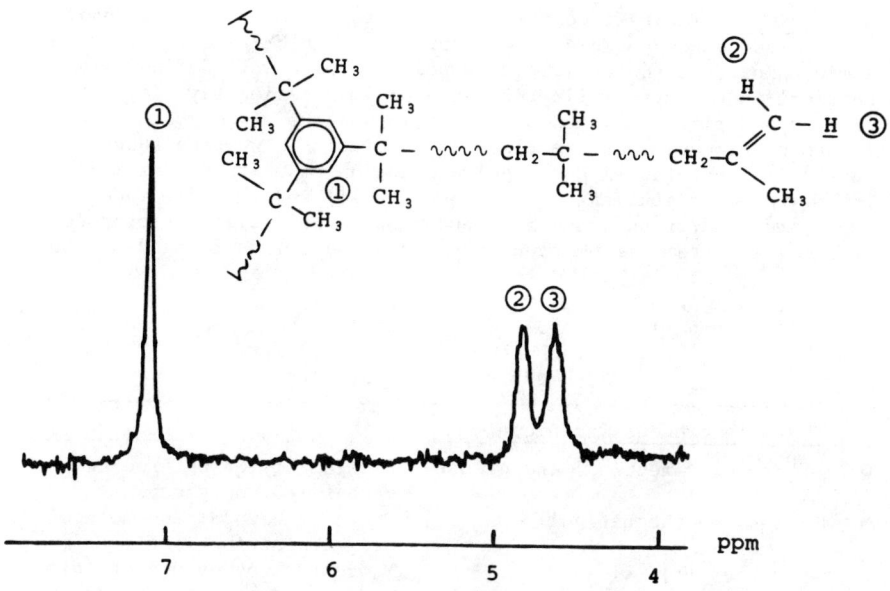

Figure 3. ^{1}H NMR Spectrum of Product 4a of Table I Obtained after Dehydrochlorination

containing unfired or once-fired end groups would not yield theoretical $\overline{M}_w/\overline{M}_n$ values even if [M] and [I] would remain constant during the polymerization; and, similarly, unfired or once-fired end groups must be absent in products exhibiting theoretical $\overline{M}_w/\overline{M}_n$ values. Thus the absence of unfired and once-fired end groups in polymers obtained in semicontinuous experiments with trinifer is indicated by the close to theoretical $\overline{M}_w/\overline{M}_n$ values obtained and independently by ^{1}H NMR spectroscopy.

Conclusions. Table II compares some characteristics of the semicontinuous and batch techniques. According to ^{1}H NMR and GPC data the semicontinuous technique produces polymers with well defined symmetrical end groups, while the batch polymerization may yield once-fired and unfired chain ends. The $\overline{M}_w/\overline{M}_n$ values of polymers obtained in semicontinuous runs are close to theoretical, whereas those harvested in conventional batch polymerization exhibit broader molecular weight distributions due to changing [I] and [M], presence of once-fired and/or unfired chain ends, and insufficient reactor control. The latter circumstance may even result in bimodal distributions.

Table II. Comparison between Conventional Batch and Semicontinuous Inifer Techniques for the Preparation of End-Reactive Polyisobutylenes

Semicontinuous Technique	Batch Technique
Well-defined symmetrical end groups	Possibility of unfired and/or once-fired end groups
$\overline{M}_w/\overline{M}_n$ close to theoretical	Broad or bimodal and molecular weight distribution
$\overline{M}_n$ control by [M]/[I]	$\overline{M}_n$ control by nature of inifer (C_I), very reactive or unreactive inifer unsuitable
100% conversion of I and M	100% conversion of both I and M only in specific cases
Reactor control easily maintained	Reactor control difficult, danger of heat jump on rapid BCl$_3$ introduction

The molecular weights of polymers formed in semicontinuous experiments are controlled by the concentration ratio [M]/[I] in the feed, while in batch polymerizations the molecular weights are controlled mostly by the reactivity of the inifer (i.e., the value of the chain transfer constant C_I rendering too reactive or unreactive inifers unsuitable). In batch polymerizations complete depletion of I and M is possible only in specific cases, while in semicontinuous runs both I and M are completely consumed and constant [M] and [I] are maintained. The semicontinuous technique facilitates satisfactory reactor control, while with the conventional batch technique reactor control is difficult, (i.e., heat jump on BCl$_3$ introduction) which is a major reason for broad or multimodal molecular weight distributions. Evidently the semicontinuous technique is a superior method for the preparation of symmetrical end-reactive polyisobutylenes by the inifer method than conventional batch polymerizations. Efforts are in progress to extend these studies to continuous polymerizations.

Acknowledgments

This material is based upon work supported by the National Science Foundation under grant DMR-81-20964.

Literature Cited

1. Kennedy, J. P., Smith, R. A.: *J. Polym. Sci.*, Polym. Chem., 18, 1523 (1980).
2. Kennedy, Chang, V.S.C., Smith, R. A., Ivan, B.: *Polym. Bull.*, 1, 575 (1979).
3. Ivan, B., Kennedy, J. P., Chang, V.S.C.: *J. Polym. Sci.*, Polym. Chem., 18, 317 (1980).
4. Kennedy, J. P., Ivan, B., Chang, V.S.C.: Urethane Chemistry and Applications, ACS Symposium Series 172, 383 (1981).
5. Kennedy, J. P., Guhaniyogi, S. C., Percec, V.: *Polym. Bull.*, 9, 27 (1983).
6. Fehervari, A., Kennedy, J. P., Tudos, F.: *J. Macromol. Sci.,-Chem.* A15, 215 (1981).
7. Tessier, M., Marechal, E.: *Polym. Bull.*, 10, 152 (1983).
8. Kennedy, J. P., Ross, L. R., Lackey, J. E., Nuyken, O.: *Polym. Bull.*, 4, 67 (1981).
9. Santos, R., Fehervari, A., Kennedy, J. P.: *J. Polym. Sci.*, Polym. Chem., in press.
10. Pask, S. D., Nuyken, O., Vischer, A., Walter, M.: IUPAC 6th International Symposium on Cationic Polymerization and Related Processes, Ghent, Belgium 1983.

RECEIVED February 6, 1985

12

Functionalization of Polymeric Organolithium Compounds
Synthesis of Macromolecular Ketones and Telechelic Amines

RODERIC P. QUIRK[1], WEI-CHIH CHEN, and PAO-LUO CHENG

Michigan Molecular Institute, Midland, MI 48640

>The amination and solid-state carbonation of polymeric
>organolithium compounds have been investigated. α,ω-Di-
>lithiumpolystyrene was aminated in 80% yield to form
>α,ω-diaminopolystyrene using the reagent generated
>from methoxyamine and methyllithium. The pure diamine
>was isolated by column chromatography. Solid-state
>carbonation of poly(styryl)lithium with high purity,
>gaseous carbon dioxide generated the dimeric ketone
>in >90% yield.

The alkyllithium-initiated, anionic polymerization of vinyl and diene monomers can often be performed without the incursion of spontaneous termination or chain transfer reactions (1). The non-terminating nature of these reactions has provided methods for the synthesis of polymers with predictable molecular weights and narrow molecular weight distributions (2). In addition, these polymerizations generate polymer chains with stable, carbanionic chain ends which, in principle, can be converted into a diverse array of functional end groups using the rich and varied chemistry of organolithium compounds (3).

The application of these functionalization reactions to polymers has been catalogued in the anionic polymer review literature (4-6). Unfortunately, many of the reported applications of these functionalization reactions to anionic chain-ended polymers have not been well characterized (7). In order to exploit these functionalization reactions to their potential, well-defined procedures for quantitative chain end functionalization must be available.

We have previously reported the results of careful investigations of the solution carbonation (8) and oxidation (9) of polymeric organolithium compounds. These studies have been extended to the investigation of solid-state carbonation reactions and these results are reported herein. In addition, a new method has been developed for the synthesis of telechelic polymers with primary amine end-group

[1] Current address: Institute of Polymer Science, The University of Akron, Akron, OH 44325

0097-6156/85/0282-0139$06.00/0
© 1985 American Chemical Society

functionality using the aminating reagent generated from methoxyamine and methyllithium (10,11).

Experimental

Styrene, benzene, and tetrahydrofuran were purified as described previously (8,11). Solutions of sec-butyllithium (Lithium Corporation of America, 12.0 wt % in cyclohexane) and methyllithium (Alfa, 1.45 M in ether) and lithium naphthalene were analyzed using the double titration procedure with 1,2-dibromoethane (12). Lithium naphthalene was prepared in tetrahydrofuran from lithium metal and a 25 mole % excess of sublimed naphthalene at -25°C using standard high vacuum procedures. Sealed ampoules of lithium naphthalene were stored in liquid nitrogen.

sec-Butyllithium-initiated polymerizations were carried out at 30°C in all glass, sealed reactors using break-seals and standard high vacuum techniques (2). α,ω-Dilithiumpolystyrene was prepared using high vacuum techniques with lithium naphthalene as initiator in a tetrahydrofuran/benzene (69/300) solvent mixture. The solution of a α,ω-dilithiumpolystyrene was stored at -78°C.

Amination (11) and solution carbonation (8) reactions were carried out as described previously. For solid-state carbonations, a benzene solution of poly(styryl)lithium was freeze-dried on the vacuum line followed by introduction of high-purity, gaseous carbon dioxide (Air Products, 99.99% pure). Analysis and characterization of polymeric amines (11) and carboxylic acids (8) were performed as described previously. Benzoyl derivatives of the aminated polystyrenes were prepared in toluene/pyridine (2/1. v/v) mixtures with benzoyl chloride (Aldrich, 99%).

Number-average molar masses were determined using a vapor pressure osmometer (VPO) (Hitachi 117 Molecular Weight Apparatus) at 54.8±0.1°C in toluene (Fisher Scientific, certified A.C.S.) which was distilled from freshly crushed CaH_2. The VPO apparatus was calibrated with pentaerythritol tetrastearate (Pressure Chemical). Gel permeation chromatographic (GPC) analyses were performed in tetrahydrofuran by HPLC (Perkin-Elmer 601 HPLC) using six μ-Styragel columns (10^6, 10^5, 10^4, 10^3, 500, and 100 Å) after calibration with standard polystyrene samples.

The concentrations of aminated chain ends were determined by titrating polymer samples dissolved in 100 ml of a 1/1 (v/v) mixture of chloroform and glacial acetic acid with standardized $HClO_4$ in glacial acetic acid using a Corning Model 10 pH meter with a calomel glass electrode (13). Elemental analyses were performed by Organic Microanalysis (P.O. Box 41838, Tucson, Arizona 85717). The reported results are in good agreement with the calculated values which were derived from $\overline{M}_n$ values (see Table I).

Results and Discussion

Amination. The synthesis of polymers with primary amine end-group functionality has been a challenge because the primary amine group can undergo rapid chain transfer and termination reactions with carbanionic chain ends (14). Schulz and Halasa (15) used a phenyllithium initiator with a bis(trimethylsilyl)-protected amine group to prepare amine-terminated polydienes. Nakahama and coworkers (16,17)

have utilized the reaction of polymeric anions with a trimethylsilyl-
protected imine to prepare amine-terminated polymers. Beak and Kokko
(10) recently described a method for the direct amination of simple
organolithium compounds using the reagent generated from methoxyamine
and methyllithium in hexane-diethyl ether (eq. 1).

$$RLi \xrightarrow[2) \text{ } H_2O]{1) \text{ } CH_3ONH_2/CH_3Li} RNH_2 \quad (1)$$

We have previously reported results of the application of these procedures for the amination of poly(styryl)lithium (11). After an
investigation of a variety of procedures, a 92% yield of poly(styryl)-
amine was obtained using two equivalents of the aminating reagent in
a THF/Et$_2$O/hexane mixture of -78°C followed by slow warming to -15°C
and quenching in methanol (eq. 2).

$$PSLi \xrightarrow[\substack{2) \text{ } -15°C \\ 3) \text{ } CH_3OH}]{1) \text{ } CH_3ONH_2/CH_3Li \\ -78°C} PSNH_2 \quad (2)$$

$\overline{M}_n = 2000$

It was of interest to explore the possibility of preparing α,ω-di-
aminopolystyrene using this aminating procedure.

α,ω-Dilithiumpolystyrene was prepared from styrene and lithium
naphthalene in a benzene/THF (300/69 v/v) mixture (eq. 3).

$$2n+2 \text{ Styrene} + 2Li^+Naph^{\overline{\cdot}} \rightarrow \underset{\underset{C_6H_5}{|}}{LiCHCH_2}\underset{\underset{C_6H_5}{|}}{\text{[}CHCH_2\text{]}_n}\underset{\underset{C_6H_5}{|}}{\text{[}CH_2CH\text{]}_n}\underset{\underset{C_6H_5}{|}}{CH_2CHLi} \quad (3)$$

<u>1</u>

The lithium naphthalene was prepared in THF from lithium metal and a
25 mole % excess of naphthalene to minimize formation of the dilith-
ium naphthalene dianion (18,19). A sample of α,ω-dilithiumpolysty-
rene was quenched with methanol for molecular weight characterization
(see Table II).

The amination of α,ω-dilithiumpolystyrenes was effected using
two equivalents of the reagent generated from methyllithium and
methoxyamine in a benzene/THF (80/20, v/v) mixture with the results
shown in eq. 4.

$$\underline{1} \xrightarrow[1) \text{ } -78°C \text{ } 2) \text{ } -15°C]{CH_3ONH_2/CH_3Li} \xrightarrow{CH_3OH} \quad (4)$$

$$H_2NCHCH_2\underset{\underset{C_6H_5}{|}}{\text{[}CHCH_2\text{]}_n}\underset{\underset{C_6H_5}{|}}{\text{[}CH_2CH\text{]}_n}\underset{\underset{C_6H_5}{|}}{CH_2CHNH_2}$$
$\underset{C_6H_5}{|}$

80%

+ PSNH + PSH

8% 12%

Table I. Elemental Analyses for α,ω-Diaminopolystyrene and the Corresponding Dibenzoyl Derivative

Sample	% C Calc.	% C Obs.	% H Calc.	% H Obs.	% N[a] Calc.	% N[a] Obs.
α,ω-Diaminopolystyrene	92.04[b]	92.25	7.71[b]	7.82	0.25[b]	0.10
Dibenzoyl-α,ω-diaminopolystyrene	91.82[c]	91.96	7.63[c]	7.84	0.25[c]	0.10

[a] The observed nitrogen content was close to the level of detection for this method (±0.06%).

[b] Calculated for $\overline{M}_n$=10952.

[c] Calculated for $\overline{M}_n$=11160.

Table II. Molecular Weight Characterization of α,ω-Diaminopolystyrene

Analytical Method	$\overline{M}_n$
GPC[a]	14,000
End-Group Titration	11,000
VPO[b]	12,500
Stoichiometry[c]	11,400
Elemental Analyses	11,000

[a] Dibenzoyl derivative. Based on polystyrene calibration.

[b] $\overline{M}_n$(VPO) = 12,000 for the polystyrene precursor.

[c] $\overline{M}_n$ calculated from the ratio of gm of monomer to one-half the moles of initiator (Naph⁻ Li⁺).

A variety of procedures were utilized to analyze this reaction mixture and to characterize α,ω-diaminopolystyrene. Thin layer chromatographic analysis using toluene as eluent exhibited three spots with R_f values of 0.85, 0.09, and 0.05 which corresponded to polystyrene, poly(styryl)amine and α,ω-diaminopolystyrene (see Figure 1). Pure samples of each of these products were obtained by silica gel column chromatography of the crude reaction mixture initially using toluene as eluent [for polystyrene and poly(styryl)amine] followed by a methanol/toluene mixture (5/100 v/v) for the diamine. Size-exclusion chromatography could not be used to characterize the diamine since no peak was observed for this material, apparently because of the complication of physical adsorption to the column packing material. Therefore, the dibenzoyl derivative (eq. 5) was prepared and used for most of the analytical characterizations.

$$H_2N-[PS]-NH_2 \xrightarrow[\text{pyridine}]{2C_6H_5COCl} C_6H_5CONH-[PS]-NHCOC_6H_5 \qquad (5)$$

The results of the various methods used to determine the molecular weight and degree of functionality of the diamine are shown in Table II. All of this evidence is consistent with the conclusion that the purified, isolated diamine possesses high difunctionality. When the diamine was chain extended with terephthaloyl chloride, a polymer product with a symmetrical molecular weight distribution was obtained (D.P. ≅ 10). Inexact stoichiometry is most probably responsible for the limited extent of chain extension.

Carbonation. The carbonation of polymeric organolithium compounds using carbon dioxide is one of the most useful functionalization reactions. However, there are special problems associated with the carbonation of polymeric organolithium compounds. For example, Wyman, Allen and Altares (20) reported that the carbonation of poly(styryl)lithium in benzene with gaseous carbon dioxide produced only a 60% yield of carboxylic acid; the acid was contaminated with significant amounts of the corresponding ketone (dimer) and tertiary alcohol (trimer) as shown in eq. 6. A recent, careful, detailed investigation of the carbonation of polymeric organolithium compounds has

$$PSLi \xrightarrow[\text{2) H}^+]{\text{1) CO}_2(g)} PSCO_2H + (PS)_2CO + (PS)_3COH \qquad (6)$$
$$\phantom{PSLi \xrightarrow[\text{2) H}^+]{\text{1) CO}_2(g)}} 60\% \quad\ \ 28\% \qquad 12\%$$

been reported (8). It was postulated that the association of polymeric organolithium compounds (21-25) promotes intra-aggregate coupling to form the dimeric ketone product as shown in eqs. 7,8.

$$(PSLi)_2 + CO_2 \longrightarrow (PSCO_2Li)(PSLi) \qquad (7)$$
$$\text{Associated}$$

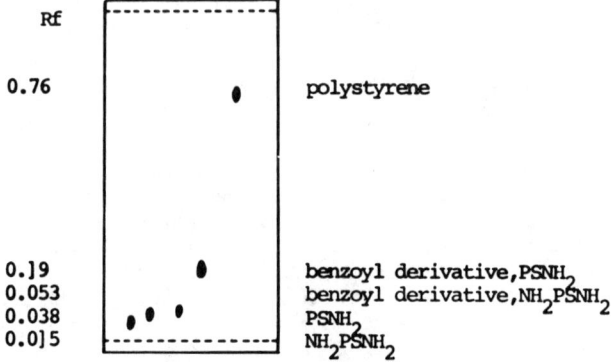

Figure 1. TLC of $PSH, PSNH_2, PS(NH_2)_2$ and derivatives.

$$(\text{PSCO}_2\text{Li})(\text{PSLi}) \rightarrow \text{PS}-\underset{\underset{\text{OLi}}{|}}{\overset{\overset{\text{OLi}}{|}}{\text{C}}}-\text{PS} \xrightarrow{\text{H}_2\text{O}} (\text{PS})_2\text{CO} \qquad (8)$$

Since Lewis base additives and basic solvents such as tetrahydrofuran are known to deaggregate polymeric organolithium compounds, (21,23,26) it was postulated that ketone formation would be minimized in the presence of sufficient tetrahydrofuran to effect dissociation of the aggregates. In complete accord with these predictions, it was found that the carbonation of poly(styryl)lithium (eq. 9), poly(isoprenyl)-lithium, and poly(styrene-b-isoprenyl)lithium in a 75/25 mixture (by volume) of benzene and tetrahydrofuran occurs quantitatively to produce the carboxylic acid chain ends (8).

$$\text{PSLi} \xrightarrow[\text{C}_6\text{H}_6/\text{THF}]{\text{CO}_2(\text{g})} \xrightarrow{\text{H}_3\text{O}^+} \text{PSCO}_2\text{H} \qquad (9)$$

(75/25)   100%

If association of the chain ends promotes coupling to form the ketone (eqs. 7,8), it was anticipated that ketone formation would be favored by conditions which maximize chain-end association. In general, higher degrees of association of organolithium compounds are favored by higher concentrations, lower temperatures, and the absence of Lewis bases (3). Our approach to maximize polymeric organolithium chain-end association was to freeze dry benzene solutions of polymeric organolithium compounds. The direct carbonation of a freeze-dried sample of poly(styryl)lithium ($\overline{M}_n$=4.2x10^3) produced the corresponding ketone in greater than 90% yield. The only other product was the unfunctionalized homopolymer, polystyrene, as shown in eq. 10.

$$(\text{PSLi})_n \xrightarrow[\text{2) H}_3\text{O}^+]{\text{1) CO}_2(\text{g})} (\text{PS})_2\text{CO}+\text{PSH} \qquad (10)$$

freeze-dried   90%   10%

No carboxylic acid functionality was detected either by thin-layer chromatographic analyses or by end-group titration. Therefore, procedures are now available to control the carbonation of polymeric organolithium compounds to efficiently produce either the carboxylated chain ends or the corresponding ketone dimer.

Literature Cited

1. Szwarc, M., "Carbanions, Living Polymers and Electron Transfer Processes," Interscience Publishers: New York, 1968.
2. Morton, M.M.; Fetters, L.J., Rubber Chem. Tech., 1975, 48, 359.
3. Wakefield, B.J., "The Chemistry of Organolithium Compounds"; Pergamon Press: Oxford, England, 1974.
4. Morton, M.M.; Fetters, L.J., Macromol. Rev., 1967, 2, 71.
5. Fetters, L.J., J. Polym. Sci., Part C, 1969, 26, 1.
6. Bywater, S., Prog. Polym. Sci., 1974, 4, 27.

7. Young, R.N.; Quirk, R.P.; Fetters, L.J., Adv. Polym. Sci., 1984, 56, 1.
8. Quirk. R.P.; Chen, W.-C., Makromol. Chem., 1982, 183, 2071.
9. Quirk, R.P.; Chen, W.-C., J. Polym. Sci.: Polym. Chem. Ed., 22, 2993 (1984).
10. Beak, P.; Kokko, B.J., J. Org. Chem., 1982, 47, 2822.
11. Quirk, R.P.; Cheng, P.-L., Polym. Preps. Amer. Chem. Soc., Div. Polym. Chem., 1983, 24(2), 461.
12. Gilman, H.; Cartledge, F.K., J. Organomet. Chem., 1964, 2, 447.
13. Fritz, J.S., Acid-Base Titration in Nonaqueous Solvents; Allyn and Bacon: Boston, 1973; Chapter 2.
14. Higginson, W.C.E.; Wooding, N.S., J. Chem. Soc., 1952, 760.
15. Schulz, D.N.; Halasa, A.F., J. Polym. Sci.: Polym. Chem. Ed., 1977, 15, 2401.
16. Hirao, A.; Hattori, I.; Sasagawa, T.; Yamaguchi, K.; Nakahama, S., Makromol. Chem., Rapid Commun., 1982, 3, 59.
17. Hattori, I.; Hirao, A.; Yamaguchi, K.; Nakahama, S.; Yamazaki, N., Makromol. Chem., 1983, 184, 1355.
18. Hsieh, H.L., J. Organomet. Chem., 1967, 7, 1.
19. Smid, J., J. Am. Chem. Soc., 1965, 87, 655.
20. Wyman, D.P.; Allen, V.R.; Altares, T., J. Polym. Sci., Part A, 1964, 2, 4545.
21. Morton, M.M.; Fetters, L.J.; Pett, R.A. and Meier, J.F., Macromolecules, 1970, 3, 327.
22. Worsfold, D.J.; Bywater, S., Macromolecules, 1972, 5, 393.
23. Fetters, L.J.; Morton, M.M., Macromolecules, 1974, 7, 552.
24. Al-Jarrah, M.M.F.; Young, R.N., Polymer, 1980, 21, 119.
25. Hernandez, A.; Semel, J.; Broecker, H.-C.; Zachmann, H.G.; Sinn, H., Makromol. Chem., Rapid Commun., 1980, 1, 75.
26. Morton, M.M.; Fetters, L.J., J. Polym. Sci., Part A, 1964, 2, 3311.

RECEIVED February 8, 1985

Free-Radical Ring-Opening Polymerization
Use in Synthesis of Reactive Oligomers

WILLIAM J. BAILEY, BENJAMIN GAPUD, YIN-NIAN LIN, ZHENDE NI, and SHANG-REN WU

Department of Chemistry, University of Maryland, College Park, MD 20742

>A new method was developed for the synthesis of reactive, end-capped oligomers which involved copolymerization of a vinyl monomer with a cyclic monomer which can undergo free radical ring-opening copolymerization followed by hydrolysis of the resulting copolymer. Since the free radical ring-opening polymerization makes possible the introduction of functional groups, such as esters, carbonates, amides, and thioesters, into the backbone of an addition polymer, hydrolysis of the copolymer will give oligmers terminated with various combinations of hydroxyl, amino, thiol, and carboxyl groups. For example, copolymerization of 2-methylene-1,3-dioxepane and styrene (r_1=0.021 and r_2=22.6) produced a copolymer containing 10 mole-percent of ester-containing units with 100% ring opening. Hydrolysis of this copolymer gave an oligomer of styrene endcapped with a hydroxyl group on one end and a carboxyl group on the other end. Similarly copolymerization of ethylene with the cyclic ketene acetal gave a series of biodegradable copolymers which, upon hydrolysis, gave oligomers of ethylene capped with a hydroxyl and a carboxyl group.

Even though functionally terminated oligomers are commercially important, most oligomers are made by ionic addition polymerization or condensation reactions rather than the convenient and inexpensive free radical process. One of the few exceptions is the hydroxyterminated polybutadiene produced by a free radical process by Arco Chemical Company. Also, the polymerization of butadiene and sulfur by a free radical mechanism that involves a ring-opening of the S_8 ring followed by reduction of the resulting polysulfide groups gives a mercapto-terminated polymer which has found limited use (1). Since it was shown that free radical ring-opening polymerization (2) made it possible to introduce functional groups, such as esters (3), carbonates (4), thioesters (5), and amides (6), into the backbone of an addition polymer, it was reasoned that simple hydrolysis would produce the desired oligomers that could be terminated with various combinations of hydroxyl, amino, thiol, and carboxylic acid groups.

0097-6156/85/0282-0147$06.00/0
© 1985 American Chemical Society

The fact that examples in the literature of free radical ring-opening polymerization are quite rare is rather surprising in view of the fact that the ionic ring-opening polymerization of heterocyclic compounds, such as ethylene oxide, tetrahydrofuran, ethylenimine, β-propiolactone and caprolactam, as well as the Ziegler-Natta metathesis ring-opening polymerization of cyclic olefins, such as cyclopentene and norbornene, are quite numerous. The few examples of free radical ring-opening polymerization that are reported in the literature include derivatives of vinylcyclopropane (7,8), o-xylylene dimer (9), derivatives of bicyclo[1.1.0]butane (10), and elemental sulfur (11).

This is probably related to the fact that simple unstrained five- or six-membered carbocyclic rings do not undergo radical ring opening readily, and in fact the open-chain radicals have been shown to undergo ring closure to the corresponding unstrained cyclic radical. For example, Butler and Angelo (12) found that diallyldimethylammonium bromide would undergo inter-intramolecular polymerization to produce a soluble polymer. Apparently the reaction is kinetically controlled to form the five-membered ring rather than the thermodynamically favored six-membered ring.

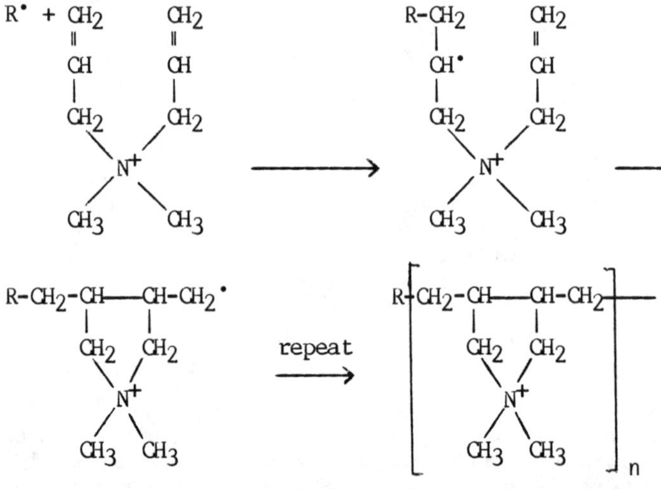

The course of some of these ring-opening and ring-closing polymerizations can be explained by the recent data of Maillard, Forrest, and Ingold (13). They studied the transformations in the cyclopropylmethyl and the cyclopentylmethyl series by electron spin resonance. In the case of the three-membered radical the reaction involves ring opening since the energy is favorable (5.94 kcal) and the rate of reaction is very high (1.3×10^8 s^{-1}). In the case of the five membered ring system the reaction proceeds in the direction of ring-closure since the energetics of that reaction is favorable (7.8 kcal) and the rate of the ring closure is also moderately high (1.0×10^5 s^{-1}). In an effort to find some mechanism other than relief of strain or aromatization to promote free radical ring-opening polymerization, we reasoned that since a carbon-oxygen double bond is 40-50 kcal more stable than a carbon-carbon double bond, the introduction of an oxygen atom into an unsaturated cyclic monomer appeared to be a possible solution.

Free Radical Ring-Opening Polymerization

For this reason, a reinvestigation of the cyclic ketene acetal, 2-methylene-1,3-dioxolane (I), that had been prepared by McElvain and Curry (14) was undertaken. Although McElvain and Beyerstedt (15) reported that benzoyl peroxide had no appreciable effect on diethyl ketene acetal, no such study was reported (14) for the 2-methylene-1,3-dioxolane (I). The synthesis was carried out as follows (6):

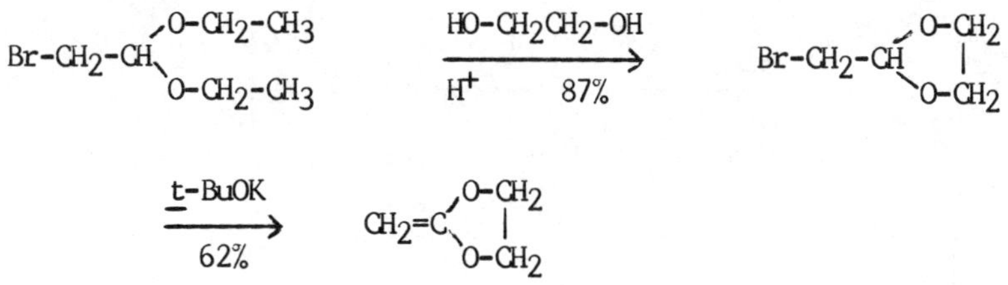

$$Br-CH_2-CH\begin{matrix}O-CH_2-CH_3\\O-CH_2-CH_3\end{matrix} \xrightarrow[H^+]{HO-CH_2CH_2-OH} \xrightarrow{87\%} Br-CH_2-CH\begin{matrix}O-CH_2\\|\\O-CH_2\end{matrix}$$

$$\xrightarrow[62\%]{t\text{-BuOK}} CH_2=C\begin{matrix}O-CH_2\\|\\O-CH_2\end{matrix}$$

I

Treatment of this monomer with benzoyl peroxide gave a high molecular weight polyester by a free radical ring-opening polymerization which can be rationalized by the accompanying scheme. The structure of the polyester IV was established by analysis and hydrolysis as well as infrared and NMR spectroscopy.

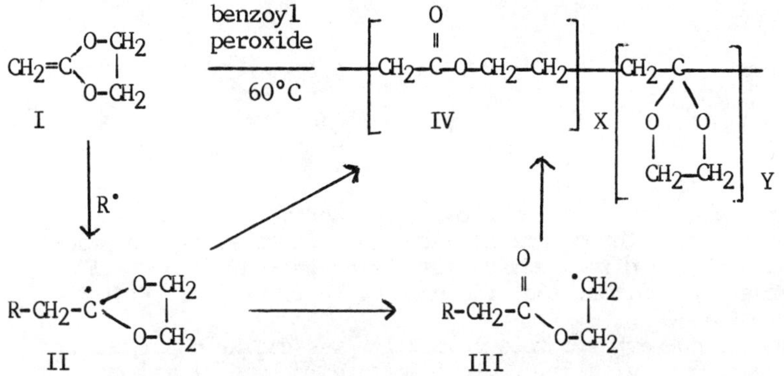

$$CH_2=C\begin{matrix}O-CH_2\\|\\O-CH_2\end{matrix} \xrightarrow[60°C]{\text{benzoyl peroxide}} \left[-CH_2-\overset{O}{\underset{\|}{C}}-O-CH_2-CH_2-\right]_{IV} \left[CH_2-C\begin{matrix}\\/\ \backslash\\O\ \ O\\|\ \ |\\CH_2-CH_2\end{matrix}\right]_X Y$$

I

$$\downarrow R^\bullet$$

$$R-CH_2-\overset{\bullet}{C}\begin{matrix}O-CH_2\\|\\O-CH_2\end{matrix} \longrightarrow R-CH_2-\overset{O}{\underset{\|}{C}}\begin{matrix}\ ^\bullet CH_2\\|\\O-CH_2\end{matrix}$$

II   III

At 60°C only 50% of the rings were opened and at 120°C, 87% of the rings were opened (6); high dilution also favored the extent of ring opening. There was a competition between the direct addition of the intermediate radical II and its ring opening to the radical III. An alternative method of analysis of the extent of ring opening was the basic hydrolysis of the copolymer IV, which cleaved the ester groups but left the cyclic ketals intact. Copolymerization of styrene and I gave a copolymer containing mostly ring-opened plus some non-ring-opened units.

In a search for other cyclic acetals that would undergo quantitative ring opening even at room temperature we prepared the seven-membered ketene acetal, 2-methylene-1,3-dioxepane (V), which underwent essentially complete ring opening at room temperature. This process can be considered a major breakthrough in polymer chemistry which makes possible the quantitative introduction of an ester group in the backbone of an addition polymer.

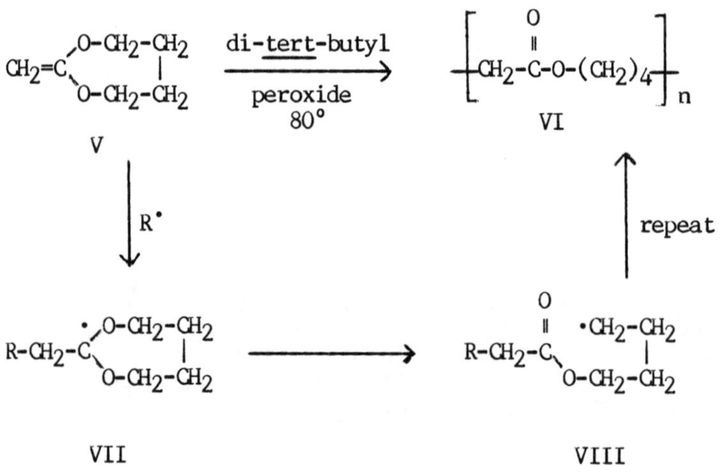

$$CH_2=C\begin{smallmatrix}O-CH_2-CH_2\\|\\O-CH_2-CH_2\end{smallmatrix} \xrightarrow[80°]{\text{di-tert-butyl peroxide}} \left[CH_2-\overset{O}{\overset{\|}{C}}-O-(CH_2)_4\right]_n$$

V ………………………………………………………… VI

$$V \downarrow R^\bullet \qquad\qquad \uparrow \text{repeat}$$

$$R-CH_2-\overset{\bullet}{C}\begin{smallmatrix}O-CH_2-CH_2\\|\\O-CH_2-CH_2\end{smallmatrix} \longrightarrow R-CH_2-\overset{O}{\overset{\|}{C}}\begin{smallmatrix}\bullet CH_2-CH_2\\|\\O-CH_2-CH_2\end{smallmatrix}$$

VII ……………………………………………………… VIII

Models show that the seven-membered ring increases the steric hindrance in the intermediate free radical VII to eliminate practically all of the direct addition and also introduces a small amount of strain so that the ring opening to the radical VIII is accelerated.

The seven-membered ketene acetal V was easily copolymerized with styrene, 4-vinylanisole, vinyl acetate, ethylene, and vinyl chloride, to give copolymers with ester groups in the main chain, all with quantitative ring opening. For example, by the use of a large amount of styrene and a small amount of the ketene acetal V, followed by hydrolysis, an oligomer of styrene was produced that was capped with a hydroxyl group and a carboxylic acid group.

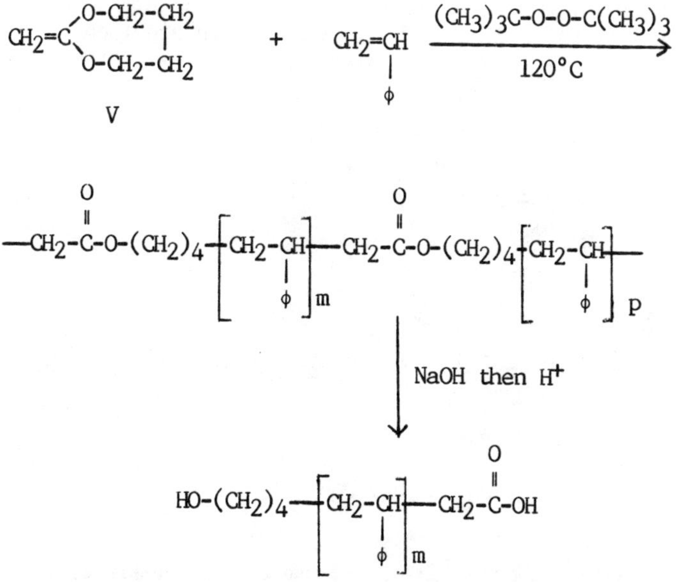

$$CH_2=C\diagup\begin{matrix}O-CH_2-CH_2\\ |\\ O-CH_2-CH_2\end{matrix} \quad + \quad \begin{matrix}CH_2=CH\\ |\\ \phi\end{matrix} \quad \xrightarrow[120°C]{(CH_3)_3C-O-O-C(CH_3)_3}$$

V

$$-CH_2-\overset{O}{\overset{\|}{C}}-O-(CH_2)_4-\left[CH_2-\underset{\phi}{\overset{|}{CH}}\right]_m-CH_2-\overset{O}{\overset{\|}{C}}-O-(CH_2)_4-\left[CH_2-\underset{\phi}{\overset{|}{CH}}\right]_p-$$

$$\downarrow \text{NaOH then } H^+$$

$$HO-(CH_2)_4-\left[CH_2-\underset{\phi}{\overset{|}{CH}}\right]_m-CH_2-\overset{O}{\overset{\|}{C}}-OH$$

For the copolymerization of the ketene acetal V with styrene, r_1 is 0.021 and r_2 is 22.6 at 120°C. With a mixture containing about 80% V and 20% styrene, a copolymer containing 90 mole-% styrene and 10 mole-% ester-containing units was obtained. The hydrolysis of this copolymer gave an oligomer of styrene containing an average of about nine styrene units end capped with the hydroxyl and carboxylic acid groups. Thus a very general method has been developed for the synthesis of a wide variety of oligomers with any desired molecular weight range. Of course, since the copolymers are random, the molecular weight distribution of the oligomers is quite broad. However, these oligomers should prove quite useful for the synthesis of polyurethanes and block polyesters.

In an effort to produce a biodegradable addition polymer, the 2-methylene-1,3-dioxepane (V) and ethylene were copolymerized at 120°C for 30 minutes at a pressure of 1800 psi to give a low conversion of copolymers with ester-containing units varying from 2.1 to 10.4 mole-%. Although most synthetic polymers are nonbiodegradable since they have not been on the earth long enough for microorganisms or enzyme systems to have evolved to utilize them as food, polyesters that are relatively low molecular weight and rather low melting are biodegradable (16). This observation is related to the fact that poly(β-hydroxybutyric acid) occurs widely in nature and many micro-organisms use this polyester to store energy in the same way that animals use fat. On the other hand, no synthetic addition polymer was known that was readily biodegradable. However, the ethylene copolymers containing the ester groups in the backbone were in fact biodegradable with the copolymers containing the high amount of ester groups being rapidly degraded and the copolymers containing only 2.1% comonomer only slowly degraded (17). Apparently there are enzymes in the micro-organisms that are capable of hydrolyzing the ester linkages in the ethylene copolymer to produce the oligomers with terminal carboxylic acid groups; these oligomers are then degraded as analogs of fatty acids by the normal metabolic processes.

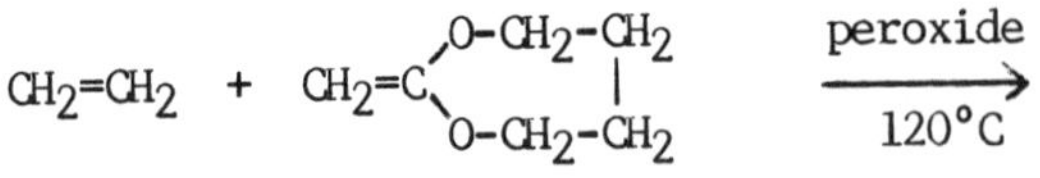

$$CH_2=CH_2 + CH_2=C\begin{matrix}O-CH_2-CH_2\\ |\\ O-CH_2-CH_2\end{matrix} \xrightarrow[120°C]{peroxide}$$

V

$$-CH_2-\overset{O}{\overset{\|}{C}}-O-(CH_2)_4-[CH_2-CH_2]_m-CH_2-\overset{O}{\overset{\|}{C}}-O-(CH_2)_4-[CH_2-CH_2]_n-$$

$$\xrightarrow[then\ H^+]{OH^-} HO-(CH_2)_4-[CH_2-CH_2]_m-CH_2-\overset{O}{\overset{\|}{C}}-OH$$

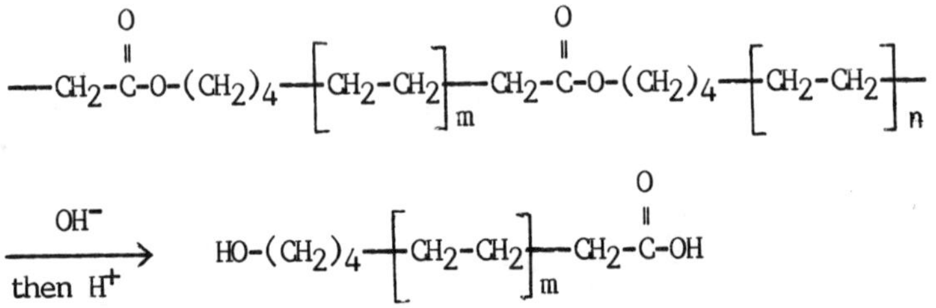

When the ethylene-2-methylene-1,3-dioxepane copolymer was hydrolyzed, it gave oligomers which were capped with a hydroxyl group at one end and a carboxylic acid group at the other. For material with an ester-containing unit content of 2.1 mole-%, the value of n was approximately 47 and for material with a content of 10.4 mole-%, the value of n was approximately 9. The copolymers with 6 or less mole-% of the ester-containing units had melting points in excess of 90°C.

Related work had shown that the nitrogen analogs of the cyclic ketene acetals were readily synthesized and would polymerize with essentially 100% ring opening. For this reason their copolymerization with a variety of monomers was undertaken (6).

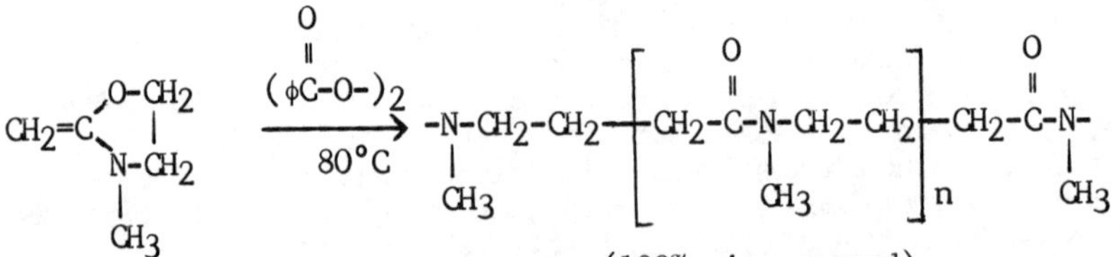

$$CH_2=C\begin{matrix}O-CH_2\\ |\\ N-CH_2\\ |\\ CH_3\end{matrix} \xrightarrow[80°C]{(\phi C-O-)_2} -N-CH_2-CH_2-\begin{bmatrix}\overset{O}{\overset{\|}{C}}H_2-C-N-CH_2-CH_2\\ |\\ CH_3\end{bmatrix}_n -CH_2-\overset{O}{\overset{\|}{C}}-N-\\ |\\ CH_3$$

IX

(100% ring opened)

Thus the amide linkage is sufficiently more stable than the ester group to greatly favor the ring opening. Copolymerization of IX with styrene proceeded with essentially quantitative ring opening (6). In the case of the copolymer with styrene and IX, the copolymer was readily hydrolyzed to give an oligomer of styrene capped with an aminomethyl group and a carboxylic acid group.

Also in a related study the sulfur analog of the cyclic ketene acetal X was prepared and polymerized. However, since the resulting thioester is apparently higher in energy than the ordinary ester and therefore retards the extent of ring opening, even at 120°C only 45% of the rings were opened. Nevertheless, copolymerization of X with styrene gave a copolymer containing some thioester groups and hydrolysis of this copolymer gave an oligomer capped with a mercaptan and a carboxylic acid group.

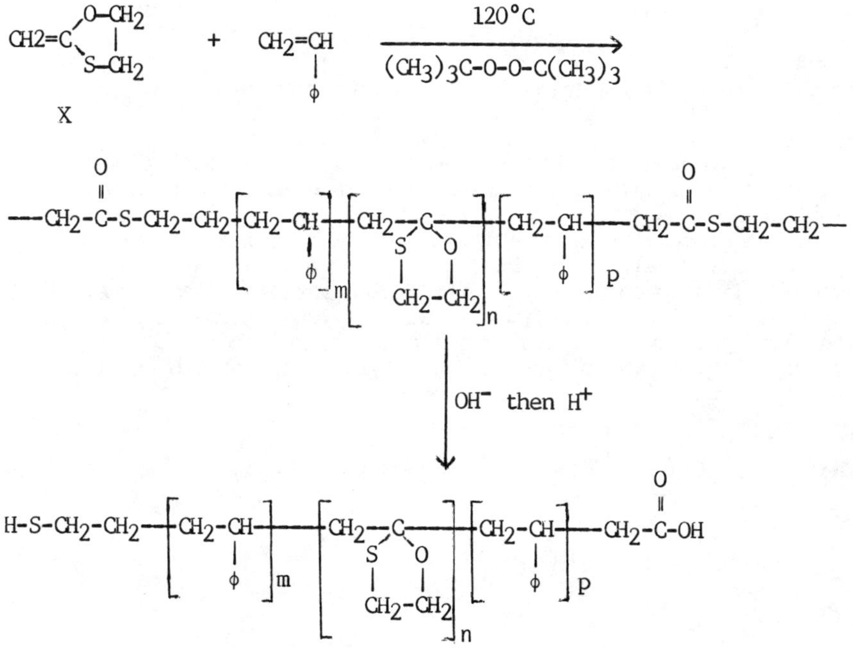

$$CH_2=C\underset{S-CH_2}{\overset{O-CH_2}{\diagup}} \quad + \quad CH_2=CH\underset{\phi}{|} \quad \xrightarrow[(CH_3)_3C-O-O-C(CH_3)_3]{120°C}$$

X

$$---CH_2-C(=O)-S-CH_2-CH_2-[CH_2-CH(\phi)]_m-[CH_2-C(S,O)(CH_2-CH_2)]_n-[CH_2-CH(\phi)]_p-CH_2-C(=O)-S-CH_2-CH_2---$$

↓ OH⁻ then H⁺

$$H-S-CH_2-CH_2-[CH_2-CH(\phi)]_m-[CH_2-C(S,O)(CH_2-CH_2)]_n-[CH_2-CH(\phi)]_p-CH_2-C(=O)-OH$$

Finally, the unsaturated spiro ortho carbonates were been shown to undergo double ring opening in a free radical polymerization to introduce a carbonate group in the backbone of an addition polymer (19).

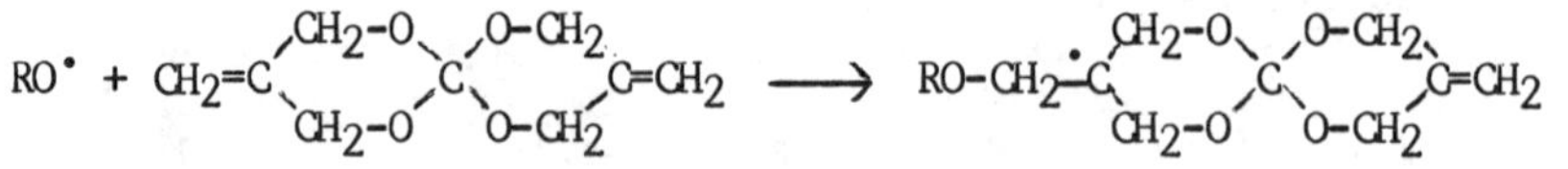

$$RO^{\bullet} + CH_2=C\underset{CH_2-O}{\overset{CH_2-O}{<}}C\underset{O-CH_2}{\overset{O-CH_2}{>}}C=CH_2 \longrightarrow RO-CH_2-\overset{\bullet}{C}\underset{CH_2-O}{\overset{CH_2-O}{<}}C\underset{O-CH_2}{\overset{O-CH_2}{>}}C=CH_2$$

IX

$$RO-CH_2-C\underset{CH_2-O}{\overset{CH_2\ \overset{\bullet}{O}}{<}}C\underset{O-CH_2}{\overset{O-CH_2}{>}}C=CH_2 \longrightarrow RO-CH_2-C\underset{CH_2-O}{\overset{=CH_2\ \ O}{<}}C\underset{{\overset{\bullet}{O}}-CH_2}{\overset{O-CH_2}{>}}C=CH_2$$

$$\xrightarrow{\text{repeat}} RO{\left[\!\!-CH_2-\overset{\overset{CH_2}{\|}}{C}-CH_2-O-\overset{\overset{O}{\|}}{C}-O-CH_2-\overset{\overset{CH_2}{\|}}{C}-CH_2-O-\!\!\right]}_n$$

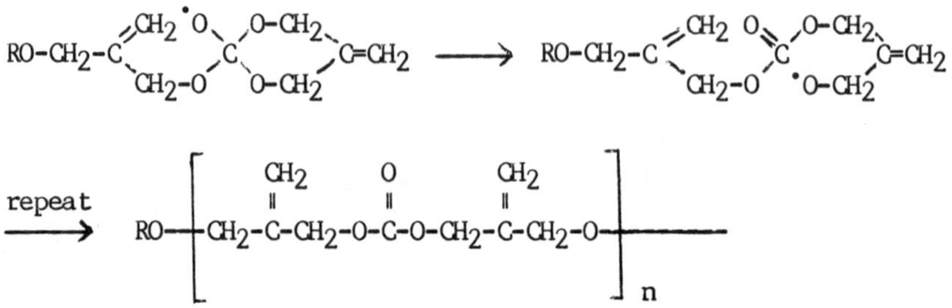

Apparently, the driving force for the ring opening is the relief of the strain in the spiro system and the formation of the stable carbonate double bond. The double ring opening is probably a concerted process from the initial radical addition product to the open-chain radical. Even though the spiro compound XI is an allyl monomer, it does copolymerize with a wide variety of comonomers. For example, XI will copolymerize with styrene to give a copolymer containing carbonate groups in the main polymer chain (20). Hydrolysis gives the oligomeric polystyrene capped with reactive hydroxyl groups (2).

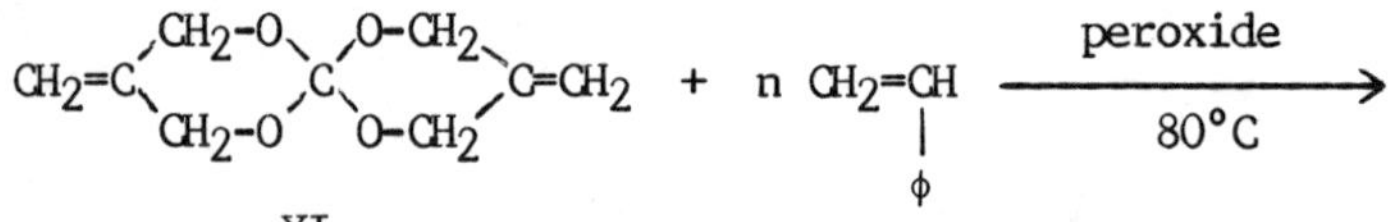

$$CH_2=C\underset{CH_2-O}{\overset{CH_2-O}{<}}C\underset{O-CH_2}{\overset{O-CH_2}{>}}C=CH_2 + n\ CH_2=\underset{\phi}{CH} \xrightarrow[80°C]{\text{peroxide}}$$

XI

$$-CH_2-\overset{\overset{CH_2}{\|}}{C}-CH_2-O-\overset{\overset{O}{\|}}{C}-O-CH_2-\overset{\overset{CH_2}{\|}}{C}-CH_2-O{\left[\!\!-CH_2-\underset{\phi}{CH}-\!\!\right]}_n CH_2-\overset{\overset{CH_2}{\|}}{C}-CH_2-O-\overset{\overset{O}{\|}}{C}-O-CH_2-\overset{\overset{CH_2}{\|}}{C}-CH_2-O-$$

$$\Big\downarrow OH^-$$

$$HO-CH_2-\overset{\overset{CH_2}{\|}}{C}-CH_2-O{\left[\!\!-CH_2-\underset{\phi}{CH}-\!\!\right]}_n CH_2-\overset{\overset{CH_2}{\|}}{C}-CH_2-OH$$

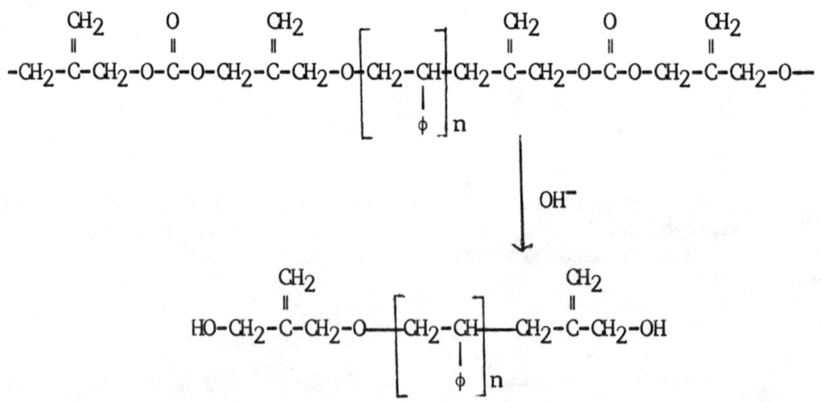

Thus, it is possible by simple free radical copolymerization with monomers that undergo ring opening followed by hydrolysis of the resulting copolymer to produce a variety of oligomers of any desired average molecular weight capped with a choice of reactive end groups.

New Chain Transfer Agents from Ketene Acetals

Since the cyclic ketene acetal V will undergo free radical polymerization to produce an ester group, a study was undertaken to see if

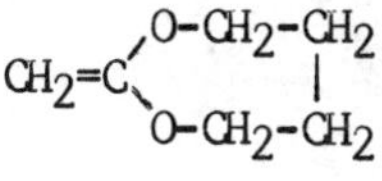

$$CH_2=C\begin{matrix}O-CH_2-CH_2\\ |\\ O-CH_2-CH_2\end{matrix}$$

V

$$CH_2=C\begin{matrix}O-CH_2-CH_3\\ \\ O-CH_2-CH_3\end{matrix}$$

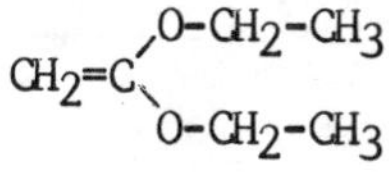

XII

the open chain ketene acetal XII would also form an ester by chain transfer by involving addition-elimination. Johnson, Barnes, and McElvain (21) reported that peroxide treatment had no appreciable effect on XII, but the criteria that they used for that determination is not clear. We have verified that treatment of XII with a peroxide does not result in a high molecular weight polymer because the monomer undergoes a degradative chain transfer reaction. When an equimolar mixture of styrene and diethyl ketene acetal (XII) was heated at 140°C in the presence of cumene hydroperoxide, a co-oligomer of styrene and the acetal XII was obtained at a 29% conversion. An elemental analysis indicated the oligomer consisted of 90 mole-% styrene and spectral studies indicated that the oligomer was capped with a carboethoxy group at one end. The ketene acetal units appear to be about equally divided between the copolymerized units and the end-capped chain transfer units. Thus it appears that XII is less reactive in the elimination process than is the 2-methylene-1,3-dioxolane (I) which undergoes cleavage to an extent of over 90% at 140°C.

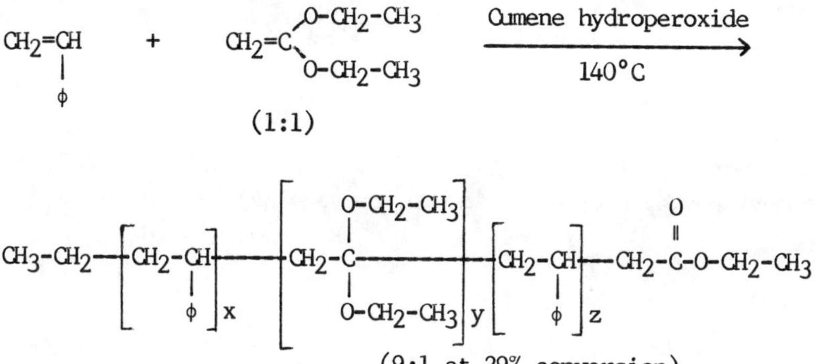

$$CH_2=CH\atop|\atop\phi \quad + \quad CH_2=C\begin{matrix}O-CH_2-CH_3\\ \\ O-CH_2-CH_3\end{matrix} \quad \xrightarrow[140°C]{\text{Cumene hydroperoxide}}$$

(1:1)

$$CH_3-CH_2-\left[\begin{matrix}CH_2-CH\\ |\\ \phi\end{matrix}\right]_x \left[\begin{matrix}O-CH_2-CH_3\\ |\\ CH_2-C\\ |\\ O-CH_2-CH_3\end{matrix}\right]_y \left[\begin{matrix}CH_2-CH\\ |\\ \phi\end{matrix}\right]_z CH_2-\overset{O}{\overset{\|}{C}}-O-CH_2-CH_3$$

(9:1 at 29% conversion)

Since the diethyl ketene acetal (XII) appears to be a moderately effective chain transfer agent as well as a comonomer, a search was made to find a ketene acetal that would be a more efficient chain transfer agent. It was reported earlier that the introduction of a phenyl group into the 2-methylene-1,3-dioxolane ring system would so stabilize the ring-opened free radical that the 4-phenyl-2-methylene-1,3-dioxolane (XIII) would undergo 100% ring opening even at room temperature (22).

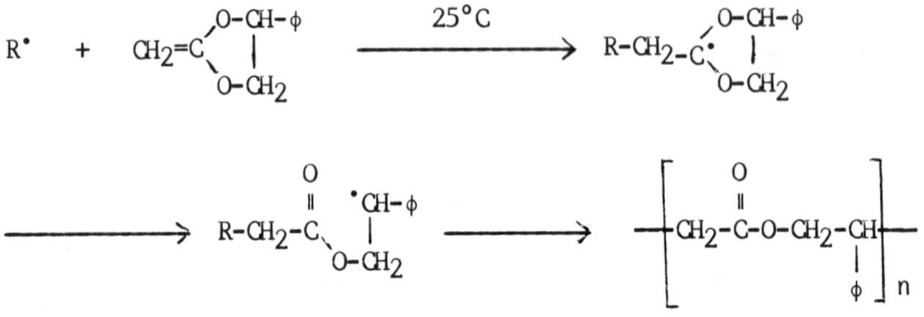

$$R^\bullet + CH_2=C\begin{smallmatrix}O-CH-\phi\\|\\O-CH_2\end{smallmatrix} \xrightarrow{25°C} R-CH_2-\overset{\bullet}{C}\begin{smallmatrix}O-CH-\phi\\|\\O-CH_2\end{smallmatrix}$$

$$\longrightarrow R-CH_2-\overset{O}{\overset{\|}{C}}\begin{smallmatrix}\bullet CH-\phi\\|\\O-CH_2\end{smallmatrix} \longrightarrow \left[-CH_2-\overset{O}{\overset{\|}{C}}-O-CH_2-\overset{}{\underset{\phi}{CH}}-\right]_n$$

On this basis it was reasoned that a benzyl group in a ketene acetal should greatly increase the extent of cleavage during polymerization and, therefore, should increase the efficiency of chain transfer. That in fact is what occurred when an equimolar mixture benzyl methyl ketene acetal (XIV) and styrene was heated at 120°C in the presence of di-<u>tert</u>-butyl peroxide; an oligomer with 80% styrene units and capped with a carbomethoxy group was obtained.

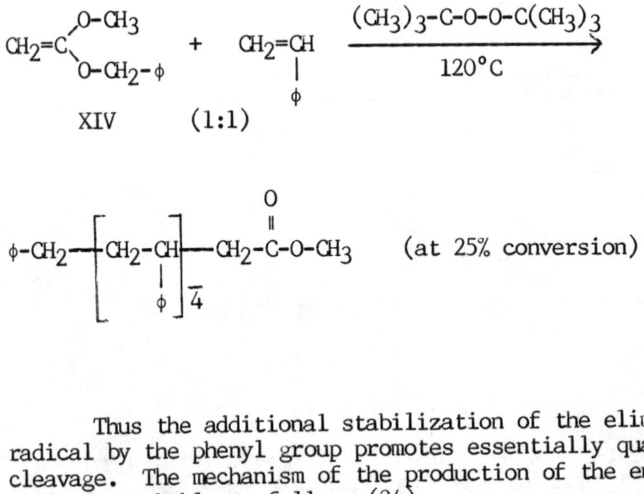

$$CH_2=C\begin{smallmatrix}O-CH_3\\ \\O-CH_2-\phi\end{smallmatrix} + CH_2=\overset{}{\underset{\phi}{CH}} \xrightarrow[120°C]{(CH_3)_3-C-O-O-C(CH_3)_3}$$

XIV   (1:1)

$$\phi-CH_2-\left[CH_2-\underset{\phi}{CH}\right]_{\overline{4}}-CH_2-\overset{O}{\overset{\|}{C}}-O-CH_3 \quad \text{(at 25% conversion)}$$

Thus the additional stabilization of the eliminated free radical by the phenyl group promotes essentially quantitative cleavage. The mechanism of the production of the end-capped oligomer is probably as follows (24):

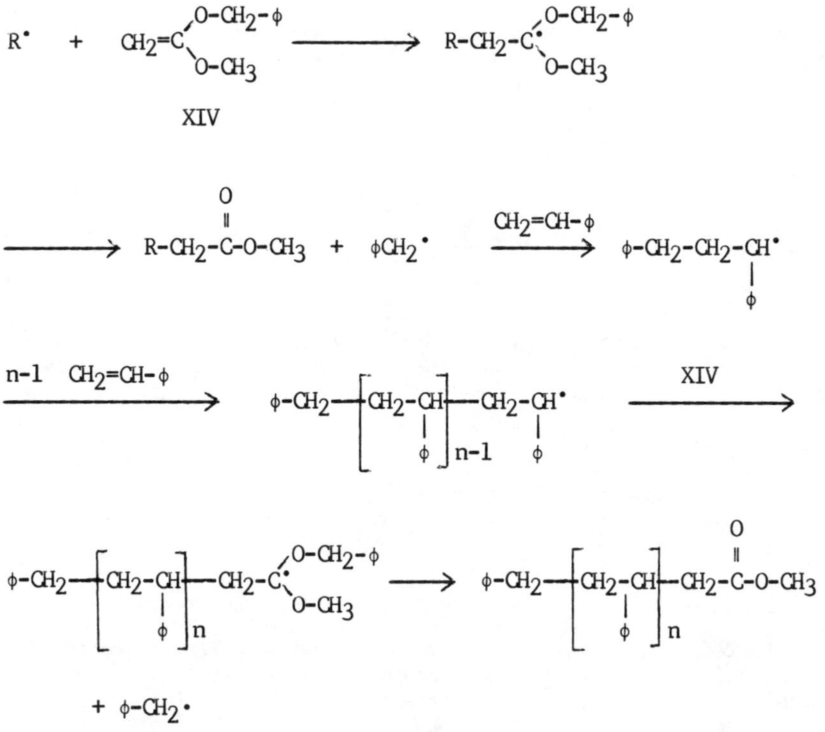

$$R^{\bullet} + CH_2=C\begin{matrix}O-CH_2-\phi\\O-CH_3\end{matrix} \longrightarrow R-CH_2-\overset{\bullet}{C}\begin{matrix}O-CH_2-\phi\\O-CH_3\end{matrix}$$

XIV

$$\longrightarrow R-CH_2-\overset{O}{\underset{\|}{C}}-O-CH_3 + \phi CH_2^{\bullet} \xrightarrow{CH_2=CH-\phi} \phi-CH_2-CH_2-\overset{\bullet}{\underset{\phi}{CH}}$$

$$\xrightarrow{n-1\ CH_2=CH-\phi} \phi-CH_2-\left[CH_2-\underset{\phi}{CH}\right]_{n-1}-CH_2-\underset{\phi}{\overset{\bullet}{CH}} \xrightarrow{XIV}$$

$$\phi-CH_2-\left[CH_2-\underset{\phi}{CH}\right]_n-CH_2-\overset{\bullet}{C}\begin{matrix}O-CH_2-\phi\\O-CH_3\end{matrix} \longrightarrow \phi-CH_2-\left[CH_2-\underset{\phi}{CH}\right]_n-CH_2-\overset{O}{\underset{\|}{C}}-O-CH_3$$

$$+ \phi-CH_2^{\bullet}$$

Although there are other unsaturated compounds that will undergo addition-elimination with free radicals, the benzyl ketene acetal XIV appears to be the most active double bond as far as rate of addition is concerned and the most efficient as far as regards to the extent of elimination is concerned. A comparison with the list of chain transfer agents listed in the Polymer Handbook (23) indicated that only the sulfur compounds appear to be more effective than XIV. Hydrolysis of the end-capped oligomer gives a macromer that is terminated with a carboxylic acid group.

Since functional oligomers were of prime interest, an effort was made to find a way to utilize the chain transfer properties of the ketene acetals to give oligomers that are end-capped at both ends with funtional groups without hydrolysis. For this reason di(p-hydroxymethylbenzyl) ketene acetal (XV) was synthesized. Preliminary studies show that copolymerization of XV with styrene gives an oligomer of styrene with hydroxyl-containing groups at both ends.

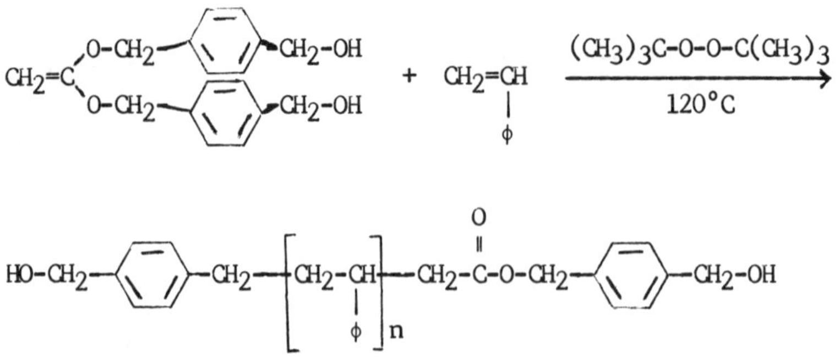

$$CH_2=C\begin{matrix}O-CH_2\\O-CH_2\end{matrix}\diagdown\bigcirc\diagdown\begin{matrix}CH_2-OH\\CH_2-OH\end{matrix} + CH_2=CH\underset{\phi}{|} \xrightarrow[120°C]{(CH_3)_3C-O-O-C(CH_3)_3}$$

$$HO-CH_2-\bigcirc-CH_2-\left[CH_2-\underset{\phi}{CH}-CH_2-\overset{O}{\underset{\|}{C}}-O-CH_2-\bigcirc-\right]_n CH_2-OH$$

By an extension of this general procedure, it appears that ketene acetals can be used effectively to produce oligomers with a variety of end groups by free radical processes (24, 25).

Acknowledgment

The authors are grateful for support for this research to the Polymer Program of the National Science Foundation, the Frasch Foundation, and the Goodyear Tire and Rubber Company.

Literature Cited

1. Weinstein, A. H.; Constanza, A. J.; Meyer, G. E. Fr. Patent 1,434,167, 1966; Chem. Abstr. 1967, 66, 76764k.
2. Bailey, W. J.; Chen, P. Y.; Chiao, W. -B.; Endo, T.; Sidney, L.; Yamamoto, N.; Yamazaki, N.; Yonezawa, K. In "Contemporary Topics in Polymer Science"; Shen, M., Ed.; Plenum Publishing: New York, 1979; Vol. 3, p. 29.
3. Bailey, W. J.; Ni, Z.; Wu, S. -R. J. Polym. Sci., Polym. Chem. Ed. 1982, 20, 2420.
4. Endo, T.; Bailey, W. J. J. Polym. Sci., Polym. Chem. Ed. 1975, 13, 2525.
5. Sidney, L.; Shaffer, S. E.; Bailey, W. J. Am. Chem. Soc., Div. Polym. Chem., Preprints 1981, 22(2), 373.
6. Bailey, W. J.; Yamazaki, N. Unpublished Results.
7. Takahashi, T. J. Polym. Sci. 1968, A6, 403.
8. Cho, I.; Ahn, K. D. J. Polym. Sci., Polym. Letters Ed. 1977, 15, 751.
9. Errede, L. A. J. Polym. Sci. 1961, 49, 253.
10. Hall, Jr., H. J.; Ykman, P. Macromol. Rev. 1976, 11, 1.
11. Tobolsky, A. V.; Eisenberg, A. J. Am. Chem. Soc. 1959, 81, 780.
12. Butler, G. B.; Angelo, R. J. J. Am. Chem. Soc. 1957, 79, 3128.
13. Maillard, B.; Forrest, D.; Ingold, K. U. J. Am. Chem. Soc. 1976, 98, 7024.
14. McElvain, S. M.; Curry, M. J. J. Am. Chem. Soc. 1948, 70, 3781.
15. Beyerstedt, F.; McElvain, S. M. J. Am. Chem. Soc. 1938, 58, 529.

16. Potts, J. E.; Clendinning, R. A.; Ackart, W. B.; Niegisch, W. D. In "Polymers and Ecological Problems"; Guillet, J., Ed.; Plenum Press: New York; 1973, p. 61.
17. Bailey, W. J. Proc. Third International Conference on Advances in the Stabilization and Controlled Degradation of Polymers, Lucerne, Switzerland, June 1, 1981, p. 12.
18. Sidney, L.; Shaffer, S. E.; Bailey, W. J. Am. Chem. Soc., Div. Polym. Chem., Preprints 1981, 22(2), 373.
19. Endo, T.; Bailey, W. J. J. Polym. Sci., Polym. Letters Ed. 1975, 13, 193.
20. Bailey, W. J. Kobunshi 1981, 30(5), 331.
21. Johnson, P. R.; Barnes, H. M.; McElvain, S. M. J. Am. Chem. Soc. 1940, 62, 964.
22. Bailey, W. J.; Wu, S. -R.; Ni, Z. Makromol. Chem. 1982, 183, 1913.
23. Young, L. W. In "Polymer Handbook"; 2nd Ed.; Brandrup, J.; Immergut, E. H., Eds., Wiley-Interscience: New York; 1975, p. II-57.
24. Bailey, W. J.; Gapud, B.; Lin, Y. -N.; Ni, Z.; Wu, S. -R. Am. Chem. Soc., Div. Polym. Chem., Preprints 1984, 25(1), 142.
25. Bailey, W. J.; Endo, T.; Gapud, B.; Lin, Y. -N.; Ni, Z.; Pan, C. -Y.; Shaffer, S. E.; Wu, S. -R.; Yamazaki, N.; Yonezawa, K. J. Macromol. Sci. - Chem. 1984, A21(8,9), 979.

RECEIVED February 28, 1985

14

Reactive Difunctional Siloxane Oligomers
Synthesis and Characterization

ISKENDER YILGÖR, JUDY S. RIFFLE[1], and JAMES E. MCGRATH

Department of Chemistry, Polymer Materials and Interfaces Laboratory, Virginia Polytechnic Institute and State University, Blacksburg, VA 24061

>Synthesis and characterization of well-defined, α,ω-terminated difunctional siloxane oligomers are discussed. Detailed procedures on the preparation of primary amine- and hydroxy-terminated oligomers are given. Control of the average molecular weight ($\bar{M}n$) and also the possible variations in the backbone structure and composition are explained. The effect of these variations on the physical, thermal and chemical properties of the resulting materials are discussed. Characterization of these oligomers by FT-IR, NMR and UV spectroscopy, potentiometric titration and DSC are summarized.

Organosiloxane based segmented elastomers have been described in the literature over the past twenty five years(1-5). The main interest in these type of block or segmented copolymers arises mainly due to the unique properties of the organosiloxane segments, which are quite different than those of conventional rubbery hydrocarbon polymers. In general polyorganosiloxanes display very good low temperature flexibility (Tg as low as -123°C), good thermal-oxidative stability, ozone and UV resistance, relatively high gas permeabilities, good biocompatibility and excellent electrical properties. Moreover, they are non-polar and have very low solubility parameters (δ ≈7.5-9.5) and interesting surface properties. Therefore when incorporated they can provide all these interesting properties to the resulting materials.

Siloxane containing block or segmented copolymers can be synthesized either by living anionic polymerization of the cyclic organosiloxane trimers with appropriate vinyl monomer(1,4,6) or by step-growth or condensation copolymerization of preformed α,ω-difunctional siloxane oligomers with conventional difunctional

[1] Current address: Thoratec Laboratories, Inc., 2023 Eighth Street, Berkeley, CA 94710

monomers or oligomers(1,5,7,8). Some investigations on the synthesis of siloxane containing graft copolymers by free radical copolymerization of vinyl-terminated polydimethylsiloxane oligomers with styrene or methyl methacrylate have also been performed(9). During the past several years we have been investigating the synthesis of reactive, α,ω-difunctional siloxane oligomers, their use in the production of a variety of segmented copolymers(5,10) and in the rubber modification of epoxy networks(11). In this paper we will principally discuss our efforts on the synthesis of various functionally terminated siloxane oligomers, their purification and structural and physical characterization.

The general structure of an α,ω-difunctional siloxane oligomer is shown in Scheme 1. In this structure X,R,Y and n can be varied and accordingly, the resulting oligomer can have a wide range of

Scheme 1.

General Structure of an α,ω-Difunctional Siloxane Oligomer

$$X-R-\underset{\underset{CH_3}{|}}{\overset{\overset{CH_3}{|}}{Si}}-O\left[\underset{\underset{Y}{|}}{\overset{\overset{Y}{|}}{Si}}-O\right]_n\underset{\underset{CH_3}{|}}{\overset{\overset{CH_3}{|}}{Si}}-R-X$$

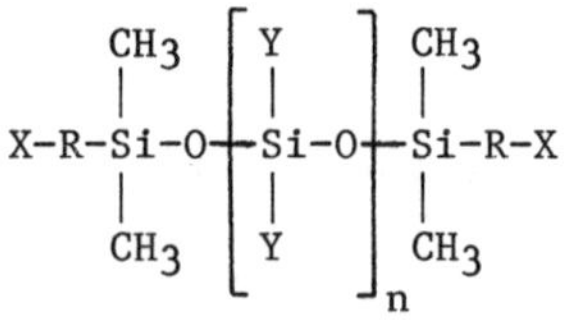

Variables: X, R, Y, n

properties such as chemical reactivity, molecular weight, thermal and physical behavior, etc. A brief summary of these possible variations is given in Scheme 2, which shows the functional end groups, molecular weights and backbone structures that have been synthesized in our laboratories and elsewhere(13,14).

As expected, the terminal functional groups mainly determine the reactivity of these siloxane oligomers towards other reactants. The variations in the backbone composition have critical effects on the glass transition temperature, solubility parameter, thermal stability and surface behavior of the resulting oligomers(12,13). In addition, if the following groups are incorporated into the

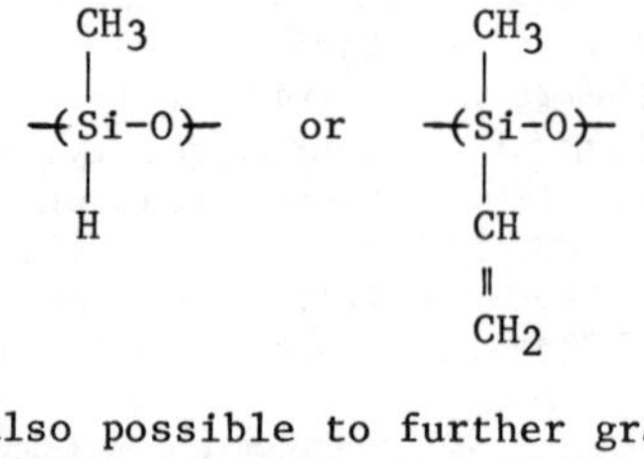

$$-\!\!\left(\underset{\underset{H}{|}}{\overset{\overset{CH_3}{|}}{Si}}-O\right)\!\!-\quad \text{or} \quad -\!\!\left(\underset{\underset{\underset{CH_2}{\|}}{\overset{\overset{CH_3}{|}}{CH}}}{\overset{\overset{CH_3}{|}}{Si}}-O\right)\!\!-$$

backbone, it is also possible to further graft or crosslink the resulting material through pendent (-Si-H) and/or (-Si-CH=CH$_2$) linkages. The average degree of polymerization (n) mainly affects the morphology and phase separation behavior of the copolymers derived from respective siloxane oligomers. This in turn determines

Scheme 2.

Possible Variations in the Structure and Composition of Siloxane Oligomers

---

X: Functional End Groups

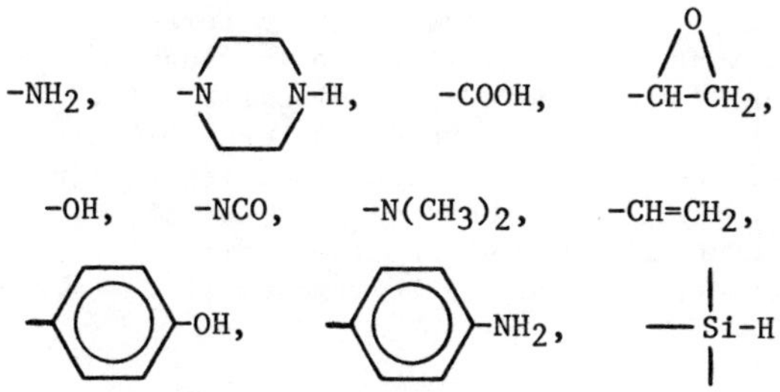

$-NH_2$, $-N\diagup\diagdown N-H$, $-COOH$, $-\underset{\diagdown O \diagup}{CH-CH_2}$,

$-OH$, $-NCO$, $-N(CH_3)_2$, $-CH=CH_2$,

$-\bigcirc-OH$, $-\bigcirc-NH_2$, $-\overset{|}{\underset{|}{Si}}-H$

R:

$-(CH_2)_{3-5}$, Chemical Bond

Y: Backbone Composition

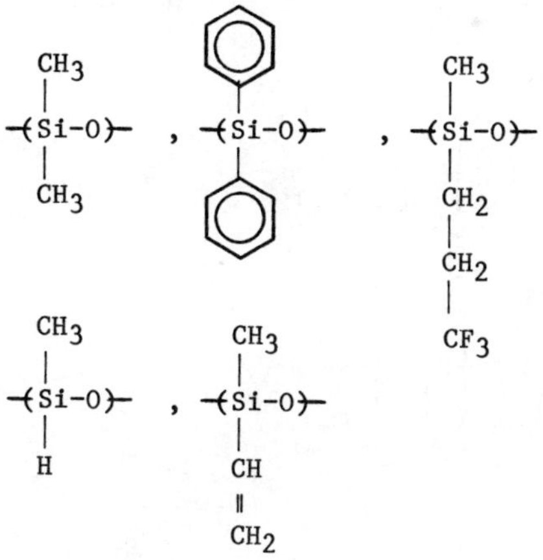

$-(\overset{CH_3}{\underset{CH_3}{|}}Si-O)-$, $-(\overset{\bigcirc}{\underset{\bigcirc}{|}}Si-O)-$, $-(\overset{CH_3}{\underset{CH_2 CH_2 CF_3}{|}}Si-O)-$

$-(\overset{CH_3}{\underset{H}{|}}Si-O)-$, $-(\overset{CH_3}{\underset{CH=CH_2}{|}}Si-O)-$

or their combinations

---

n: Degree of Polymerization

0-150 Higher

the thermal, mechanical and solution properties, surface activity and processibility of the resulting multiphase materials.

As it is well known, the (Si-O) bond in organosiloxanes may be considered to be polar or partially (~50%) ionic.(12) Therefore, it can be cleaved by the attack of strong acids or bases. This is the main rationale behind the "equilibration" route to the synthesis of a wide variety of functionally terminated siloxane oligomers(12-14) from cyclic siloxanes and α,ω-difunctional disiloxanes as shown in Scheme 3.

These type of reactions are generally named as "equilibration" or "redistribution" reactions due to the nature of the processes. During the reactions the catalyst can only cleave the (Si-O) bonds in the cyclic or linear species including that of the "end blocker" and growing chains. However, the (Si-R) or (R-X) bonds are stable. Therefore at the end of the reactions the linear oligomers are functionally terminated and the minority (10-15%) cyclic side products are nonfunctional. After elimination of catalyst, the cyclic side products can usually be removed from the system by

Scheme 3.

Synthesis of Functionally Terminated Siloxane Oligomers

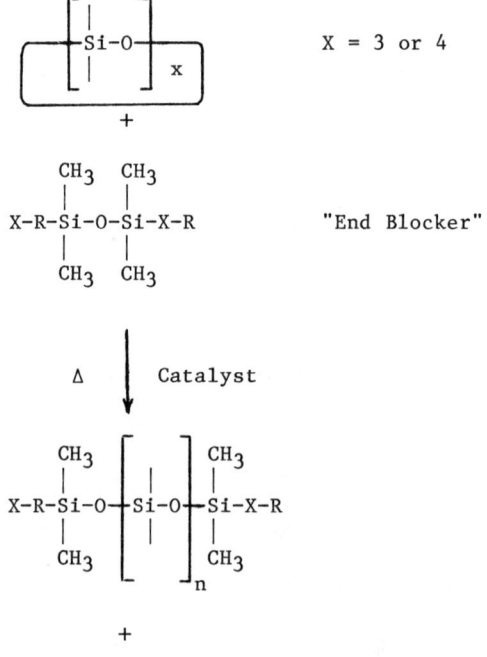

vacuum distillation at elevated temperature. The number average molecular weight of the final product is determined by the initial ratio of cyclic monomers to the end blocker. Backbone composition can be dictated by the variety and proportion of the cyclic monomers present in the original reaction mixture.

In this paper we will discuss the synthesis of α,ω-hydroxybutyl terminated polydimethylsiloxane oligomers by cationic routes and α,ω-aminopropyl terminated poly(dimethyl-diphenyl)siloxane oligomers by anionic methods respectively. Detailed procedures for the synthesis of aminopropyl, carboxypropyl and glycidoxypropyl terminated polydimethylsiloxane oligomers have already been described elsewhere(11).

EXPERIMENTAL

Materials

Octamethyltetrasiloxane (D_4) and octaphenyltetrasiloxane ($D_4"$) were generously supplied by General Electric Co., Silicone Products Division. 1,3-Bis(4-hydroxybutyl)tetramethyldisiloxane and 1,3-Bis(γ-aminopropyl)tetramethyldisiloxane were purchased from Silar Laboratories and Petrarch respectively. Trifluoroacetic acid was a product of Aldrich. All other chemicals were reagent grade and used without further purification.

Synthesis of α,ω-Hydroxybutyl Terminated Polydimethylsiloxane Oligomers

Depending on the desired average molecular weight of the final oligomer, calculated amounts of D_4 and "end blocker" were charged into a 3-necked round bottom Pyrex reactor fitted with an argon inlet, thermometer and condenser. Stirring was provided by a magnetic bar. Then, trifluoroacetic acid (a sufficient quantity to esterify hydroxyl end groups as well as 10-15% by weight excess) was added as the catalyst. The mixture was heated to 60°C and maintained at this temperature for 48 hours. At the end of this period the excess trifluoroacetic acid was extracted with distilled water until the aqueous extract was neutral. Then the reaction mixture was stripped off under high vacuum to remove the cyclic side products. At this point the respective oligomers were isolated in the form of their trifluoroacetic esters. These ester groups were converted to alcohols by a weak base catalyzed hydrolysis reactions. The ester form of the oligomer was dissolved in THF and an equal volume of 5% aqueous sodium carbonate solution was added. Next, the system was heated to 40-45°C, for 48 hours, while being stirred. When the system was cooled down the organic and aqueous phases separated out. The product was recovered by the evaporation of THF. Hydrolysis reactions were followed by FT-IR spectroscopy.

Synthesis of α,ω-Aminopropyl Terminated Poly(dimethyl-diphenyl)-siloxane Oligomers

Depending on the desired levels of diphenylsiloxane units and the molecular weight of the final product, calculated amounts of D_4, $D_4"$ and 1,3-bis(γ-aminopropyl)tetramethyldisiloxane were introduced into a round bottom, 3-necked flask fitted with a condenser, thermometer and argon inlet. Later 0.01-0.1% by weight of finely ground potassium hydroxide was added and the system was heated to 160°C. The mixture was heterogeneous initially but in about 30 minutes the potassium hydroxide reacted to generate a homogeneous and viscous system. The reaction was allowed to continue for 24 hours. At the end of this period the reaction was cooled to 25°C and the catalyst was neutralized with alcoholic hydrochloric acid and extracted with distilled water. The product was dissolved in methylene chloride, dried with magnesium sulfate, filtered, rotavapped and stripped under high vacuum. Molecular weights, diphenylsiloxane contents and thermal behavior of the resulting oligomers were determined.

Characterization of the Products

The structural characterization of the oligomers included FT-IR (Nicolet MX-1), NMR (Varian EM390) and UV (Perkin Elmer 552) spectroscopy. Molecular weights were determined by end group analysis and vapor pressure osmometry. Hydroxyl end groups were analyzed by phthalic anhydride-pyridine method (15) and amine end groups were determined by potentiometric titration with standard hydrochloric acid. Thermal characterization of the products were obtained by Perkin Elmer DSC-2, under nitrogen atmosphere using a heating rate of 10°C/min.

RESULTS AND DISCUSSION

α,ω-difunctional reactive siloxane oligomers are very versatile starting materials for the synthesis of numerous segmented block copolymers and modification of network structures. The flexible polymerization and copolymerization chemistry of siloxanes with acid and base catalysis (12-14) leads to the synthesis of a wide variety of oligomers having different molecular weights, backbone compositions and functionalities. With all these variations, it is possible to design and synthesize siloxane oligomers with a wide range of well defined chemical, physical and mechanical properties.

In our laboratories we have been working on the synthesis of various α,ω-difunctional siloxane oligomers for several years(11). A brief summary of these studies is given in Scheme 2, which shows various end groups, molecular weights and backbone structures that have already been synthesized. Here we are going to discuss our studies on the synthesis and characterization of α,ω-hydroxybutyl terminated polydimethylsiloxane and α,ω-aminopropyl terminated poly(dimethyl-diphenyl)siloxane oligomers structures of which are given in Scheme 4 respectively.

Synthesis and Characterization of Hydroxybutyl Terminated Siloxane Oligomers

In their earlier studies on the synthesis of α,ω-hydroxyl terminated siloxane oligomers Marvel and co-workers (21) have utilized sulfuric acid as the catalyst in the equilibration of cyclic tetramer (D_4) and 1,3-bis(4-hydroxylbutyl)tetramethyldisiloxane (DSX). However the molecular weights of the oligomers obtained were several times higher than the expected values. They have attributed this to the dehydrating action of the sulfuric acid which resulted in the loss of end-group functionality. They were unable to improve the equilibration by varying the reaction solvent, temperature or time. As an alternate route they have also tried to reduce the carboxyl terminated polysiloxane with lithium aluminum hydride to produce the corresponding diol, however in this case infrared spectrum of the compound showed Si-H absorption peak at 2120 cm^{-1} which indicated an appreciable amount of cleavage in the siloxane linkages. As a result they were unable to obtain clean, α,ω-difunctional hydroxy terminated siloxane oligomers.

In this study hydroxybutyl terminated siloxane oligomers were synthesized by the trifluoroacetic acid (TFAA) catalyzed equilibrations of D_4 and 1,3-bis (4-hydroxybutyl)tetramethyl-disiloxane (DSX), in bulk, at 60°C. In this system, under these reaction conditions the catalyst, TFAA also reacts with the (OH) end groups to from respective esters, so for effective equilibrations a 10-15% by weight excess of TFAA is necessary. At the end of the reactions (OH) groups are regenerated by the hydrolysis of ester groups under slightly basic conditions using Na_2CO_3 in a mixed solvent system of THF and distilled water. These reactions were followed by FT-IR spectroscopy by observing the decrease and finally

Scheme 4.

Structures of the Siloxane Oligomers Synthesized

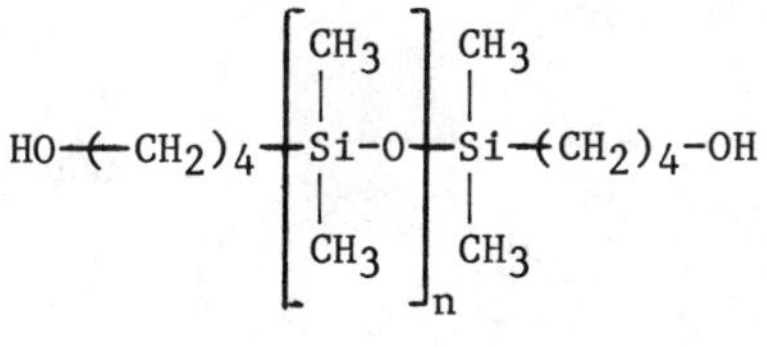

$$HO\text{-}(CH_2)_4\text{-}\left[\begin{array}{c}CH_3\\|\\Si\text{-}O\\|\\CH_3\end{array}\right]_n\begin{array}{c}CH_3\\|\\Si\text{-}(CH_2)_4\text{-}OH\\|\\CH_3\end{array}$$

α,ω-Bis(γ-hydroxybutyl)polydimethylsiloxane

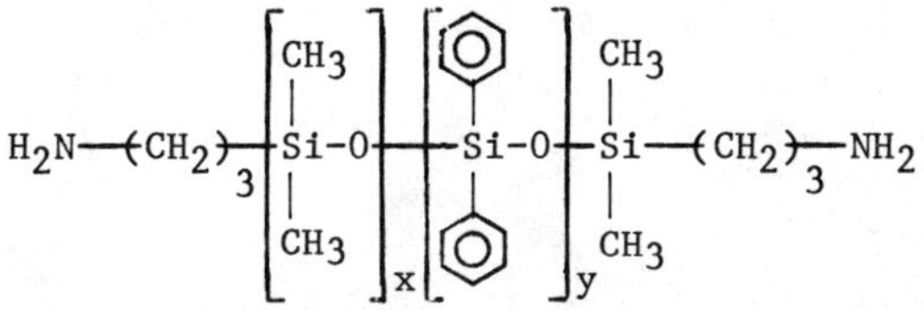

$$H_2N\text{-}(CH_2)_3\text{-}\left[\begin{array}{c}CH_3\\|\\Si\text{-}O\\|\\CH_3\end{array}\right]_x\left[\begin{array}{c}Ph\\|\\Si\text{-}O\\|\\Ph\end{array}\right]_y\begin{array}{c}CH_3\\|\\Si\text{-}(CH_2)_3\text{-}NH_2\\|\\CH_3\end{array}$$

α,ω-Bis(γ-aminopropyl)poly(dimethyldiphenyl)siloxane

the disappearance of the carbonyl stretching around 1800 cm^{-1} and the increase in (-OH) absorption band around 3300 cm^{-1}. The respective FT-IR spectra for the oligomer having Mn @ 2000, before and after, the complete hydrolysis are shown in Figure 1. It has been observed that hydrolysis with dilute aqueous Na_2CO_3 at 40-45° for 48 hours is sufficient to convert all ester groups to alcohols without cleaving any siloxane bonds and without changing the molecular weight distribution of these oligomers, as evidenced by

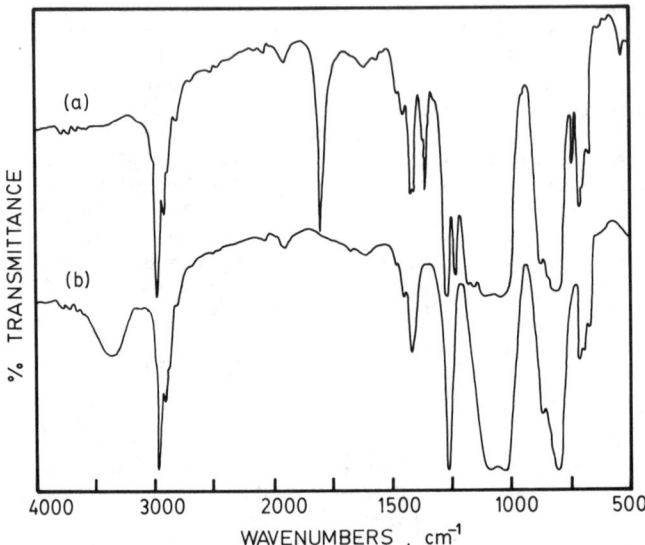

Figure 1. FT-IR Spectra of α,ω-Hydroxybutyl terminated polydimethylsiloxane oligomer (Mn ~1000)
(a) Trifluoroacetate ester
(b) Alcohol obtained after hydrolysis

GPC studies. The ^{1}H-NMR spectrum of the hydroxy terminated oligomer with Mn @ 1000 is given in Figure 2. The position of the peaks, as marked on the spectrum and the relative ratio of integrations, confirm the formation of the predicted oligomer structure.

Table I provides a summary of the experimental data and the results obtained. As mentioned earlier, the final molecular weight of the oligomers are dictated by the initial ratio of D_4 to DSX, the "end blocker". As expected, Mn values are usually slightly lower than that aimed for. This is no doubt due to the formation of the cyclic side products. The latter are removed from the system by vacuum distillation and are known to be present in such equilibrations at a typical level of around 10 percent by weight. The molecular weights of the stripped products obtained by VPO and end group analysis are in very good agreement.

Table I

Synthesis of α,ω-Bis(4-hydroxybutyl)polydimethylsiloxane Oligomers

Sample No.	D_4 (g)	DSX (g)	$\overline{M}n$ (g/mole) Stoichiometry	$\overline{M}n$ Obtained VPO	End Gr. Analy.
1	72.10	27.86	1000	910	940
2	86.00	13.93	2000	1650	1720
3	93.03	6.97	4000	3480	3550
4	95.36	4.64	6000	5200	5400

Base catalyzed equilibrations of hydroxyl terminated siloxane oligomers can also be achieved, however these systems can sometimes be complicated by the attack of base to the (OH) end groups to form $[(CH_2)_x\ O^-\ X^+]$ type species, which may also cleave the siloxane bonds. This generally results in the loss of terminal functionality in the oligomers produced, which is of course not desirable.

Synthesis of Aminopropyl Terminated Poly(dimethyl-diphenyl)siloxane Oligomers

Acid and base catalyzed coequilibration reactions of D_4 and $D_4"$ have been studied in the literature by several workers (16-18). These studies were directed towards either the synthesis of very high molecular weight "modified" silicone rubbers, or to the analysis of reaction kinetics (13, 16-19). There has been no systematic studies in the open literature on the synthesis and characterization of functionally terminated, low molecular weight (dimethyl-diphenyl)-siloxane oligomers which can be used in the preparation of segmented block copolymers. It is known that the incorporation of diphenyl-siloxane units generally disrupts the low temperature crystallization of polydimethylsiloxane resins and also increases their thermal and radiation stability (12,13). Glass temperature (Tg) and solubility parameter values of the resulting polymers are also raised accordingly, depending on the level of diphenylsiloxane incorporation.

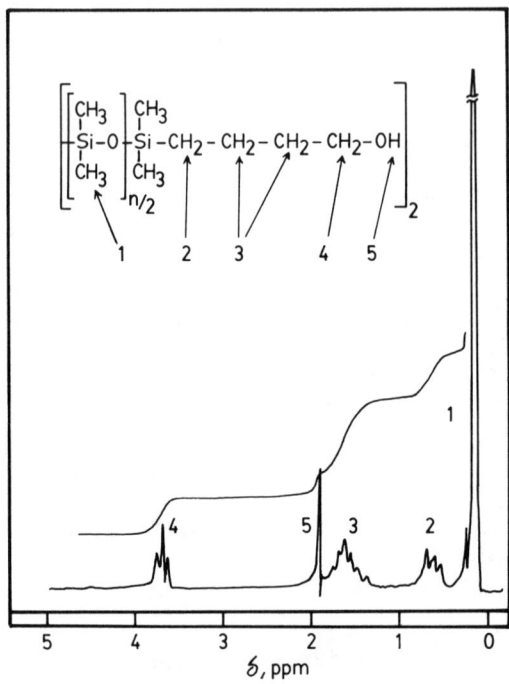

Figure 2. ^{1}H-NMR spectrum of α,ω-Hydroxybutyl terminated polydimethylsiloxane oligomer (Mn ~1000)

In this study our aim was to systematically synthesize α,ω-aminopropyl terminated (dimethyl-diphenyl)siloxane oligomers having low molecular weights (1000-3000 g/mole) and to subsequently analyze the composition and thermal behavior of the copolymers. Some of these oligomers have later been used in the synthesis of various segmented urea or imide type copolymers and in the modification of epoxy networks, which have been discussed elsewhere (5,11).

Table II provides a summary of the results on the characteristics of aminopropyl terminated poly(dimethyl-diphenyl)siloxane oligomers synthesized. These reactions were conducted in bulk at 160°C with KOH as the initiator. As can be seen from Table II the stoichiometric number average molecular weights sought and obtained are in very good agreement. The level of diphenylsiloxane incorporation was determined by UV spectroscopy. There is no absorption of dimethylsiloxane backbone in the spectral range of 240 to 280 nm. On the other hand, phenyl groups absorb very strongly over these wavelengths (Figure 3). For quantitative analysis we have used the absorption peak at 270 nm as the reference. Chloroform was used as the solvent for the UV measurements. Standard mixtures of D_4 and $D_4"$ were used for the calibration. It is clear from Table II that the level of diphenylsiloxane charged and incorporated shows almost a one to one correspondence. The slight difference may be due to the residual diphenyl containing cyclic species, which are difficult to remove because of their very high boiling points (17).

Table II

Characteristics of Aminopropyl Terminated Poly(Dimethyldiphenyl)siloxane Oligomers

Sample No.	$\overline{M}_n$ (g/mole)		% Diphenylsiloxane		T_g (°C)
	Stiochiometry	Obtd.(a)	Charged	Incorp.	
1	1900	1770	0	0	-123
2	2000	1780	5.0	5.5	-119
3	2000	1660	10.0	11.9	-115
4	1250	1380	16.0	15.8	-112
5	2000	1950	20.0	21.5	-105
6	2000	1990	40.0	43.4	- 79
7	2000	2150	56.0	59.6	- 49
8	2000	2330	73.0	78.3	- 35

(a)Titrated value

Glass transition temperatures of the resulting oligomers increase with increasing levels of diphenylsiloxane present in the system as expected. This is entirely consistent with the idea that dimethyl and diphenyl units are randomly distributed along the

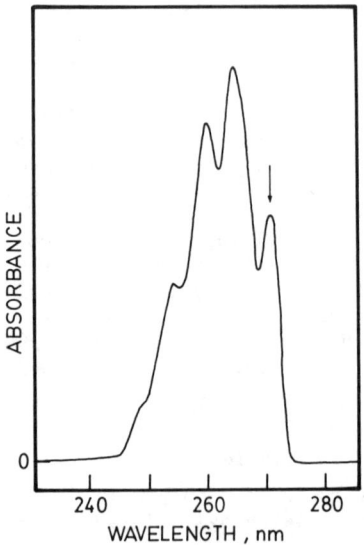

Figure 3. Typical UV absorption spectrum of D_4'' or D_4''/D_4 blend or a poly(dimethyl-diphenyl)siloxane oligomer ($CHCl_3$ solution)

oligomer backbone. This is an expected behavior in view of the long equilibration times, since the simultaneous formation and scission of siloxane bonds throughout the reactions should tend to randomize the sequence distribution. However it is known that $D_4"$ is more reactive than D_4 in base catalyzed reactions due to the stabilization considerations of siloxanolate anion by the phenyl substituents (12,13). Therefore it may be possible to obtain "block-like" distributions in these oligomers if the reaction conditions (i.e. type and concentration of catalyst, reaction time and temperature) can be adjusted properly. Our future work is proceeding in this direction. Alternatively, we have also investigated lower equilibration temperatures where the diphenyl tetramer is less soluble. "Blocky" sequences could also be achieved in this situation.

Conclusions

We have demonstrated the synthesis and characterization of reactive, α,ω-difunctional siloxane oligomers with hydroxyl or primary amine silicon-carbon linked end groups by using acid or base catalyzed equilibration reactions. We have also shown the effect of backbone composition on the thermal behavior of resulting oligomers. It is clear that, the Tg of the dimethyl-diphenyl oligomers can be varied from -123°C to -35°C by increasing the level of incorporation of diphenylsiloxane units, while maintaining or changing the number average molecular weight. This is a very effective tool for the design and the synthesis of a wide variety of siloxane oligomers suitable for specific needs.

In addition we are also investigating the synthesis and characterization of reactive, difunctional (trifluoropropyl, methyl)siloxane containing oligomers. Also very recently, we have been able to fractionate various functionally terminated siloxane oligomers into very narrow fractions by using the supercritical fluid extraction techniques (20). This is a very important step in the production of α,ω-difunctional reactive siloxane oligomers with narrow molecular weight distributions.

Literature Cited

1. A. Noshay and J. E. McGrath, "Block Copolymers: Overview and Critical Survey", Academic Press, New York (1977).
2. J. B. Plumb and J. H. Atherton, in "Block Copolymers", Ed. D. C. Allport and W. H. Janes, Halstead Press, New York (1972), Ch. 6.
3. M. Morton, A. Rembaum and E. E. Bostick, J. Appl. Polym. Sci., 8, 2707 (1964).
4. E. E. Bostick, in "Block Copolymers", Ed. S. W. Aggarwal, Plenum Press, New York (1972) p. 237.
5. J. E. McGrath, et al., Polym. Prepr., 24(2), 35, 39, 47, 78, 80 (1983).
6. J. C. Saam, D. J. Gordon and S. Lindsey, Macromolecules, 3, 1 (1970).
7. H. A. Vaughn, Jr., J. Polym. Sci., Part B, 7, 569 (1969).

8. M. Matzner, A. Noshay, L. M. Robeson, C. N. Merriam, R. Barlay, Jr. and J. E. McGrath, Appl. Polym. Symp., 22, 143 (1973).
9. Y. Kawakami, Y. Miki, T. Tsuda, R. A. N. Murty and Y. Yamashita, Polym. J., 14(11), 913 (1982).
10. J. S. Riffle, Ph.D. Thesis, VPI & SU, Blacksburg, VA (1981).
11. J. S. Riffle, I. Yilgor, C. Tran, G. L. Wilkes, J. E. McGrath and A. K. Banthia in "Epoxy Resins II", Ed. R. S. Bauer, ACS Symp. Ser. No: 221, Ch. 2 (1983).
12. M. C. Voronkov, V. P. Mileshkevich and Yu A. Yuzhelevskii, "The Siloxane Bond", Consultants Bureau, New York, 1978.
13. W. Noll, "Chemistry and Technology of Silicones", Academic Press, New York, 1968.
14. C. Eaborn, "Organosilicon Compounds", Butterworths Publ. Lim., London, 1960.
15. "Analysis of Silicones", Ed. A. Lee Smith, John Wiley, New York (1974), p. 135.
16. K. A. Andrianov, B. G. Zavin and G. F. Sablina, Polym. Sci. USSR, 14, 1294 (1972).
17. Z. Laita and M. Jelinek, Polym. Sci. USSR, 6, 342 (1964).
18. R. L. Merker and M. J. Scott, J. Polym. Sci., 43, 297 (1960).
19. K. A. Andrianov et al., Polym. Sci. USSR, 12, 1436 (1970).
20. I. Yilgor, J. E. McGrath and V. Krukonis, Polym. Bull., 12(6), 499, (1984).
21. K. Kojima, C. R. Gore and C. S. Marvel, J. Polym. Sci., A-1, 4(9), 2325, (1966).

RECEIVED February 19, 1985

15

Synthesis of Poly(phenylene Oxides) by Electrooxidative Polymerization of Phenols

EISHUN TSUCHIDA, HIROYUKI NISHIDE, and TOSHIHIKO MAEKAWA

Department of Polymer Chemistry, Waseda University, Tokyo 160, Japan

>Anodic oxidation of phenols gave the corresponding poly(1,4-phenyleneoxide)s by selecting the electrolysis conditions to prevent passivation of the electrode. The mechanism of the polymerization was discussed based on electrochemical measurements. Applications of the electro-oxidative polymerization were also described.

It is well known that 2,6-dimethylphenol is oxidatively polymerized to poly(2,6-dimethyl-1,4-phenyleneoxide) with a copper amine complex as catalyst in the presence of oxygen at room temperature (Eq. 1) (1). This polymerization can also proceed by anodic oxidation, accompanying evolution of an equivalent of hydrogen as shown in Equation 2.

Recently much research has been made to coat electrodes with thin polymer films by electro-oxidative polymerization of phenols (2). The formed thin and uniform poly(phenyleneoxide) films on electrode are interesting because of their electric and electrochemical properties. Figure 1 shows a typical cyclic voltammogram for the oxidation of 2,6-dimethylphenol at a platinum electrode in

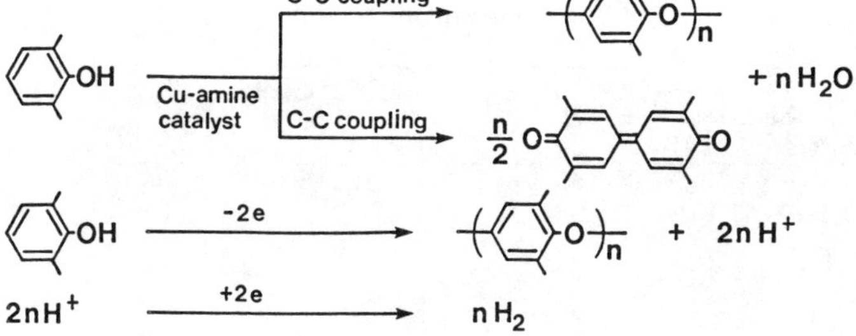

0097-6156/85/0282-0175$06.00/0
© 1985 American Chemical Society

alkaline methanol solution, where the potential sweep was cycled between 0 and 1.0 V. A rapid current decrease during the second sweep indicates that the electro-oxidation of the phenol produced an insulating film which adhesively coated the platinum electrode surface. The Fourier-transform IR spectrum and the ESCA analysis of the surface of the electrode indicated the formation of poly(2,6-dimethyl-1,4-phenyleneoxide) (3).

However the formation of thin polymer film on the electrode, i.e. passivation of the electrode, resulted in cessation of the polymerization, which restricted the electro-oxidation as a polymerization procedure. The electro-oxidative polymerization as a method of producing poly(phenyleneoxide)s had not been reported except in one old patent, in which a copper-amine complex was added as an electron-mediator during the electrolysis (4). The authors recently found that phenols are electro-oxidatively polymerized to yield poly-(2,6-disubstituted phenyleneoxide)s, by selecting the electrolysis conditions: This electro-oxidative polymerization is described in the present paper.

Electro-Oxidative Polymerization of 2,6-Disubstituted Phenols
---

The electrolysis apparatus for the polymerization is illustrated in Figure 2, which is characterized by a single cell without a partition membrane between the electrodes. In poor solvents of poly(phenyleneoxide)s such as methanol and acetonitrile, the polymer was deposited on the electrode, i.e. passivation of the electrode occured. Dichloromethane, nitrobenzene, and hydroquinone dimethyl ether were selected as the solvents because both the polymer and a supporting electrolyte dissolved in them and they were relatively stable under electrolysis conditions.

Electro-oxidative polymerization of 2,6-disubstituted phenols is listed in Table I, with the polymerizations catalyzed by the copper-pyridine complex and oxidized by lead dioxide. 2,6-Dimethylphenol was electro-oxidatively polymerized to yield poly(2,6-dimethylphenyleneoxide) with a molecular weight of 10000, as was attained by other polymerization methods. The NMR and IR spectra were in complete agreement with those measured for the other polymerization

Table I. Electro-oxidative Polymerization of 2,6-Disubstituted Phenols

substituents of phenols		yield (wt%) of poly(phenyleneoxide)		
R_2	R_6	electro-[1] lysis	Cu-py cat.[2]	PbO_2 ox.[3]
CH_3-	CH_3-	83	85	54
C_6H_5-	C_6H_5-	38	46	86
CH_3O-	CH_3O-	23	0	32
Cl-	Cl-	98	0	0
CH_3-	H-	39	86	38
H-	H-	78	0	0

1) $[(C_2H_5)_4NBr]/[phenol]$: 2, in CH_2Cl_2/CH_3OH (4/1), 8 F/mol, 10 mA/cm^2
2) [Cu-pyridine]/[phenol]: 0.01 in benzene
3) $[PbO_2]$/[phenol]: 2.5 in benzene

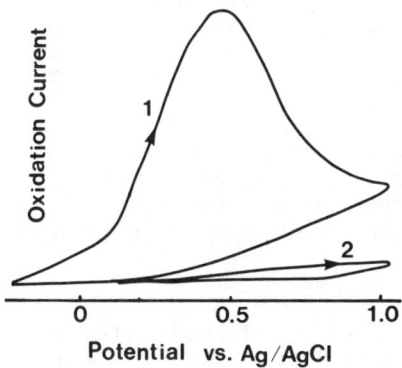

Figure 1. Cyclic voltammogram for the oxidation of 10 mM 2,6-dimethylphenol at a pltinum electrode in alkaline methanol.

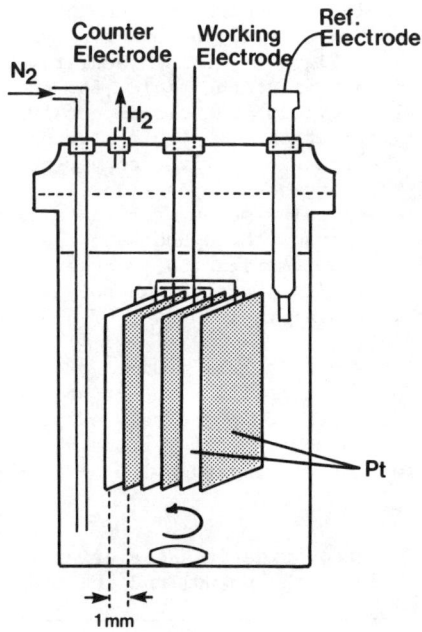

Figure 2. Electrolysis apparatus for the polymerization of phenols.

methods. The electro-oxidation was also effective for 2,6-diphenyl-, 2,6-dichloro-, 2,6-dimethoxy-, and 2-methylphenol to produce the corresponding polymers. Poly(2,6-dimethoxyphenyleneoxide) was formed by the electro-oxidation as well as by the oxidation with lead dioxide, while 2,6-dimethoxyphenol with the copper-pyridine catalyst gave only 3,3',5,5'-tetramethoxydiphenoquinone through C-C coupling reaction (Eq. 1). For 2,6-dichlorophenol the polymer was produced in high yield by the electro-oxidation, while the dichlorophenol reacted very slowly with the copper-pyridine catalyst and lead dioxide even at 60°C and did not at room temperature.

In Table I, one notices also that phenol is electro-oxidatively polymerized, which does not occur by the other methods.

Mechanism of the Polymerization

The following questions on the electro-oxidative polymerization arose. First, why various phenol derivatives were smoothly polymerized which could not occur by the oxidation with the copper catalyst or lead dioxide. Secondly, why the activated phenol was reacted preferentially through C-O coupling to form the poly(phenyleneoxide). The mechanism of the electro-oxidative polymerization is discussed below by using the example of 2,6-dimethylphenol.

The study of the molecular weight of the intermediate course is an effective method for the classification of polymerization as chain or stepwise reaction. In Figure 3, the molecular weight of the obtained polymer is plotted against the yield, for the oxidative polymerization of dimethylphenol with the copper catalyst and for the electro-oxidative polymerization. The molecular weight rises sharply in the last stage of the reaction for the copper-catalyzed polymerization. This behavior is explained by a stepwise growth mechanism. On the other hand, the molecular weight for the electro-oxidative polymerization remains constant throughout the reaction course. That is, the electro-oxidative polymerization proceeds seemingly through a chain reaction mechanism. (In practice the polymerization proceeds heterogeneously only in the diffusion layer of the electrode. The details are described below.)

The polymerization starting from a dimer or an oligomer is also an effective method for the classification of polymerization as chain or stepwise reaction. For the copper-catalyzed polymerization, the poly(phenyleneoxide) was also and rather rapidly obtained from the dimer of dimethylphenol(2,6-dimethylphenyl 3,5-dimethyl-4-hydroxy-

Table II. Electro-oxidative Polymerization of 2,6-Dimethylphenol and its Dimer

dimethyl-phenol	polymn.	yield (wt%)	$M \cdot 10^{-3}$
monomer	electrolysis	83	10
	Cu-py cat.	85	23
dimer	electrolysis	5	1
	Cu-py cat.	88	25

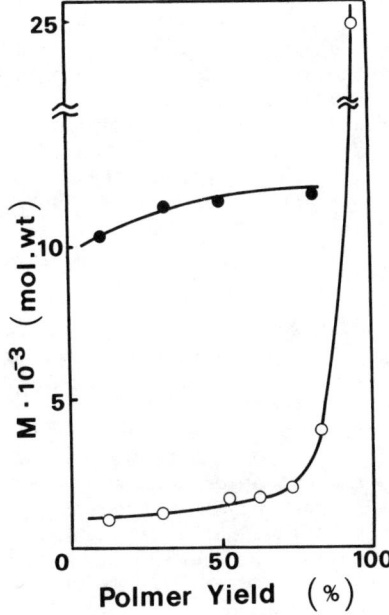

Figure 3. Relationship between the polymer yield and the molecular weight. o: Polymerization catalyzed by the copper-pyridine complex. ●: Electrolysis

phenyl ether). On the other hand, the electro-oxidative polymerization proceeded slowly, when the dimer was supplied as the starting material. This result also suggests a chain reaction profile (Table II).

The polymer yield increased with the current density at the constant electricity, although the molecular weight of the obtained polymer remained constant. The polymerization rate was related to the current density on the electrode, i.e. the concentration of the activated phenol. When the reaction solution was vigorously stirred, the polymer formation was much suppressed and yield of the diphenoquinone through C-C coupling increased. This result suggests that the polymerization occurs on the surface region of electrode or within the diffusion layer on electrode.

An ESR spectrum of the electrolysis solution of 2,4,6-tri-t-butylphenol showed the formation of phenoxy radical which was accumulated in the solution with the supplied current amount. The active species of polymerization reaction was assumed to be a phenoxy radical formed by the electrolysis.

Electrochemical Features of the Polymerization

The polymerization mechanism was also studied by electrochemical methods (e.g. see Figure 4). The cyclic voltammogram of 2,4,6-tri-t-butylphenol in alkaline methanol solution showed the reversible oxidation and corresponding reduction curve in the repeated sweeps. This means that the phenol stays at the electrode or within the diffusion layer and repeats the redox reaction. On the other hand, for 2,6-dimethylphenol only the oxidation peak was observed at 0.47 V vs. Ag/AgCl even when the scan rate was much faster than 10 V/s. Furthermore, the dimer of 2,6-dimethylphenol gave the oxidation peak more cathodic (at 0.42 V vs. Ag/AgCl). These results indicate that the oxidized and activated phenol coupled rapidly with each other to form the dimer before the electrolytic reduction of the activated phenol occurs. That is, the coupling reaction of the activated phenol takes place very rapidly at the electrode surface. The dimer, probably also an oligomer, is more easily oxidized at the electrode because of their lower oxidation peak potential than the monomer is. The oxidation peak potential observed for the polymerization of 2,6-dimethylphenol, in which passivation of the electrode was protected by selecting the solvent (Figure 4 (d)), was shifted to cathodic with the repeated sweeps. This corresponds to the progress of polymerization.

Diffusion coefficient of the phenol (D) and surface excess of the phenol on electrode (Γ) were estimated by double potential step chronocoulometry (5) using for an example the 2,4,6-tri-t-butylphenol. Tributylphenol is a simple model compound to evaluate the electrochemical values because of a lack of subsequent coupling and because it has a reversible oxidation and corresponding reduction peak at 0.23 V vs. Ag/AgCl. The plot of square root of time vs. charge passed gave a straight line. From the slope and the intercept D and Γ were calculated: $D = 2.1 \times 10^{-6}$ cm^2/s, $\Gamma = 8 \times 10^{-11}$ mol/cm^2. The D value is relatively small, and it can be estimated that the dimer and the oligomer for the polymerization system exhibit much smaller D values than the phenol monomer. This suggests that the oxidation products are accumulated on the electrode surface during the electrolysis. The Γ value indicates that the phenol concentration within

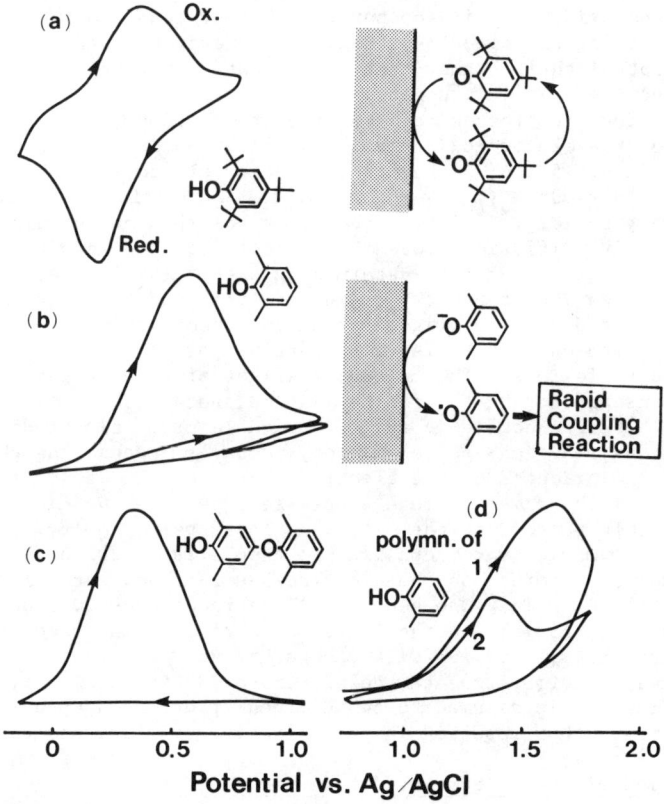

Figure 4. Cyclic voltammograms for the oxidation of 2,6-dimethylphenol, its dimer and tri-t-butylphenol at a platinum electrode. (a)-(c): in alkaline methanol, (d): in dichloromethane.

diffusion layer is ca. 100 times of that estimated from the phenol concentration in the bulk phase. This result suggests that the phenol is considerably adsorbed on the electrode surface.

Oxidation peak potentials of phenol derivatives were measured with cyclic voltammetry: 0.53, 0.47, 0.47, 0.28, and 0.77 V vs. Ag/AgCl for phenol, 2,6-dimethyl-, 2,6-diphenyl-, 2,6-dimethoxy-, and 2,6-dichlorophenol respectively. The oxidation potential of phenol and 2,6-dichlorophenol are relatively high and this high potential is one of the reasons why phenol and dichlorophenol could not be polymerized by the oxidation with copper catalyst or lead dioxide. On the other hand, for the electro-oxidative polymerization the potential can be kept slightly higher than the oxidation potential of phenols and the polymerization proceeds.

From the electrochemical measurements mentioned above, the aspects of the electro-oxidative polymerization is summarized as foloows (Figure 5). (i) The polymerization proceeds in the diffusion layer of electrode. (ii) Coupling and the polymerization reaction occurs very rapidly. (iii) The phenol is adsorbed on the electrode surface. (iv) Diffusion process of phenol from and into the bulk phase is very slow and rate-determining step. (v) Active species of the polymerization is phenoxy radical. (vi) Oxidation potential is lower for the dimer and oligomer than the phenol monomer.

The features of the electro-oxidative polymerization can be explained as follows. The molecular weight of the obtained polymer stayed constant during the polymerization, because the polymerization proceeds heterogeneously in the diffusion layer of electrode. The C-O coupling reaction is predominant, probably because the phenol is adsorbed and oriented on the electrode surface. The polymerization started from the dimer is much suppressed, because the dimer diffuses from the bulk phase into the diffusion layer very slowly.

The polymerization mechanism of phenols is described as follows. The phenol is adsorbed on the electrode surface and accumulated in the diffusion layer. The adsorbed phenol undergoes one-electron oxidation to the phenoxy radical on the electrode surface. The concentrated phenoxy radical is coupled with each other at p-position to form the dimer, and the dimer repeats the electro-oxidation and coupling. The phenoxy radical is assumed to be adsorbed or oriented upon the electrode surface thus resulted in the selective coupling reaction.

Application of the Electro-Oxidative Polymerization

The first application example is the electro-oxidative polymerization of phenol in the presence of 2,2-bis[3,5-dimethyl-4-hydroxyphenyl]-propane, which is the procedure to obtain terminally hydroxylated poly(phenyleneoxide), i.e. the oligomer contained two hydroxy groups per one molecule.

(Eq. 3)

The second example is the electro-oxidative polymerization of phenols bearing functional substituents. It is known that salicylic acid forms a stable chelate with copper ion, thus the copper catalyst is deactivated and the polymerization does not occur. On the other hand, salicylic acid was electro-oxidatively polymerized to produce the poly(phenyleneoxide) bearing carboxylic group.

(Eq. 4)

The third application is the oligomerization of phenol. By selecting solvent and supporting electrolyte, phenol is electro-oxidatively polymerized to yield poly(phenyleneoxide) as a tan-colored powder.

The NMR signal was simple one peak assigned to the phenyleneoxide structure. The IR spectrum showed that the absorption at 3400 cm^{-1}, corresponding to the hydroxy group, has disappeared, and new absorption assigned to an ether bond was observed at 1190 cm^{-1}. It has been reported by the study on oligo(phenyleneoxide) prepared by Ullmann reaction that oligo(1,2-phenyleneoxide) showed the IR absorption at 970 and 1025 cm^{-1} but the linear oligo(1,4-phenyleneoxide) did not show this absorption (6). The IR spectrum of oligo(phenyleneoxide) prepared by the electro-oxidation showed no absorption signal at both 970 and 1025 cm^{-1}: The linear 1,4-phenylene structure was suggested (Figure 6). Phenol is assumed to be adsorbed and oriented on the platinum electrode, which probably brings about the selective coupling to form oligo(1,4-phenyleneoxide).

The thermal stability of the oligo(phenyleneoxide) was much higher than that of oligo(2,6-dimethylphenyleneoxide) with the same molecular weight. The poly(1,4-phenyleneoxide) structure is expected to show interesting properties.

Conclusions are as follows. (i) Various phenol derivatives can be smoothly oxidized to yield poly(phenyleneoxide)s. (ii) The 1,4-phenyleneoxide structure is predominant in the polymer. (iii) The properties of oligo(phenyleneoxide) and the polymer with two terminal hydroxyls should be interesting.

Experimental

3 Pairs of platinum plate (2 x 5 cm) were set in a cell with 1 mm spacing as the working and the auxiliary electrode (Figure 2). Reference electrode was Ag/AgCl. The solution (50 ml) of phenol (0.005 mol) and electrolyte (0.01 mol) such as tetraethylammonium bromide and tetraethylammonium perchrolate was kept under nitrogen atmosphere in the cell. The electrolysis was carried out with constant potential or current density (10 mA/cm^2) which was supplied by a potentiogalva-

Figure 5. Scheme for the electro-oxidative polymerization of phenols.

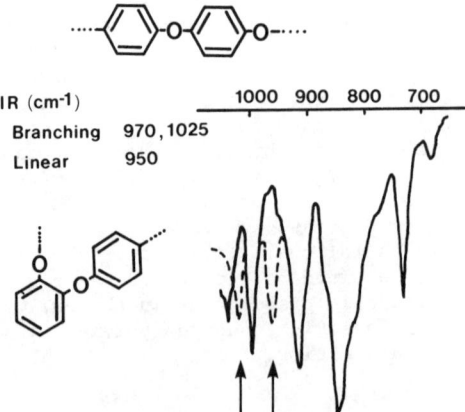

Figure 6. IR spectrum of poly(1,4-phenyleneoxide) obtained by the electrolysis of phenol.

nostat. During the electrolysis hydrogen gas was evolved from the cathode. After the electrolysis, the reaction mixture was washed with water to extract the supporting electrolyte. The concentrated organic layer was slowly poured into methanol. White or pale brown powder was collected by filtration, washed with methanol, and dried in vacuo.

The NMR signals of the polymerization product agreed with those assigned to the poly(phenyleneoxide) structures shown in Equation 2. The IR spectrum of the polymerization product showed that the absorption at 3400 cm^{-1}, corresponding to the hydroxyl group, has disappeared, and new absorption assigned to an ether bond was observed as follows.

Poly(2,6-dimethyl-1,4-phenyleneoxide): ^{1}H NMR(CDCl$_3$) δ = 2.10 (CH$_3$) 6.90(benzene H), IR 1180 cm^{-1}(ν_{C-O-C}), poly(2,6-diphenyl-1,4-phenyleneoxide); ^{1}H NMR δ = 6.70, 7.65(benzene H), IR 1170 cm^{-1} (ν_{C-O-C}), poly(2,6-dichloro-1,4-phenyleneoxide): ^{1}H NMR δ = 7.3 (benzene H) IR 1180 cm^{-1}(ν_{C-O-C}), poly(2,6-dimethoxy-1,4-phenyleneoxide): ^{1}H NMR δ = 3.67(OCH$_3$), 7.10(benzene H) IR 1210 cm^{-1}, 1195 cm^{-1} (ν_{C-O-C}), poly(2-methyl-1,4-phenyleneoxide): ^{1}H NMR δ = 2.18(CH$_3$), 6.60, 7.00, 7.20(benzene H), IR 1200 cm^{-1}(ν_{C-O-C}), poly(2-hydroxycarbonyl-1,4-phenyleneoxide): IR 1190 cm^{-1} (ν_{C-O-C}), 1710 cm^{-1}(ν_{COOH}) carboxylic acid residue determined by titration: 0.006 equivalent/g, poly(1,4-phenyleneoxide); ^{1}H NMR δ = 7.2(benzene H), IR 1190 cm^{-1} (ν_{C-O-C}), co-oligomer of phenol and bisphenol: ^{1}H NMR δ = 7.10 (benzene H), 2.30(CH$_3$), 1.80(C(CH$_3$)$_2$), IR 1190 cm^{-1}(ν_{C-O-C}), 1359, 1378 cm^{-1}(geminal dimethyl δ$_{C-H}$).

The refered oxidative polymerization of phenols with the copper-pyridine catalyst was carried out as in lit. (7,8).

Acknowledgments

The authors thank the Ministry of Education, Science and Culture Japan, for the Grant-in-Aid for Scientific Research.
This work derives from a collaborative effort with Prof. F. C. Anson (California Institute of Technology) that is supported by the Japan-U.S. Cooperative Science Program.

Literature Cited

1. Hay, A. S. J. Polymer Sci. 1962, 58, 581-591.
   Hay, A. S. Adv. Polymer Sci. 1967, 4, 496.
2. Pham, M. C.; Hachemi, A.; Dubois, J. E. J. Electroanal. Chem. 1984, 161, 199-204 and the references therein.
   Ohnuki, Y.; Ohsaka, T.; Matsuda, H.; Oyama, N. J. Electroanal. Chem. 1983, 158, 55.
3. Tsuchida, E.; Nishide, H.; Maekawa, T. Polymer J. in contribution.
4. Anson, F. C.; Osteryoung, R. A. J. Chem. Edu. 1983, 60, 293.
   Anson, F. C. Anal. Chem. 1966, 38, 54.
5. Borman, F. H. Willem. U.S. Patent 3 335 075, 1964.
6. Van Dort, H. M.; Hoefs, C. A. M.; Magre, E. P.; Schopf, A. J.; Yntema, K. Eur. Polymer J. 1968, 4, 275.
7. Tsuchida, E.; Nishide, H.; Nishiyama, T. Makromol. Chem. 1968, 4, 275.
8. Van Dort, H. M.; De Jonge, C. R. H. H.; Mijs, W. J. J. Polymer Sci. 1968, C-22, 431.

RECEIVED February 19, 1985

16

Coupling and Capping Reactions on Poly(2,6-dimethyl-1,4-phenylene Oxide)

DWAIN M. WHITE and GEORGE R. LOUCKS

Corporate Research and Development Center, General Electric Company, Schenectady, NY 12301

> The phenolic hydroxyl endgroups in poly(2,6-dimethyl-1,4-phenylene oxide) react with various mono, di- and trifunctional acylating (or related) reagents under phase transfer catalysis conditions to produce capped and/or coupled products in high yields. Acetic anhydride, even at low concentrations, acetylated the hydroxyl endgroups almost quantitatively. Bifunctional reagents (e.g., isophthaloyl chloride) coupled the polymer producing a viscosity increase consistent with a two-fold increase in molecular weight. Trifunctional reagents (e.g., phosphorous oxychloride) produced a branched polymer and tripled the molecular weight. Typical increases in intrinsic viscosity of the polymer with trifunctional coupling were from 0.25 dl/g to 0.50 dl/g. With bifunctional poly(phenylene oxide)s, even larger molecular weight increases were attained. Quaternary ammonium and phosphonium halides were used as the phase transfer catalysts. For effective coupling, high-shear mixing and high concentrations of polymer and base were used.

The terminal phenolic groups of oligomers of poly(2,6-dimethyl-1,4-phenylene oxide), $\underline{1}$, ($\underline{1}$, $\underline{2}$) react efficiently with multifunctional coupling reagents to form polymers with increased molecular weights. A high degree of coupling has been attained by adding a stoichiometric quantity of a coupling reagent such as phthaloyl chloride under anhydrous conditions to the preformed anion of the phenolic hydroxyl end group of the polymer.($\underline{3}$) A more convenient procedure which utilizes phase transfer catalysis (PTC) ($\underline{4-6}$) and does not require anhydrous conditions or exact stoichiometric ratios of reactants is described in this report. The reaction is of the general type:

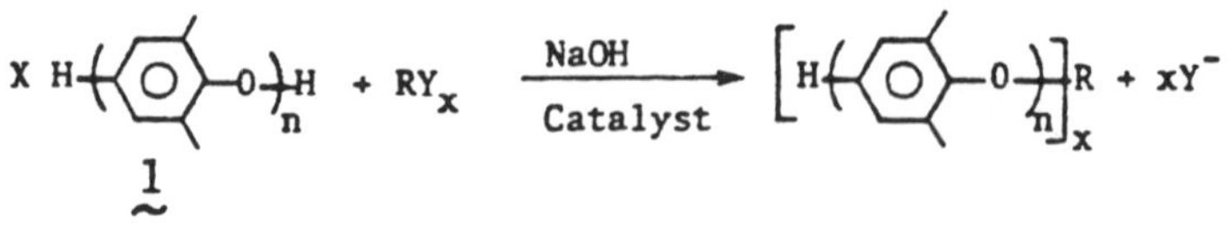

$$X\ H{+\!\!\langle O\rangle\!\!-\!\!O+\!\!}_n H\ +\ RY_x\ \xrightarrow[\text{Catalyst}]{\text{NaOH}}\ \Big[H{+\!\!\langle O\rangle\!\!-\!\!O+\!\!}_n R\Big]_x\ +\ xY^-$$

1

and is applicable to monofunctional (where RY_x is a capping reagent) or polyfunctional (RY_x is a coupling reagent) reagents. The reaction can also be used to prepare block copolymers. If RY_x is an oligomeric species, block copolymers result. If another polymer with hydroxyl end groups is present with the poly(phenylene oxide) during the coupling reaction, cross-coupled products result.

Results and Discussion

The PTC reaction is carried out by adding the capping or coupling reagent to a vigorously stirred mixture of a concentrated solution of low molecular weight 1 which contains a quaternary ammonium halide catalyst and a 50% aqueous solution of sodium hydroxide. In many cases the reaction is over within several minutes after the coupling reagent is added and the polymer can be isolated by precipitation with methanol.

Analyses of the products of several typical reactions after isolation by methanol precipitation are presented in Table I.

Table I. Analysis of Capped and Coupled 1

RY_x	[η] (dl/g) Initial	Prod.	OH Absorbance Initial	Prod.
None	0.33	0.33	0.180	0.180
Acetic Anhydride	.49	.49	.092	.000
Isophthaloyl Chloride	.33	.57	.180	.022
BPA-bischloroformate	.28	.54	.267	.017
Phosphorous Oxychloride	.33	.72	.180	.016

Capping with acetic anhydride (10% based on polymer weight) results in a disappearance of the phenolic hydroxyl end groups of the polymer. The absorbance of the end groups was measured in the infrared at 3610 cm^{-1} in carbon disulfide in a cell with a 1 cm pathlength. By comparing the spectrum of the initial polymer and the isolated product with the spectrum of a hydroxyl-free polymer sample, greater than 99% capping is found for the PTC reaction. With less acetic anhydride (1.25% based on polymer weight) 97% of the hydroxyl groups are capped. For high molecular weight polymer ([η] 0.49 dl/g in CHCl$_3$ at 25°C, no change in intrinsic viscosity, molecular weight distribution or glass transition temperature is noted as a result of the capping reaction.

The coupling reactions result in marked increases in the molecular weight of the polymer. These increases are measured by intrinsic viscosity determinations. The changes in viscosity for several coupling reagents are listed in Table I. Isophthaloyl chloride and the bischloroformate from bisphenol-A produce 70 to 90% increases in the intrinsic viscosity while the trifunctional reagent (phosphorous oxychloride) produces an increase of 120%. These changes in intrinsic viscosity correspond approximately to doubling and tripling the molecular weight for bifunctional and trifunctional coupling, respectively. The high degree of capping of the hydroxyl groups that accompanies the coupling is also shown in Table I by the 90% decrease in hydroxyl absorbance.

The second and third reagents in Table I are solids. They are added to the reaction mixture (containing $\underline{1}$, Adogen 464, and aqueous sodium hydroxide) as solids either at once or in several portions over a short period of time. The gradual dissolution of the solids into the reaction mixture makes this mode of addition equivalent to slow addition of a solution of the coupling reagent (the method used with liquid coupling reagents). The reaction between 1 and RY_x is sufficiently fast that the coupling reaction is favored over competing hydrolysis reactions. Furthermore, all of the functionality Y of RY_x reacts before an excess of RY_x accumulates. This is an important feature since an excess of RY_x at this stage can lead to capping of too many of the polymer chains with $-RY_{x-1}$ groups and not leave enough phenolic endgroups to react with the capped chains to form the coupled product. Once the rapid coupling is complete, the accumulation of additional coupling reagent has no effect on the reaction. It slowly hydrolyzes and is removed during workup.

Coupling reactions are favored over hydrolysis reactions under PTC conditions when a catalyst such as Adogen 464 (tricaprylmethyl ammonium chloride) is used. The effectiveness of the catalyst may be due to its high solubility in the organic phase where the key steps of the reaction are believed to take place. Expressions (1)-(4) summarize these steps. The catalyst transfers hydroxide into the organic phase (step 1, Y = halide) to form the ammonium phenoxide (Step 2) which reacts in Step 3 with the coupling reagent. The product still contains reactive functionality and reacts with additional oligomeric ammonium phenoxide (Step 4) until all of the Y groups are displaced.

$$R'_4N^+Y(\text{org}) + OH^-_{(\text{aq})} \longrightarrow R'_4N^+OH(\text{org}) + Y^-_{(\text{aq})} \quad (1)$$

$$R'_4N^+OH^-_{(\text{org})} + H \left(\underset{n}{\underset{}{}\text{⟨O⟩}-O} \right) H \longrightarrow H \left(\underset{n}{\underset{}{}\text{⟨O⟩}-O} \right)^- NR'^+_4 + H_2O \quad (2)$$
$$A$$

$$A + RY_x \longrightarrow H \left(\underset{n}{\underset{}{}\text{⟨O⟩}-O} \right) RY_{(x-1)} + R'_4N^+Y^- (\text{org}) \quad (3)$$
$$B$$

$$B + \underset{\sim}{2} \longrightarrow \left[H \left(\underset{n}{\underset{}{}\text{⟨O⟩}-O} \right) \right]_2 RY_{(x-2)} + R'_4N^+Y^- (\text{org}) \quad (4)$$

While Steps 3 and 4 take place, Step 1 is repeated, replenishing the supply of $R'_4N^+OH^-$.

The coupling is carried out with a high concentration (50% w/w) of sodium hydroxide in water and with high-shear mixing which ensures a large interfacial area to aid in transport of hydroxide across the interface. The quantity of NaOH on a weight basis is relatively small, however, since only two to five times the quantity of phenolic end groups is required. The polymer concentration is kept high (20 to 35% w/v) to offset the low molar endgroup concentration that results from the relatively high molecular weight of the polymer. Temperatures (often 50°) were sufficient to solubilize the polymer but not so high as to cause decomposition of the catalyst. With the conditions described here the catalyst can be used at a concentration of ca. 0.1% (i.e., 0.5% w/w based on polymer) and provide rapid coupling rates with many coupling reagents. Both Adogen 464 and cetyltrimethylammonium bromide are effective catalysts at this concentration.

The bifunctional polymer 2 (2) that is formed when oligomers of 1 are heated with 3,3',5,5'-tetramethyl-4,4'-diphenoquinone (a byproduct of the synthesis of 1) reacts with a bifunctional coupling reagent to produce a coupled product

with an increase in molecular weight that is greater than two-fold. Gelation occurs when the coupling reagent is trifunctional but can be avoided if a sufficient quantity of monofunctional polymer is also present. Examples of such coupling reactions on mono- and difunctional polymer mixtures are presented in Table II.

Table II. Coupling Reactions on Mixtures of 1 and 2

Coupling Reagent	Wt. of Reagent (g)[a]	Intrinsic Viscosity (dl/g)	
		Initial Pol.	Product
None (Control)	–	.31	.32
Isophthaloyl Chloride	0.23	.31	.60
4,4'-Biphenyl Disulfonyl dichloride	.39	.31	.70
Toluene-2,4-diisocyanate	.19	.31	.47
2,3-Bis(bromomethyl)-quinoxaline	.35	.31	.53
α,α-Dibromo-p-Xylene	.29	.31	.55
Methylene bromide	7.5	.31	.75
Methylene chloride	4.0	.31	.49
Cyanuric chloride	.18	.24	.90
1,3,5-Benzene-Tricarboxylic acid chloride	.55	.24	.79

(a) For 10g polymer

Methylene bromide can function as the coupling reagent if it is used in an excess. This unusual coupling reaction succeeds, presumably, because the intermediate α-bromo ether, **3**, reacts much more rapidly with the phenoxide endgroup of another polymer than methylene bromide does to produce the formal linked product, **4**.

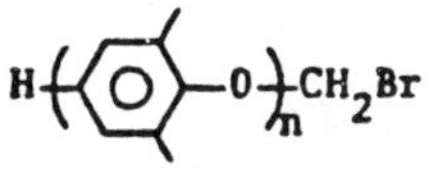

$$H\text{-}(\phi\text{-}O)_n\text{-}O\text{-}CH_2Br \qquad H\text{-}(\phi\text{-}O)_n\text{-}O\text{-}CH_2\text{-}(O\text{-}\phi)_n\text{-}H$$

3 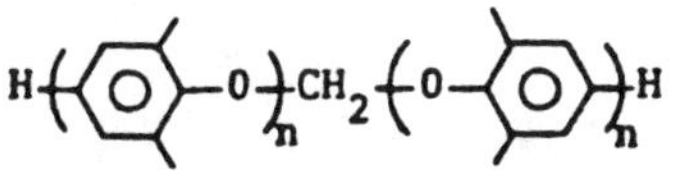 **4**

Methylene chloride also acts as a coupling reagent but is less effective than methylene bromide. Examples of coupling material containing structure **2** are shown in Table II.

Block copolymers containing poly(phenylene oxide) and a second polymer can be prepared with the PTC coupling system in various ways. In one method, oligomers of the second polymer which contain acid chloride endgroups are used as the coupling reagent. When the oligomeric coupling reagent is bifunctional, reaction with **1** produces an ABA block copolymer while reaction with **2** produces an (AB)$_n$ block copolymer. Typical polymeric coupling reagents are **5** (**8**) and **6** (**3**).

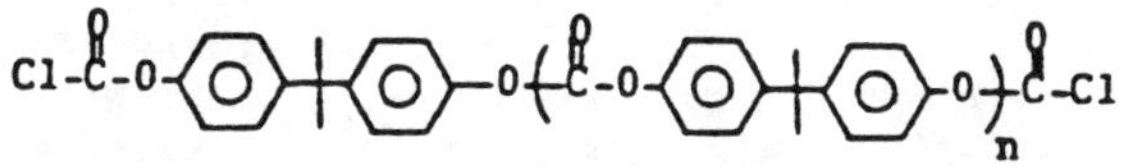

$$Cl\text{-}\overset{O}{C}\text{-}O\text{-}\phi\text{-}\phi\text{-}O\text{-}(\overset{O}{C}\text{-}O\text{-}\phi\text{-}\phi\text{-}O)_n\overset{O}{C}\text{-}Cl$$

5

$$Cl\text{-}\overset{O}{C}\text{-}\phi\text{-}\overset{O}{C}\text{-}(O\text{-}\phi\text{-}\phi\text{-}O\text{-}\overset{O}{C}\text{-}\phi\text{-}\overset{O}{C})_n\text{-}Cl$$

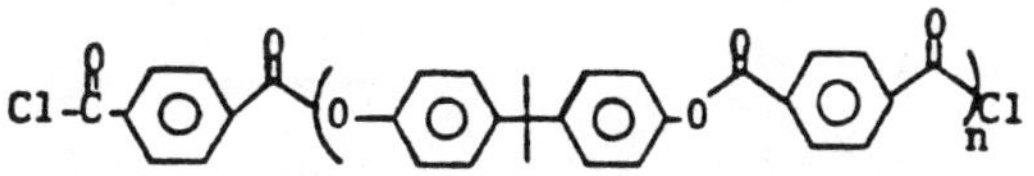

6

Care must be taken with such oligomeric coupling reagents because of the possibility of hydrolysis of backbone ester or carbonate groups during the coupling reaction. Reagent **6** and the corresponding block in the product are particularly susceptible to cleavage by sodium hydroxide in the presence of the catalyst. However, high molecular weight product can be obtained if the addition time of the reagent is short and the reaction mixture is acidified immediately after the addition step is finished.

A second method for preparing block copolymers is a cross-coupling process. A low molecular weight coupling reagent is added to a mixture of poly(phenylene oxide) and a second homopolymer with phenolic hydroxyl endgroups such as **7** (**9**) or **8** (**10**) in the presence of sodium hydroxide and catalyst.

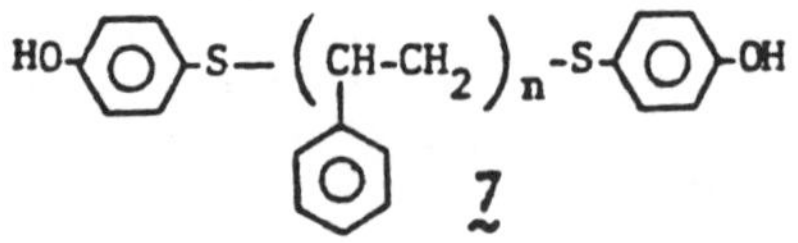

HO—⟨O⟩—S—(CH-CH₂)ₙ—S—⟨O⟩—OH
         |
         ⟨O⟩
         7

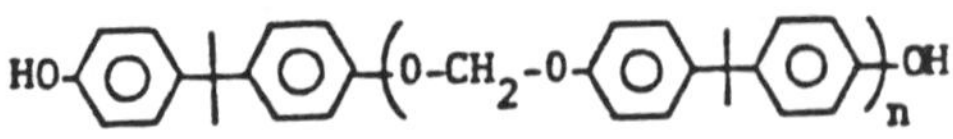

HO—⟨O⟩—⟨O⟩—(O-CH₂-O—⟨O⟩—⟨O⟩)ₙ—OH
         8

The resultant copolymer from **1** and **8** has two glass transition temperatures (T_g's at 97° and 202°) which indicates two phases are present. In contrast, the copolymer from **1** and **7** has only a single T_g value which is consistent with the compatibility between 1 and polystyrene homopolymers.

Evidence for the block-type nature of the copolymers is provided by an analysis based on the ability of **1** to form a complex with methylene chloride that is insoluble in methylene chloride. (**11**) When each of the coupled copolymers described above is dissolved in methylene chloride, a precipitate forms and can be isolated by filtration. Both **1** and the second polymer are present in the soluble fraction and both are present in the precipitate. Since **1** normally precipitates quantitatively from methylene chloride solution and the other polymers remain soluble, the coupled products must be block copolymers.

Experimental

The solvents and coupling reagents were reagent grade materials. The tricaprylmethylammonium chloride (Adogen 464) was not purified further. Intrinsic viscosities were measured in chloroform at 25°. Infrared spectra for hydroxyl analysis were measured on 2.5% solutions in carbon disulfide (vs. carbon disulfide) in a cell with a 1.00 cm path length. The absorbance at 3610 cm^{-1} was subtracted from a similar spectrum of **1** which had been quantitatively acetylated. Gel permeation chromatographic measurements were performed on a Waters Associates HPLC with μ-Styragel columns (10^5, 10^4, 10^3, and 500ANG) in chloroform with a polystyrene calibration.

Low molecular weight polymers were prepared by Hay's method (**12**) by using short reaction times and terminating the reactions by adding aqueous Na₃EDTA to complex the copper. The reaction mixtures were then either treated with methanol to precipitate **1** or heated at 50-55° under nitrogen for 1-1.5 hrs to discharge the color of the 3,3',5,5'-tetramethyl-4,4'-diphenoquinone and generate **2**. Polymer containing **2** was then isolated by precipitation with methanol. Evidence for the presence of structure **2** in the starting material was obtained by IR (the absorbance of a 2.5% solution at 3610 cm (1 cm path) increased 0.12 units after the heating step).

Acetic Anhydride Capping. To a solution of 1 (5.0 g, intrinsic
viscosity 0.49 dl/g) in 15 mL chlorobenzene was added 0.10g Adogen
464 and sodium hydroxide (0.10 g 50% aqueous solution). The mixture
was stirred with a vibromixer stirrer at 25° under nitrogen and 0.5 g
acetic anhydride was added. After two minutes the reaction mixture
was diluted with 20 mL toluene and the polymer was isolated by
methanol precipitation. The dried product weighed 5.0 g; intrinsic
viscosity 0.49 dl/g; the hydroxyl absorbance had decreased from an
initial value of 0.092 to 0.000. With less reagents (1.25% acetic
anhydride, 1.1% 50% aqueous NaOH and 0.50% Adogen 464 based on 1 by
weight) at 44-48°, the acetic anhydride was added over a two minute
period to the high shear zone of a paddle stirrer agitated reaction
mixture to reduce the hydroxyl absorbance of a 20% (in toluene) solution of 1 to 3% of the initial value.

Coupling Reactions with Solid Coupling Reagents. This procedure
includes a step where 1 is converted to structure 2 by heating the
polymerization reaction mixture; the procedure for the coupling step
also applies to normal 1.
   A 300 mL jacketed Waring Blender equipped with a nitrogen inlet,
thermocouple and septum port was charged with 30 mL monochloroben-
zene and 10g polymer containing 2. Water at the desired reaction
temperature was circulated through the blender jacket. Adogen 464
(0.05g) was added followed by 1.3 mL 50% aqueous NaOH. After high
speed mixing for 3 minutes, the coupling agent was added all-at-once.
High shear conditions were maintained during the addition and 2-3
minutes thereafter. The reaction mixture was diluted with toluene,
acidified with conc. HCl and precipitated with methanol. Results for
various coupling reagents are listed in Table II.
   A series of coupling reactions on unequilibrated polymer was run
using a slower addition of isophthaloyl chloride than described above
and at several catalyst levels. To a solution of polymer 1 (5.0g,
intrinsic viscosity 0.36 dl/g) in 39 mL toluene was added Adogen 464
(15 mg in 0.15 mL toluene) and sodium hydroxide (180 mg 50% aqueous
solution). The mixture was stirred under a nitrogen atmosphere and
solid isophthaloyl chloride (73 mg) was added over a 15 min period.
The polymer was precipitated by dropwise addition of methanol, then
washed with methanol and dried. The product weighed 5.0g, intrinsic
viscosity 0.57 dl/g. The infrared absorbance at 3610 cm^{-1} decreased
upon coupling from 0.180 to 0.022. The procedure was repeated at
several other Adogen 464 concentrations and in one case without the
coupling reagent. The products showed the following analyses:

Weight of Adogen 464 (mg)	Isophthaloyl Chloride (mg)	Coupled Polymer [η] (dl/g)	3610cm^{-1} absorbance
0 (control)	)	0.36	0.180
0	73	.40	.142
5	73	.56	.035
15	73	.57	.022
25	73	.50	.020
25	0	.39	.140

The reaction was repeated on a different polymer sample with the solid bis(chloroformate) of 2,2-bis(4-hydroxyphenyl)propane which resulted in the following increase in intrinsic viscosity of the polymer: 0.28 to 0.54 dl/g.

Coupling with $POCl_3$ (Vibromixer Stirrer). To a solution of polymer 1 (2.44g, intrinsic viscosity 0.33 dl/g $\overline{M}w$ 34,000, $\overline{M}n$ 17,000) in 8.2 mL chlorobenzene was added Adogen 464 (0.12 mL 10% solution in toluene) and sodium hydroxide (180 mg 50% aqueous solution). The mixture was stirred at 25° with a Vibromixer Stirrer under a nitrogen atmosphere for 30 minutes and phosphorus oxychloride (0.020g) was added over a 10 minute period with a syringe. The reaction mixture was diluted with toluene (10 mL) and the polymer isolated by methanol precipitation and dried; intrinsic viscosity 0.72 dl/g. The initial polymer had an OH infrared absorption at 3610 cm^{-1} of 0.18; the value for the coupled product was 0.016.

The reaction was repeated using one half the quantity of phosphorous oxychloride (equivalent to the estimated number of OH groups in the polymer). The intrinsic viscosity increased from 0.33 to 0.70 dl/g and the OH absorbance at 3610 cm^{-1} decreased from 0.18 to 0.015.

With a Polytron homogenizer to provide high-shear mixing for a larger scale reaction at 40°, the intrinsic viscosity of a polymer (2000g) with an initial value of 0.24 dl/g was increased to 0.51 dl/g (6 minute $POCl_3$ addition) and to 0.40 dl/g (1.5 minute $POCl_3$ addition). Similar results occurred when a Waring blender was used to provide high shear mixing for intermediate scale reactions.

Coupling with $POCl_3$ (Paddle Stirrer). To a solution of polymer 1 (15.0g, intrinsic viscosity 0.28 dl/g $\overline{M}w$ 27,000, $\overline{M}n$ 13,000) in 69 mL toluene was added 75 mg cetyltrimethylammonium bromide (CTMAB) and sodium hydroxide (1.25 mL 25% aqueous solution) at 30° under a nitrogen atmosphere. After stirring with a paddle stirrer in a 3-neck round bottom flask for ten minutes, a solution of phosphorus oxychloride (4.7 mL 2% toluene solution) was added over a 47 minute period at a constant rate using a syringe controlled with an infusion pump. The reagent was introduced into the high-shear region immediately above the outer portion of the rapidly stirring paddle. A reaction temperature between 40 and 45° was maintained throughout the addition. When addition was completed, acetic acid was added to neutralize excess base and the polymer was precipitated by slowly adding 500 mL methanol to the reaction mixture. After washing with methanol and drying at 80° (10 Torr), the polymer weighed 14.9g. The analyses for viscosity and hydroxyl end groups of the product and two others prepared under identical conditions with 150 mg and 30 mg catalyst are presented:

CTMAB (mg)	Initial Polymer		Coupled Product	
	[η] (dl/g)	3610 Abs	[η] (dl/g)	3610 Abs
150	0.28	0.20	0.60	0.01
75	0.28	0.20	.62	0.01
30	0.28	0.20	.46	0.01

Similar results were obtained with similar weights of Adogen 464 instead of CTMAB.

Coupling with Methylene Bromide. A 300 mL high speed mechanical blender equipped with nitrogen inlet, thermocouple and addition port was charged with 30 mL chlorobenzene and 10.0g of polyphenylene oxide containing structures 1 and 2 ([η] 0.31 dl/g $\bar{M}w$ 32,000, $\bar{M}n$ 16,000; OH absorbance at 3610 cm^{-1}: 0.301. Water at ~ 70°C was circulated through the blender jacket, 0.05g Adogen 464 and 1.3 mL 50% aqueous sodium hydrogen were added to the blender. After high speed mixing for 3 minutes, 2.5g methylene bromide was added all at once and high shear mixing conditions were maintained for 40 minutes. A sharp increase in viscosity occurred at about 30 minutes. The resulting viscous polymer solution was diluted with 5 volumes of toluene, transferred to a larger blender and acidified with conc. hydrochloric acid. The polymer was precipitated by addition of four volumes of methanol, collected, washed with fresh methanol and dried overnight in vacuo at 60°C. [η] 0.62 dl/g; IR absorbance at 3610 cm^{-1}: 0.016. When the reaction was repeated using methylene chloride (3.0 mL) instead of methylene bromide, the polymer had values of [η] 0.49 dl/g and OH: 0.134.

Preparation of Polyformal, 8. A solution of 45.6g (0.20 mole) of bisphenol-A in 62 mL of methylene chloride and 93 mL of N-methylpyrrolidone (NMP) was stirred vigorously under a nitrogen atmosphere with 16.0g (0.50 mole) sodium hydroxide pellets at 80°C for 5 hours. Additional methylene chloride was added and the mixture was filtered while warm. A mixture of 250 mL methanol and 250 mL acetone was added to the filtrate to precipitate the polymer. The polymer was collected on a filter and dried at 60°C and 10 Torr for 20 hours. The weight of dry polymer was 35.8g. Intrinsic viscosity: 0.21 dl/g in chloroform at 25°C; number average molecular weight from infrared analysis of hydroxyl end groups: 8150. Using either less sodium hydroxide or less methylene chloride with a similar procedure yielded lower molecular weight polyformal. Thus, with 5% less sodium hdyroxide, the polymer had an intrinsic viscosity of 0.08 dl/g in chloroform at 25°C; $\bar{M}_n$ was 2560 and T_g was 52°C. With 70% less methylene chloride: [η] 0.13 dl/g, $\bar{M}_n$ 3770.

Cross-coupling of Poly(phenylene oxide) and Polyformals. A 300 mL mini-blender was equipped with a nitrogen inlet, a thermocouple near the high shear impeller region of the blender and a port for the introduction of a coupling agent. The blender was charged with polymers 1 (5g [η] 0.24 dl/g and 8 (5g), 20 mL of chlorobenzene, 0.5 mL of a 10% (W/V) solution of Adogen 464 (methyl tricaprylammonium chloride) and 50% aqueous sodium hydroxide. The reaction medium was maintained under an inert nitrogen atmosphere at a temperature of approximately 50°C, while the reaction ingredients were premixed vigorously for 2 minutes. Isophthaloyl chloride (IPC) pellets were then added. Within 2 minutes, a viscous reaction mixture resulted which was diluted with 20mL of chlorobenzene and added to 300 milliliters of methanol with vigorous stirring. A resulting solid precipitate was filtered, washed with methanol, dried at 80°C for 20 hours under a vacuum of 10 Torr. A description of reactants and products for several polyformals is listed below:

[η] (dl/g)	g (Mn)	50% NaOH (g)	IPC (g)	Product (g)	(η) (dl/g)	Tg (°C)	
0.08	2560	2.06	1.04	8.82	0.33	81	*
.11	3440	1.66	.84	9.65	.80	97	202
.13	3770	1.56	.60	9.85	.51	96	*
.21	8150	.99	.50	9.54	.80	96	*

*Higher transition not measured.

<u>Preparation of 7.</u> Styrene (100g, 0.96 mole, freshly distilled), azobisisobutyronitrile (0.05g) and bis(4-trimethylsiloxyphenyl) disulfide (5.83g, 0.15 mole) were heated under nitrogen in a 4 oz. screwcap bottle for seven days at 50°±3°C. The clear viscous liquid was diluted with toluene and precipitated by adding to 500 mL methanol containing a solution of 2g calcium nitrate in 50 mL ethanol in a stirred blender. The polymer was filtered off, dried, and reprecipitated as above, then dried, redissolved in toluene and precipitated by dropwise addition into 3 liters of methanol containing 1 mL conc. hydrochloric acid. After filtering and drying, the polymer weighed 50.4g, exhibited an intrinsic viscosity (measured in chloroform at 25°C) of 0.20 dl/g, and an infrared OH absorbance at 3590 cm^{-1} of 0.238 (500 mg/10 mL CS_2).

<u>Cross-coupling 1 and 2 with 7.</u> A 300 mL jacketed Waring blender equipped with a nitrogen inlet--in order to provide an inert nitrogen purged reaction medium, a thermocouple--located in the high fluid shear stress reaction region, and septum port was charged with 14 mL chlorobenzene, 2 g of α,ω-(bis-4-trimethylsiloxyphenylthio)polystyrene as prepared above, and 2g of a mixture of polyphenylene oxides 1 and 2. Water at 45°C was circulated through the blender jacket. A solution (0.2 mL) containing 10% Adogen 464 in toluene, was added to the blender followed by 0.2 mL of a 50% aqueous solution of NaOH. After high speed mixing for 2 minutes, isophthaloyl chloride was added as a solid over a 2-minute period. High shear reaction conditions were maintained throughout the addition of isophthaloyl chloride and for 2 minutes thereafter. The resulting reaction mixture was diluted with toluene and slowly poured into a larger blender containing several volumes of methanol. The block polymer product was collected by filtration, washed and dried overnight in vacuo at 60°C. [η] 0.51 dl/g; OH absorbance at 3610 cm^{-1}: 0.01.

<u>Methylene Chloride Fractionation of Cross-Coupled 1, 2 and 7.</u> A sample of the block polymer (above; 0.50g) was dissolved in 10 mL of methylene chloride. The soluton was stored at 2°C for 2 days. A polymer:methylene chloride complex precipitate formed which was removed by filtration at 2°C. The precipitate was then heated at 50° to drive off the methylene chloride. The dried polymer weighed 0.43g and contained (based on IR analysis) 58% by weight of poly(phenylene oxide) and 42% by weight of polystyrene. Analysis of the filtrate after evaporation of the methylene chloride established the presence of a residue containing 17% polyphenylene oxide and 83% polystyrene. On the basis of these results, at least 72% of the initial polystyrene charged to the reaction medium was calculated as having been incorporated into an acyl-coupled polyphenylene oxide-polystyrene block polymer.

Acknowledgments. The authors would like to thank Drs. AR Gilbert and AS Hay for useful discussions; Mr. DG Keyes and Mrs. JE Croteau for valuable technical assistance.

Literature Cited

1. Hay, A. S.; Blanchard, H. S.; Endres, G. F.; and Eustance, J. W.,J.Am.Chem.Soc.,81 6335 (1959).
2. White, D. M. J. Org. Chem. 34, 297 (1969).
3. White,D. M. U.S. Patent 3,703,564 (1972).
4. Stark, C.M. J.Am.Chem.Soc.,93, 195 (1971).
5. Dockx, J. Synthesis, 443 (1973).
6. Dehmlow, E. V. Chem. Tech., 5, 210 (1975).
7. White, D. M. J. Polymer Science.: Polymer Chem. Edition, 19, 1367 (1981).
8. White, D. M. U.S. Patent 3,875,256 (1975).
9. White, D.M. U.S. Patent 4,156,764 (1979).
10. Hay, A.S. et al, J. Polymer Science, Polymer Letters Ed., 21, 449 (1983).
11. Factor, A. et al.,Polymer Letters, 7, 205 (1969).
12. Hay, A. S. U.S. Patent 4,028,341 (1977)

RECEIVED February 28, 1985

Characterization of Hydroxyl-Terminated Liquid Polymers of Epichlorohydrin

SIMON H. YU

Avon Lake Technical Center, The BFGoodrich Company, Avon Lake, OH 44012

> Hydroxyl-terminated epichlorohydrin (HTE) liquid polymers are synthesized by a cationic ring-opening polymerization of epichlorohydrin in conjunction with a glycol or water as a modifier. Cyclic oligomers were removed by extraction. HTE liquid polymers free from cyclic oligomers are characterized by gel permeation chromatography, infrared, ^{13}C and ^1H NMR and field desorption mass spectrometry, and chemical titrations. The functionality and the spectroscopic data are consistent with the general structure I for the polymer made using ethylene glycol as a modifier. One unit of modifier is incorporated into the polymer chain and the functionality of the polymer is the same as that of the modifier. The terminal hydroxyl groups are predominately secondary; there are some head-to-head and tail-to-tail linkages. The results provide structural evidence that the polymerization proceeds via a polyaddition mechanism by propagating simultaneously from both ends of the difunctional modifier.

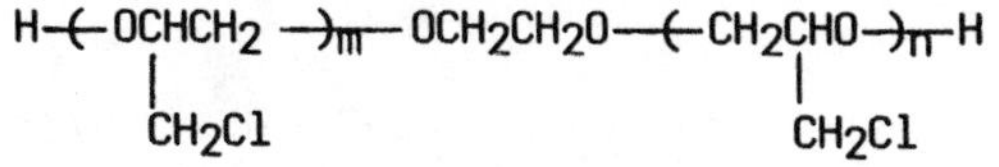

$$H\text{-}(\text{OCHCH}_2)_m\text{-}OCH_2CH_2O\text{-}(CH_2CHO)_n\text{-}H \quad I$$
$$\quad\quad\quad | \quad\quad\quad\quad\quad\quad\quad\quad\quad\quad\quad | $$
$$\quad\quad CH_2Cl \quad\quad\quad\quad\quad\quad\quad\quad CH_2Cl$$

The cationic ring-opening polymerization of epichlorohydrin in conjunction with a glycol or water as a modifier produced hydroxyl-terminated epichlorohydrin (HTE) liquid polymers (1-2). Hydroxyl-terminated polyethers of other alkylene oxides (3-4), oxetane and its derivatives (5-6), and copolymers of tetrahydrofuran (7-15) have also been reported. These hydroxyl-terminated polyethers are theoretically difunctional and used as reactive prepolymers.
Formation of cyclic oligomers is a characteristic feature of the cationic ring-opening polymerization of cyclic ethers (16-17).

Unfortunately, the presence of cyclic oligomers interferes with the determination of molecular weight and consequently functionality (10). By definition, the functionality in telechelic polymers is dependent on the colligative property of molecular weight coupled with hdyroxyl number. Therefore, the functionality of only few low molecular weight polyethers made by cationic polymerization have been determined. Furthermore, the role of glycol during the polymerization is still uncertain. The incorporation of the glycol at the head of the polymer chain via a monocation mechanism has been proposed (5-6). On the other hand, according to a newly proposed mechanism (18), the glycol is incorporated in the middle of the polymer chain by propagating simultaneously at both ends of the glycol through the polyaddition of the monomer. The expected structures of HTE polymers made in conjunction with ethylene glycol as a modifier according to these mechanisms are shown below.

Monocation Mechanism:
$$HOCH_2CH_2O-(-CH_2CHO-)_n-H$$
$$|$$
$$CH_2Cl$$

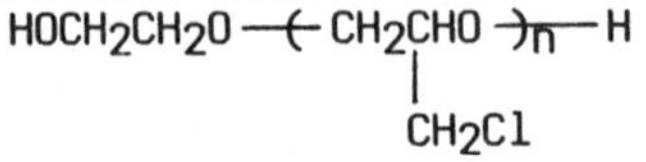

Polyaddition Mechanism:
$$H-(-OCHCH_2-)_m-OCH_2CH_2O-(-CH_2CHO-)_n-H$$
$$|\qquad\qquad\qquad\qquad\qquad |$$
$$CH_2Cl\qquad\qquad\qquad\quad CH_2Cl$$

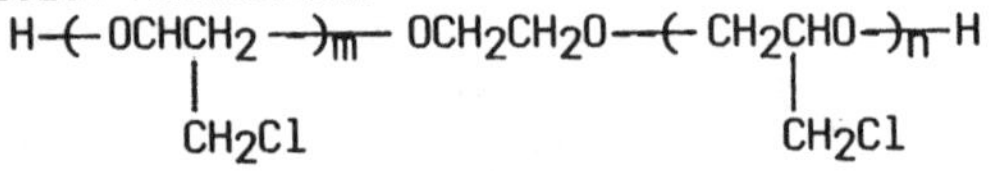

In this presentation, we report the characterization of HTE liquid polymers free from the interference of cyclic oligomers. Cyclic oligomers of the higher molecular weight HTE liquid polymers ($\bar{M}_n$>1200), including commercial Hydrin 10 liquid polymers, were removed by extraction. The functionality of the liquid polymer was then determined, and structures are proposed as determined by infrared, carbon-13 and proton NMR, and field desorption mass spectroscopic analyses.

Experimental

HTE liquid polymers were synthesized by cationic ring-opening polymerization of epichlorohydrin (ECH) in the presence of water or ethylene glycol (EG) as a modifier (1). Cyclic oligomers were removed by extraction. After extraction, the liquid polymers were essentially free from cyclic oligomers as determined by gel permeation chromatography (GPC) (Figure 1).

Hydroxyl numbers were determined by acetylation with an acetic anhydride-pyridine mixture. GPC analysis was carried out at 40°C using a Waters GPC Model 200 instrument with columns packed with Styrogel [10^2Å (4 ft), 10^3Å (2 ft), 10^4Å (2 ft), 10^4-10^5Å (4 ft), 10^6Å (4 ft), and 10^6-10^7Å (4 ft)]. THF was used as a solvent. Molecular weights were calibrated with respect to polystyrene. Molecular weights of several liquid polymers were also determined by vapor-pressure osmometry (VPO) using a Corona Wescan molecular weight apparatus. FT infrared spectra were recorded with a Nicolet 7199 spectrometer. Samples were prepared by applying a thin coat of liquid polymers on a KBr crystal. Carbon-13 NMR spectra were acquired at 20.1 MHz using a Bruker WP-80 spectrometer. Liquid polymers were examined as a 20 wt. % solution in benzene-d_6 or chloroform-d with internal tetramethylsilane reference at 30°C. Proton NMR spectra were acquired at 200.13 MHz using a Bruker WH-200 spectrometer, in chloroform-d at 30°C. Trichloroacetylisocyanate was used as a derivatizing agent (19). Field desorption mass spectra were obtained with a Varian MAT 311A mass spectrometer in the field desorption mode. The samples were dissolved in either methanol or tetrahydrofuran. The solutions were then saturated with solid lithium bromide so that the lithiated molecular ions [MLi]$^+$ were produced during analysis.

Results and Discussion

Functionality. For reactive liquid polymers, the most important property is the functionality. Functionality is determined from number average molecular weight ($\overline{M}_n$) and hydroxyl number by the following equation:

$$\overline{F}_n = \frac{\overline{M}_n}{\text{Hydroxyl Equivalent Weight}} = \frac{\overline{M}_n \times \text{Hydroxyl Number}}{56,100}$$

The hydroxyl number is defined as the milligram equivalent of KOH per gram of the polymer, where a mole of KOH is equivalent to one mole of hydroxyl group.

According to GPC analysis, lower molecular weight HTE liquid polymers ($\overline{M}_n < 1200$) synthesized by cationic ring-opening polymerization in conjunction with water or ethylene glycol are free from cyclic oligomers; higher molecular weight HTE polymers ($\overline{M}_n > 1200$) contain 10-20 wt. % of cyclic oligomers. Without the interference of cyclic oligomers, the determination of functionality of lower molecular weight HTE polymers is straightforward. They are difunctional (Table I). In contrast, higher molecular weight HTE liquid polymers containing cyclic oligomers show a bimodal molecular weight distribution by GPC. The functionality calculated from the observed $\overline{M}_n$ varies from 0.5 to 1.5. After the cyclic oligomers are removed by extaction, higher molecular weight liquid polymers no longer exhibit bimodal distribution and the $\overline{M}_n$ is nearly twice that of the observed $\overline{M}_n$ prior to the extraction; the functionality determined is $\overline{F}_n = 2.0$ (Table I). The $\overline{M}_n$ of several HTE polymers free from cyclic oligomers has also been determined by vapor pressure osmometry (VPO). For higher molecular weight HTE polymers, $\overline{M}_n$'s obtained from GPC are the same as those obtained from VPO.

Infrared spectra. An FT infrared spectrum of HTE liquid polymer ($\overline{M}_n \simeq 500$) is shown in Figure 2. All spectra of HTE polymers show characteristic absorptions: a broad band at 3530 cm^{-1} for the hydroxyl stretching, three bands at 2958, 2913, and 2875 cm^{-1} assigned to the carbon-hydrogen stretching, an extremely strong

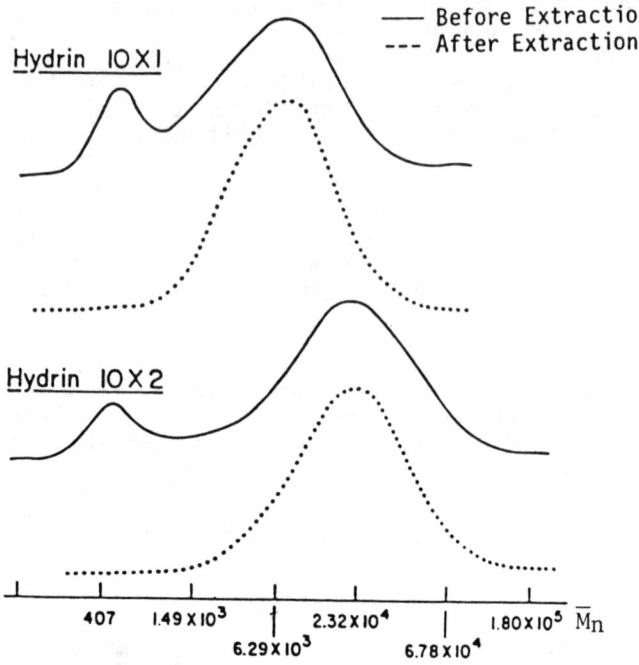

Figure 1. GPC Curves of Hydrin 10 Liquid Polymers Before and After Extraction

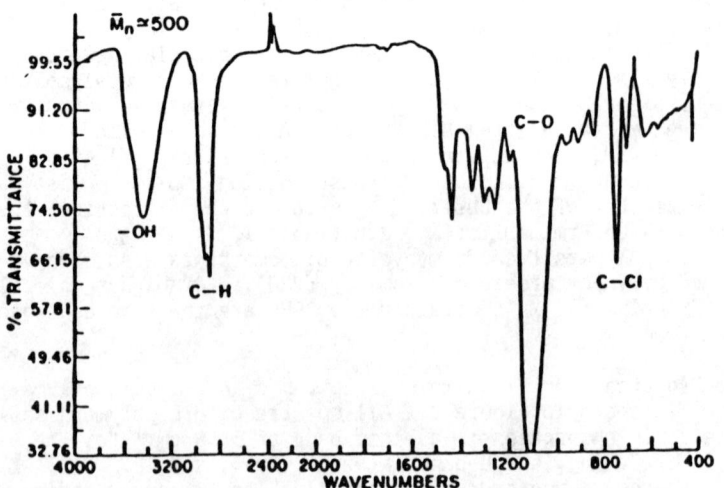

Figure 2. FT-IR Spectrum of HTE Liquid Polymer

Table I. Functionality of Hydroxyl-Terminated Epichlorohydrin Liquid Polymers[a]

	HTE 500	HTE 1000	HTE 1000A[b]
Lower molecular weight polymers containing no cyclic oligomers.			
$\bar{M}_n$ by GPC	623	1,130	1,030
OH Number	175	114	118
Functionality	1.9	2.3	2.2
$\bar{M}_n$ by VPO	527	874	--

	HTE 2000		HTE 3000		
	A	B	A	B	C
Higher molecular weight polymers before extraction.					
$\bar{M}_n$ by GPC	1,500	1,220	1,430	1,390	1,530
OH Number	56	57	38	48	34
Functionality	1.5	1.2	0.97	1.2	0.93
Higher molecular weight polymers after extraction.					
$\bar{M}_n$ by GPC	2,320	1,920	3,070	2,750	3,610
OH Number	56	62	37	45	27
Functionality	2.3	2.1	2.0	2.2	1.8
$\bar{M}_n$ by VPO	2,300	--	2,960	--	--

	Hydrin 10X1	Hydrin 10X2
Commercial polymers before extraction		
$\bar{M}_n$ by GPC	1,700	1,570
OH Number	32	11
Functionality	0.97	0.5
Commercial polymers after extraction		
$\bar{M}_n$ by GPC	4,080	11,800
OH Number	23	8.2
Functionality	2.0	1.7

a. Synthesized liquid polymers are designated as HTE followed by $\bar{M}_n$. Hydrin 10 liquid polymers are commercially available from BFG.
b. HTE 1000A was synthesized using water as a modifier, other synthesized polymers using ethylene glycol as a modifier.

Table II. Ratios of Infrared Absorbances[a]

	OH/CO	CH/CO	CCl/CO
HTE Liquid Polymer	0.0~0.7	0.38	0.37
PECH Elastomer (Hydrin 100)	0.0	0.31	0.35

a. The absorbances of OH at 3530, CH at 2875, CO at 1125, and CCl at 747 cm^{-1}.

band at 1125 cm^{-1} assigned to the single bond carbon-oxygen stretching for ether linkages in the polymer chain and a doublet at 747 and 710 cm^{-1} attributed to the carbon-chlorine stretchings for trans and gauche configuration, respectively. As the molecular weight of liquid polymers increases, the hydroxyl absorption becomes weaker. For higher molecular weight Hydrin 10X2 ($\overline{M}_n \simeq 12,000$), the hydroxyl band nearly disappears.

Table II summarized the ratios of major absorbances of HTE liquid polymers. Only the ratio of hydroxyl absorbance is sensitive to the molecular weight, whereas the other two ratios are independent of it. For comparison, the ratios of absorbances of high molecular weight elastomers (Hydrin 100) made by a coordination catalyst (R_3Al/H_2O) are also shown in Table II. No hydroxyl absorbance is detected. The two ratios of CH/CO and CCl/CO absorbances of HTE liquid polymers match well with those of elastomers and are characteristic feature of polyepichlorohydrin.

Carbon-13 NMR spectra. A carbon-13 NMR spectrum of HTE polymer ($\overline{M}_n \simeq 500$) is shown in Figure 3. The carbon-13 NMR spectra of HTE polymers show the carbon chemical shifts at 79.3 and 72.7 ppm for the backbone and the terminal methine carbons respectively, at 43.9 and 46.2 ppm for the backbone and the terminal chloromethyl carbons, respectively, and in the range of 69.7-72.0 ppm as a multiplet for the methylene carbon. It is a characteristic feature of hydroxyl-terminated polyethers that the terminal carbon exhibits a up-field shift due to the substituent effect of the hydroxyl group, whereas the β carbon(s) to the terminal hydroxyl group exhibits a down-field shift (Table III). The terminal methine carbon also suggests that the hydroxyl groups are predominantly secondary.

Table III. Substituent Effect of Hydroxyl Groups on Carbon Chemical Shifts of Polyethers

Hydroxyl-terminated Polyether	α-Carbon, ppm		β-Carbon, ppm		Reference
	Backbone	Terminal	Backbone	Terminal	
Epichlorohydrin	79.3	72.7	43.9	46.2	This study
Ethylene Oxide	70.5	61.6	70.5	72.5	21
Oxetane	68.4	60.0	31.0	33.8	24
Tetrahydrofuran	70.4	61.8	26.5	29.7	22
Propylene Oxide	75.8	66.8	17.5	19.7	23
	75.3	65.6			

The complicated pattern for the methylene carbon of the polymers indicates the presence of an irregular structure of head to head and tail to tail linkages. On the other hand, the uniformly head to tail structure of polyepichlorohydrin elastomers made by the coordination catalyst shows a doublet for the methylene carbon at 70.2 and 70.0 ppm (21). No peak corresponding to the terminal methine or chloromethyl carbons is detected in elastomers.

Proton NMR spectra. Since all protons of HTE polymers show the resonance peaks at 3.4-4.1 ppm, no additional structural information can be derived from the proton NMR spectra. However, when HTE polymers are derivatized by trichloroacetylisocyanate, the resonance peaks of methylene and methine protons adjacent to the terminal hydroxyl groups shift downfield to 4.4 and 5.1 ppm corresponding to the primary and secondary alcohols, respectively (Figure 4). The proton NMR spectra of the derivatized HTE polymers show that more than 90% of the terminal hydroxyl groups are secondary.

FD Mass spectra. A FD mass spectrum of HTE polymers synthesized in conjunction with ethylene glycol is shown in Figure 5. The numbers shown indicate the units of ECH incorporated per polymer chain. The spectra are slightly complicated by the ^{35}Cl and ^{37}Cl isotope patterns. A major series of species A with molecular weight of (62 + 92.5n) corresponding to [EG + n(ECH)] is identified. A minor series of species B with molecular weight of (18 + 92.5n) corresponding to [H_2O + n(ECH)] is also detected as impurities due to the presence of trace amount of water during the polymerization. This series of impurities can be eliminated by careful precluding water in the polymeriation system. Species B become the predominant series in HTE polymers synthesized in conjunction with water (Figure 6).

The FD mass spectra provide qualitative distribution of various species in HTE polymers. Most importantly, the spectra also provide the structural information to prove the incorporation of one unit of modifier, ethylene glycol or water, in HTE polymers. This is also the first analysis that distinguishes HTE polymers synthesized in conjunction with ethylene glycol and water. The incorporation of one unit of modifier into the polymer chain has been estimated semi-quantitatively with ^1H NMR method for the copolymerization of tetrahydrofuran and propylene oxide in conjunction with 1,4-butanediol as a modifier (7).

Proposed structures. The functionality and the spectroscopic data are consistent with the general structures as shown.

Ethylene glycol as a modifier:
$$H-(-OCHCH_2-)_m-OCH_2CH_2O-(-CH_2CHO-)_nH$$
$$\quad\quad\quad | \quad\quad\quad\quad\quad\quad\quad\quad\quad\quad\quad | $$
$$\quad\quad CH_2Cl \quad\quad\quad\quad\quad\quad\quad\quad CH_2Cl$$

Water as a modifier:
$$H-(-OCHCH_2-)_m-O-(-CH_2CHO-)_nH$$
$$\quad\quad\quad | \quad\quad\quad\quad\quad\quad\quad\quad | $$
$$\quad\quad CH_2Cl \quad\quad\quad\quad\quad CH_2Cl$$

The proposed structures are consistent with the polyaddition mechanism of cationic ring-opening polymerization of ECH in conjunction with a modifier (18). The polymer chain propagates simultaneously at both ends of the difunctional modifier through polyaddition of monomers. Consequently, one unit of modifier is incorporated into the polymer chain and the functionality of the

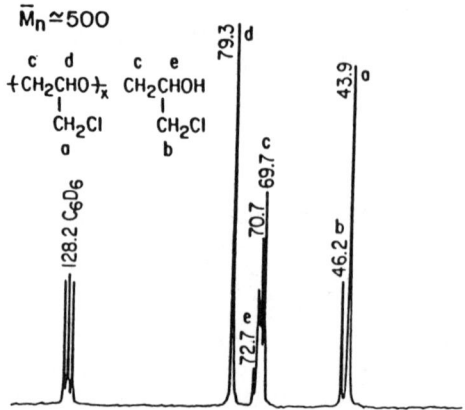

Figure 3. Carbon-13 NMR Spectrum (20.1 MHz) of HTE Liquid Polymer

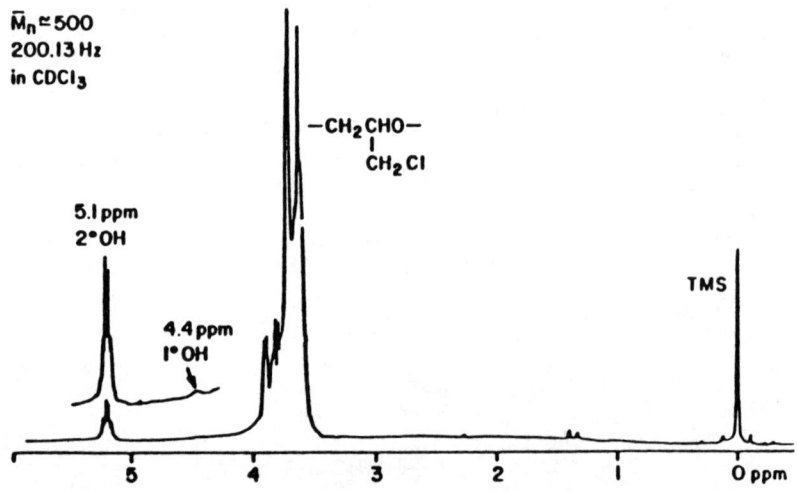

Figure 4. Proton NMR Spectrum of Derivatized HTE

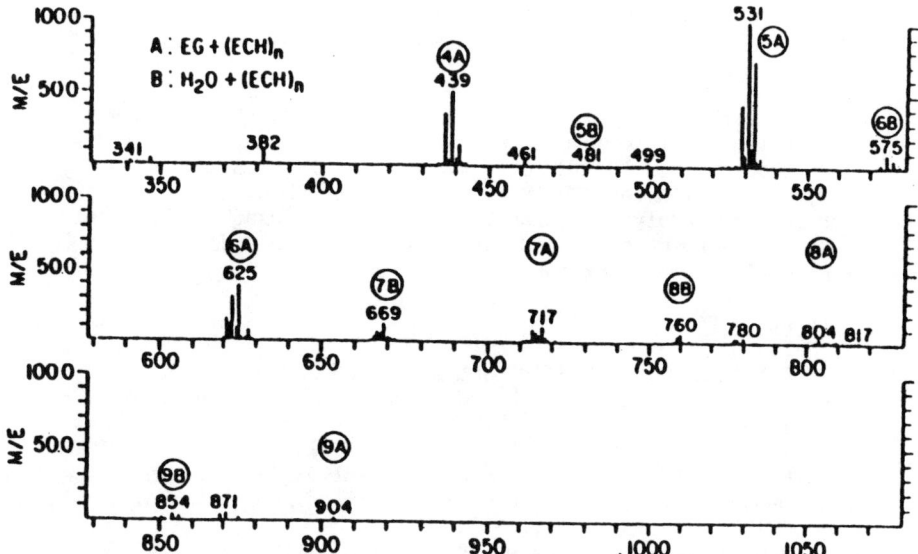

Figure 5. FD Mass Spectrum of HTE Liquid Polymer (EG as a modifier)

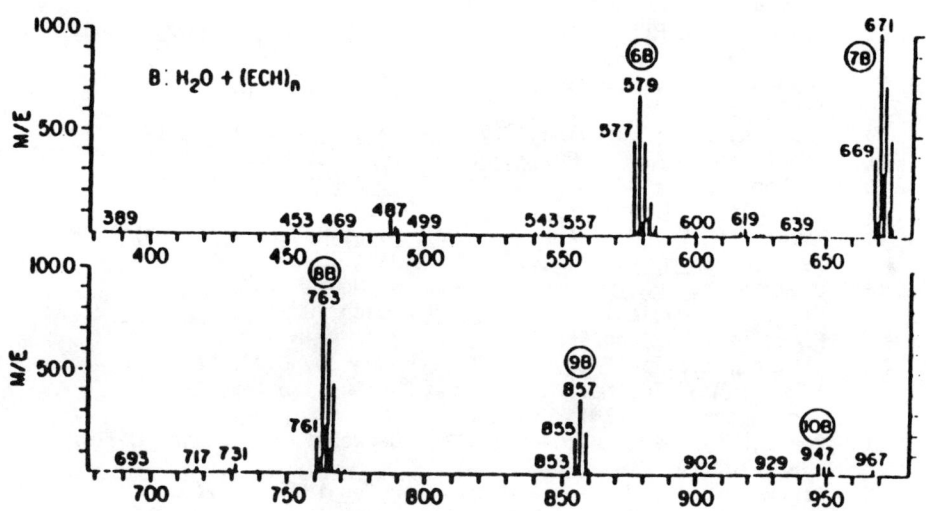

Figure 6. FD Mass Spectrum of HTE Liquid Polymer (Water as modifier)

polymer is the same as that of the modifier. The proposed structures against the monocation mechanism which suggests only one end of the difunctional modifier participates and a modifier unit is incorporated at the head of the polymer chain (5,6).

Acknowledgments

The author gratefully acknowledges the support of this work by BFGoodrich Staff Technical Services: J.L. Dorsch for ^{13}C NMR, J.P. Shockcor for ^{1}H NMR, J.W. Ryan for FT-IR, J.B. Haehn for GPC, D.J. Harmon for VPO, R.P. Lattimer for FD-MS, and M. Chesler for chemical analysis. Also the laboratory support of D.A. Versace and encouragement of D.E. Mackey and valuable discussions with M.P. Drefuss and Y. Okamoto.

Literature Cited

1. Dreyfuss, M.P. U.S. Patent 3 850 856, 1974.
2. Young, C.I.; Barber L.L. U.K. Patent Application 2 021 606, 1979.
3. Aelony, D. U.S. Patent 4 284 826, 1981.
4. Bruson, H.A.; Rose, J.S. U.S. Patent 3 269 961, 1966.
5. Manser, G.E.; Rose, D.L. "Synthesis of Energetic Polymers", SRI Project PYU 8627, 1981.
6. Manser, G.E. U.K. Patent Application 2 101 619 A, 1983.
7. Hammond, J.M.; Hooper, J.F.; Robertson, W.G.D. J. Polym. Sci. Part A-1 1971, 9, 265.
8. Alvarez, E.J.; Hornof, V.; Blanchard, L.P. J. Polym. Sci. Part A-1 1972, 10, 1895 and 2237.
9. Baijal, M.D.; Blanchard, L.P. J. Polym. Sci. Part C 1968, 157.
10. Blanchard, L.P.; Baijal, M.D. J. Polym. Sci. Part A-1 1967, 5, 2045.
11. Blanchard, L.P.; Singh, J.; Baijal, M.D. Can. J. Chem. 1966, 44, 2679.
12. Taganov, N.G.; Korovina, G.V.; Entelis, S.G. Polym. Sci. USSR 1980, 22, 1717.
13. Entelis, S.G.; Korovina, G.V. Makromol. Chem. 1974, 175, 1253.
14. Dickinson, L.A. J. Polym. Sci. 1962, 58, 857.
15. Murbach, W.J.; Adicoff, A. Ind. Eng. Chem. 1960, 52, 772.
16. Goethals, E.J. Adv. Polym. Sci. 1977, 23, 104.
17. Yamashita, Y.; Kawakami, Y. Yuki Gosei Kagake Kyokaishi 1978, 36, 183; Chem. Abstr. 1978, 89, 248431.
18. Okamoto, Y. Polym Reprints 1984, 2, No. 1, 264.
19. Groom, T.; Babiee, J.S. Jr.; Van Leuwen, G.B. J. Cellular Plastics 1974, Jan/Feb, 43.
20. Steller, K.E. in "Polyethers"; Vandenberg, E.J., Ed.: ACS SYMPOSIUM SERIES No. 6, American Chemical Society: Washington, D.C. 1975; p. 136.
21. Bayer, E.; Zheng, H.; Albert, K.; Geckeler, K. Polym. Bull. 1983, 10, 231.
22. Pruckmayr, G.; Wu, T.K. Macromolecules 1978, 11,265.
23. Mochel, V. D.; Bethea, T. W.; Futamura, S. Polymer 1979, 20,65.
24. Yu, S. H. unpublished data.

RECEIVED March 5, 1985

18

Thermal and Photo Stability of Polyarylates
Styrylpyridine-Based Polymers

HOH-JIEAR YAN and ELI M. PEARCE

Polymer Research Institute and Chemistry Department, Polytechnic Institute of New York, Brooklyn, NY 11201

Improvements in both the thermal and UV stability of polymers can be made by increasing rigidity and UV dispersive functional groups in the polymer backbone. This may be accomplished in systems containing active C=C double bonds capable of undergoing a Diels-Alder reaction and cyclization and also the Fries rearrangement. In this study, styrylpyridine based polyesters, polyacrbonates and their related model compounds were synthesized and characterized by TGA and OI to determine the effect of structure on thermal stability. In addition, the Photo-Fries rearrangement occurring during the UV irradiation of the styrylpyridine based model compound of the ester and carbonate was also investigated.

It is known that increased char yield is usually associated with improved flammability behavior (1). This can be understood if one considers that the volatile flammable products can only diffuse with difficulty through the char, and that the thermal conductivity of a porous char layer is relatively poor (2). The structure of the polymer can contribute to the amount of char formed based on the character of the functional groups present and the nature of the backbone (2,3). Ritchie (4) found that for a series of unsaturated polyesters and their copolymers, the temperatures at which carbon dioxide is eliminated was in the range of 280 to 345°C depending on the structure of the polyester. Aliphatic polyesters and their copolymers have less thermal

0097-6156/85/0282-0209$06.00/0
© 1985 American Chemical Society

stability than the polyarylates. Z. Jedlinski (5) had studied the influence of chemical structure on fourteen aromatic copolymers, containing naphthalene rings in the backbone, and found that the thermal stability increased as the number of condensed naphthalene rings and the symmetry of the copolymer chain increased. Van Krevelen (6) and Parker (7) have correlated char yield with the aromaticity of polymers. Lin and Pearce (8) have confirmed the general relationship between char yield and oxygen indices for phenolphthalein related polyesters and polycarbonates.

Polymers containing certain chromophores can absorb light to undergo photochemical reactions involving the formation of free radicals, photoionization, cyclization, intramolecular rearrangement, and fragmentation. These can cause discoloration, extraction, distortion, shrinkage, surface cracking, and electrical failures of the polymeric material. In order to prevent these effects, many polymers have been protected against photodegradation by the addition of stabilizers. These stabilizers fall into three general types (9): light screens, ultraviolet absorbers, and quenching compounds. Studies on stabilization mechanisms show that these substances must be compatible with the polymer and stable to ultraviolet irradiation and elevated temperatures (10-13). Finding stabilizers with such properties was often a difficult problem, and the synthesis of polymers capable of self-stabilization due to an inherent stabilizing group in the molecule has received considerable attention (14,15). The Fries arrangement of polyesters (14,16,17) and polycarbonates (15) to o-hydroxybenzophenones (as a common UV absorber (11)) has been studied. The reaction scheme was:

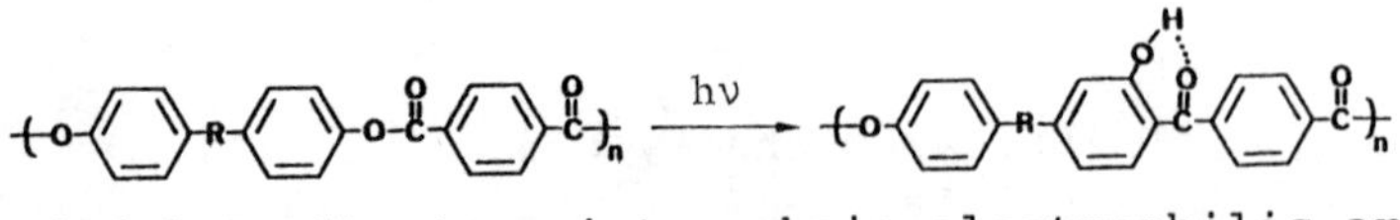

A Friedel-Crafts type inter-chain electrophilic aromatic substitution also could occur(9).

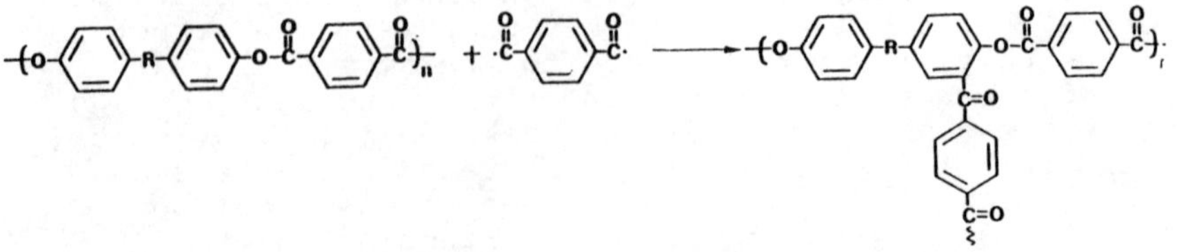

This effect decreased the formation of o-hydroxybenzophenone.

During the course of this study, styrylpyridine based polyesters and polycarbonates as well as their related model compounds were synthesized and characterized by TGA, OI, and UV irradiation to determine the effect of the structure on thermal and UV stability.

Experimental

A. Model Compounds

(I)   p-(β-2vinylpyridyl)phenyl benzoate (p-VPPS)
(II)   p,p'-Bis(β-2-vinylpyridyl)diphenyl isophthalate (p,p'-BVPDPI)
(III)   p,p'-Bis(β-2-vinylpyridyl)diphenyl terephthalate (p,p'-BVPDPT)
(IV)   p,p'-2,6-(β-2-vinylpyridyl)diphenyl dibenzoate (p,p'-2,6-VPDPDB)
(V)   p,p'-Bis(β-2-vinylpyridyl)diphenyl carbonate (p,p'-BVPDPC)

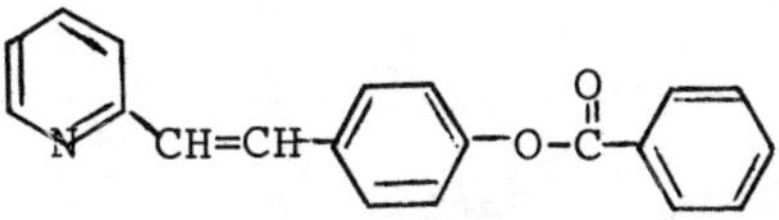

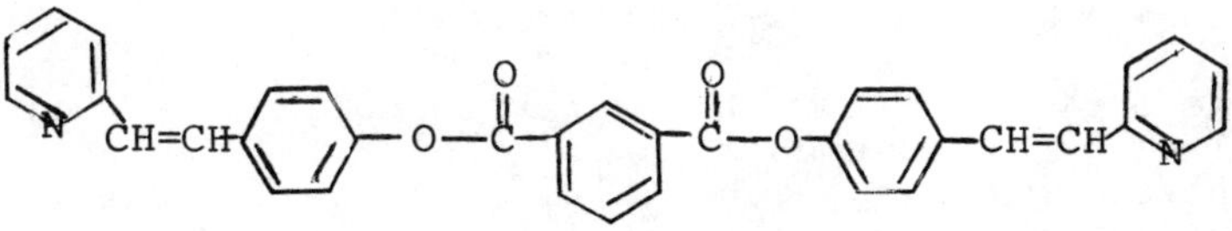

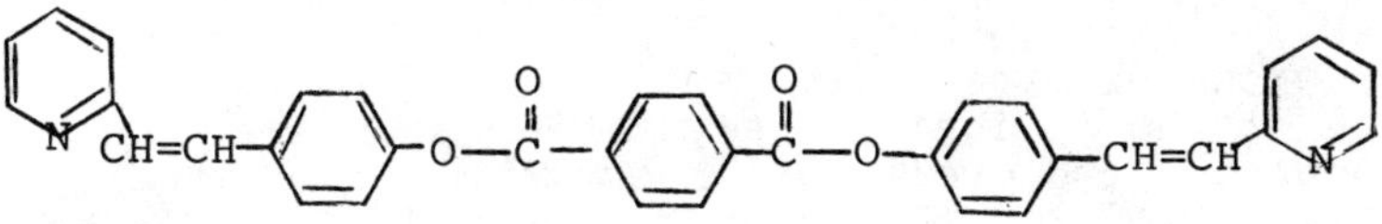

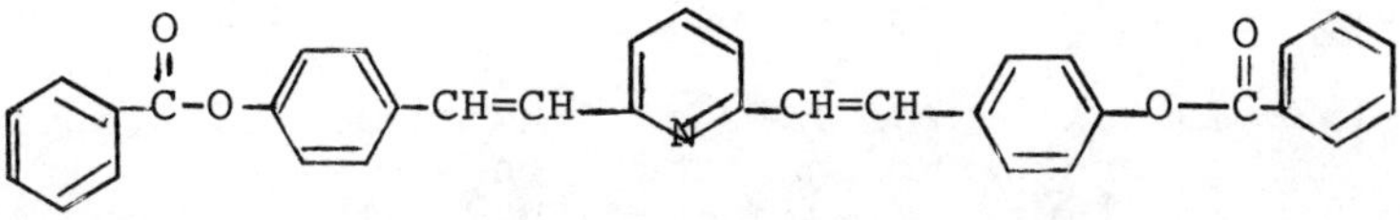

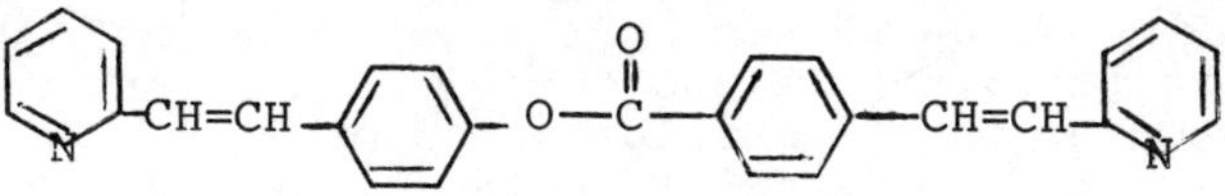

The preparation of hydroxy-terminated styrylpyridine has been described previously (18). The model esters and carbonate were prepared by an interfacial reaction. The scheme for the reaction was as follows:

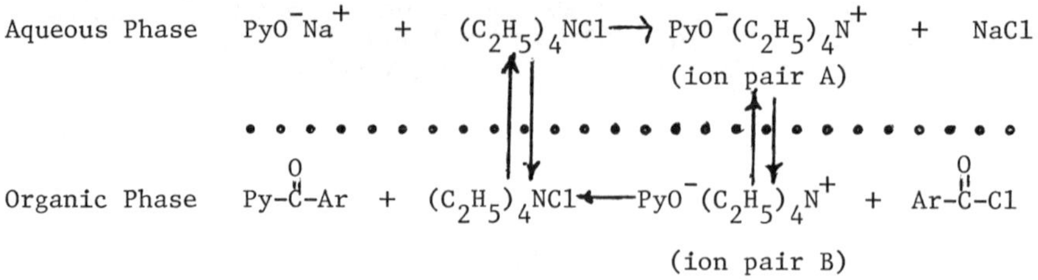

Aqueous Phase PyO^-Na^+ + $(C_2H_5)_4NCl \longrightarrow PyO^-(C_2H_5)_4N^+$ + $NaCl$

$\qquad\qquad\qquad\qquad\qquad\qquad$ (ion pair A)

• •

Organic Phase $Py-\overset{O}{\underset{\|}{C}}-Ar$ + $(C_2H_5)_4NCl \longleftarrow PyO^-(C_2H_5)_4N^+$ + $Ar-\overset{O}{\underset{\|}{C}}-Cl$

$\qquad\qquad\qquad\qquad\qquad\qquad$ (ion pair B)

where: PyO^-Na^+ : (p-hydroxystyryl)pyridine salt.

$Ar-\overset{O}{\underset{\|}{C}}-Cl$: Acyl Chloride (benzoyl chloride, terephthaloyl chloride).

Solvent : organic phase (1,2 dichloroethane).

ion pair: depends on the partition coefficient of aqueous and organic solvent.

$(C_2H_5)_4N^+Cl^-$: phase-transfer agent.

In the above reaction, the water soluble nucleophile was dissolved in an aqueous NaOH solution. The phase-transfer catalyst, $(C_2H_5)_4N^+Cl^-$, allows for the transfer of the nucleophile as an ion-pair $(PyO^-(C_2H_5)_4N^+)$ into the organic phase where later reaction with the organic reagent, Ar-CO-Cl occurred. Migration of the cationic catalyst back to the aqueous phase completed the cycle, which continued until the nucleophile, PyO^-, or the organic compound, $Ar-\overset{}{\underset{\underset{O}{\|}}{C}}-Cl$, have been completely consumed.

Synthesis of Ester

A 500 ml resin kettle was fitted with a stirrer, dropping funnel, and a condenser connected with a drying tube. All the equipment was dried and flushed with nitrogen for 10 minutes. Monohydroxy-terminated styrylpyridine (0.02 mole), NaOH (0.05 mole) and 200 ml of distilled water were placed in the resin kettle and cooled with stirring to 0°C in an ice-water bath. A solution of mono or diacid chloride (0.01 to 0.02 mole) in 150 ml of 1,2 dicholoroethane was placed in a dropping funnel. The $(C_2H_5)_4NCl \cdot H_2O$ (10 gram) a phase transfer agent, was dissolved in 50ml of 1,2 dichlorethane and poured into

the resin kettle. The rate of addition of diacid
chloride was controlled and the complete addition
required about 3 hours. After completion of the
reaction, stirring was continued for one more hour until
the reaction kettle reached room temperature. The
solids were filtered, washed twice with 20% NaOH (300 ml)
and subsequently rinsed with water. Recrystallization
was done from 1,2 dicholoroethane/ethanol (in a 4:1
volume ratio) and dried in a vacuum oven at 80°C for 2
days. The products were characterized by various methods
(Tables I and II). In the case of p,p'-2,6-VPDPDB, the
acid chloride and dihydroxy-terminated styrylpyridine
were mixed by dropping dihydroxy-terminated
styrylpyridine into the acid chloride.

Synthesis of Carbonate

A 500ml round-bottom flask was fitted with a stirrer, a
condenser with a drying tube, and a gas inlet adaptor
tube which reached the bottom of the flask. All the
equipment was dried and flushed with nitrogen for 10
minutes. Monohydroxy-terminated styrylpyridine (0.01
mole), 0.015 mole of NaOH and 200 ml of distilled water
were placed in the flask. The mixture was stirred
vigorously and cooled with ice-water. The $(C_2H_5)_4NCl \cdot H_2O$
(0.08 mole) a phase transfer agent, was dissolved in 50
ml of 1,2 dichloroethane and poured into the flask.
Phosgene gas was bubbled into the solution at a
controlled rate of 1 ml/min. After 2 minutes the
solution became cloudy and the pH of the
solution decreased as a result of the addition of
phosgene. small amount of product was obtained, and in
order to increase the yield, 10 ml of 20% NaOH was added
and phosgene was again bubbled through the solution for
30 minutes. The product was pale yellow in color, and
after washing twice with 50 ml of 20% NaOH, the color of
the material became white. Subsequently, the product was
washed generously with water and recrystallized from
ethanol. The characterization data is shown in Tables I
and II.

Polyarylates
(VI)   poly(2,4(β-vinylpyridyl)diphenyl isophthalate)
        [poly(2,4-VPDPI)]
(VII)   poly(2,4(β-vinylpyridyl)diphenyl terephthalate)
        [poly(2,4-VPDPT)]
(VIII) poly(2,6(β-vinylpyridyl)diphenyl isophthalate)
        [poly(2,6-VPDPI)]
(IX)   poly(2,6(β-vinylpyridyl)diphenyl terephthalate)
        [poly(2,6-VPDPT)]
(X)   poly(2,4(β-vinylpyridyl)diphenyl carbonate)
        [poly(2,4-VPDPC)]
(XI)   poly(2,6(β-vinylpyridyl)diphenyl carbonate)
        [poly(2,6-VPDPC)]

Table I. The Characterization of p-VPPB, p,p'-BVPDPI, p,p'-BVPDPT, pp'-2,6VPDPDB, and p,p'-BVPDPC.

Compound	Melting point (°C) (observed)	λ_{max} (nm)	log ε	H/C*	H/C**
p-VPPB	159.5-161.5	318	4.52	4.98/79.73	4.87/79.48
p,p'BVPDPI	213.2-214.5	319	4.85	4.58/77.86	4.61/77.62
p,p-BVPDPT	260.5-263.0	319	4.79	4.58/77.86	4.30/77.23
p,p'2,6VPDPDB	188.1-189.7	345	4.67	4.78/80.31	4.56/80.23
p,p'-BVPDPC	210.0-214.0	319	4.85	4.76/77.14	4.87/77.10

* : theoretical ratio of hydrogen/carbon

**: found ratio of hydrogen/carbon

Table II. The Oxygen Index and Char yield (TGA residue, 800 °C, under nitrogen) for Model Compounds and Polymers.

component	(I)	(II)	(III)	(IV)	(V)	(VI)	(VII)	(VIII)	(IX)	(X)	(XI)
char yield	1.0	34.8	35.5	54.5	53.5	31.0	54.5	51.2	18.9	53.8	54.3
oxygen index	21.5	28.0	27.5	33.2	34.6	24.7	33.0	33.9	23.2	33.9	34.9

I.   : p-(β-2-vinylpyridyl)phenyl benzoate (p-VPPB).
II. : p,p'-bis(β-2-vinylpyridyl)diphenyl isophthalate (p,p'-BVPDPI).
III. : p,p'-bis(β-2-vinylpyridyl)diphenyl terephthalate) (p,p'-BVPDPT).
IV. : poly-[2,4(β-vinylpyridyl)diphenyl isophthalate] [poly(2,4VPDPI)].
V.   : poly-[1,4(β-vinylpyridyl)diphenyl terephthalate] [poly(2,4VPDPT)].
VI. : p,p'-2,6(β-vinylpyridyl)diphenyl dibenzoate [(p,p'-2,6VPDPDB)].
VII. : poly-[2,6(β-vinylpyridyl)diphenyl isophthalate] [poly(2,6VPDPI)].
VIII.: poly-[2,6(β-vinylpyridyl)diphenyl terephthalate] [poly(2,6VPDPT)].
IX. : p,p'-bis(β-2-vinylpyridyl)diphenyl carbonate (p,p'-BVPDPC).
X.   : poly-[2,4(β-vinylpyridyl)diphenyl carbonate] [poly(2,4VPDPC)].
XI. : poly-[2,6(β-vinylpyridyl)diphenyl carbonate] [poly(2,6VPDPC)].

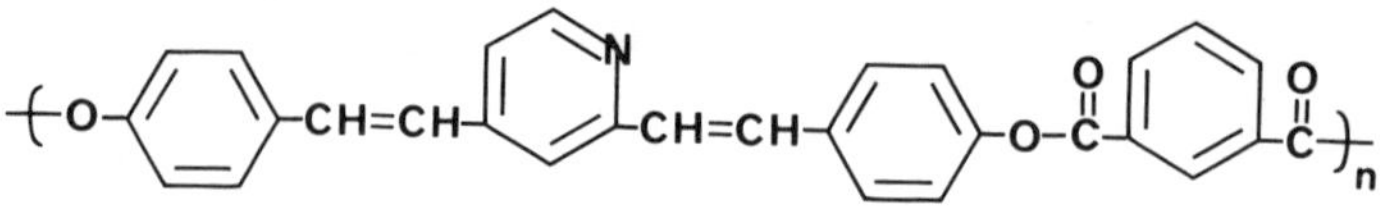

(VI)

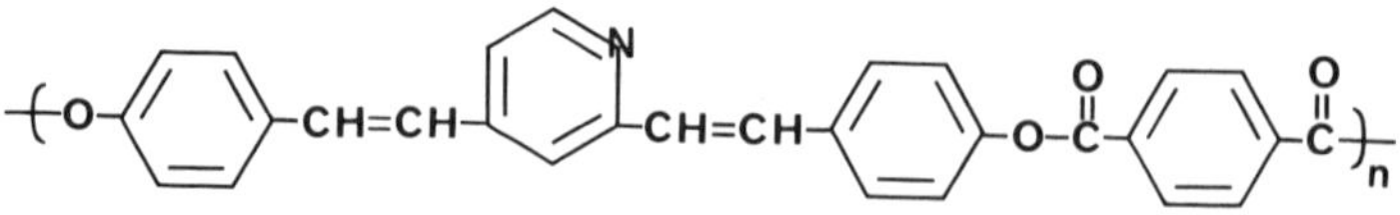

(VII)

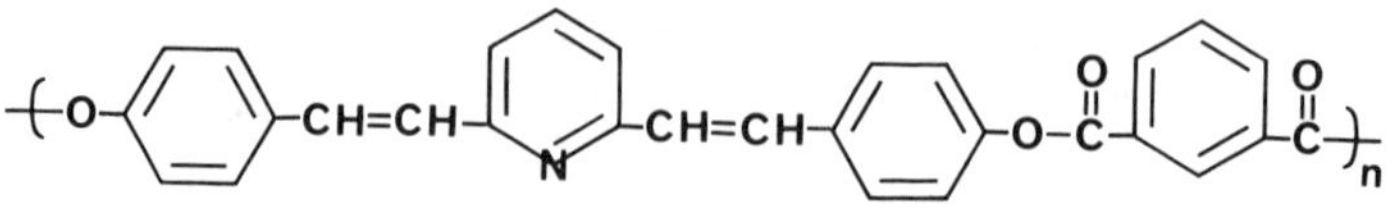

(VIII)

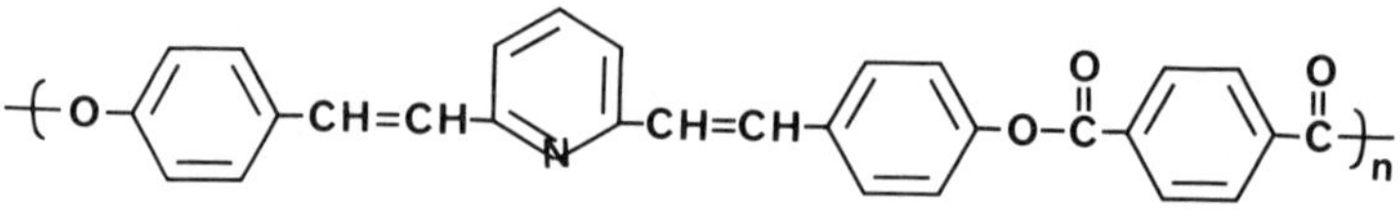

(IX)

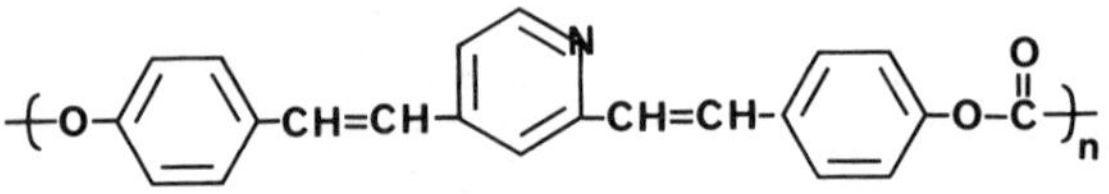

(X)

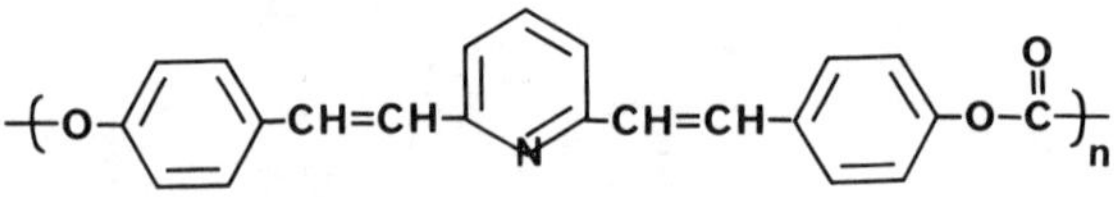

(XI)

Synthesis of Polyesters

0.01 mole of di(p-hydroxystyryl)pyridine, 0.8 gram of NaOH, and 1.5 gram of tetraethylammonium chloride monohydrate were dispersed in 200 ml of water using a high-speed blender. To this mixture, 0.01 mole of diacid chloride in 50 ml%dichloroethane was added quickly and the resulting mixture was stirred vigorously for five minutes. The polymer was rapidly precipitated on the wall of the blender. After five minutes, 250 ml of n-hexane was added and stirring was continued. The polymer was collected, washed with 100 ml portions of 20%

NaOH. The product was then washed with water and subsequently with ethanol and then dried in a vacuum oven at 80°C for 2 days. The characterization data is shown in Table II.

Synthesis of Polycarbonates

The procedure used for the preparation of polycarbonates was the same as the procedure mentioned in the synthesis of carbonate(A-2).

Results and Discussion

Flammability Characterization of Styrylpyridine Polyesters and Polycarbonates and Their Related Model Compounds

Thermal characterization data (char yields) and oxygen indices for styrylpyridine based polyesters and their model compounds are shown in Fig. 1 and Table II. An increase of char yield is generally reflected as an improved oxygen index. p,p'-BVPDPI, p,p'-BVPDPT and p,p'-2,6-VPDPDB have higher char yields and oxygen indices than p-VPPB. This was probably due to increased double bonds, pyridine, and phenyl rings in the backbone and/or occurrences of the Diels-Alder reaction with the C=C double bond in the backbone forming highly crosslinked C-C bonds. These conclusions were similar to van Krevelen's suggestion that polymers containing high aromatic ring content and/or double bonds in the polymer backbone unit usually gave greater char yields and oxygen indices than the model component for similar reasons. There is no great difference in char yields and oxygen indices of the polyesters based on isophathaloyl chloride or terephthaloyl chloride because of their similarity in chemical composition and structure.

The styrylpyridine based polycarbonates showed similar char yield and oxygen indices (Fig. 2 and Table II) to that of the polyesters.

Photo-Fries Rearrangement of Styrylpyridine Based Ester and Carbonate-UV-Spectroscopic Studies

The ultraviolet (UV) rearrangement of polyarylesters and their related model compounds have been previously studied (20,21). The chemical changes which occur during the UV irradiation of styrylpyridine based ester and carbonate were investigated. The UV spectra of the p-VPPB and p,p'-BVPDPC in 1,2-dichloroethane were monitored during the irradiation (Fig. 3 and 4). The maximum absorption for unirradiated p-VPPB was at 319 nm. After UV irradiation, the maximum peak shifted from 319 nm to 350 nm and the observed increased absorption in the

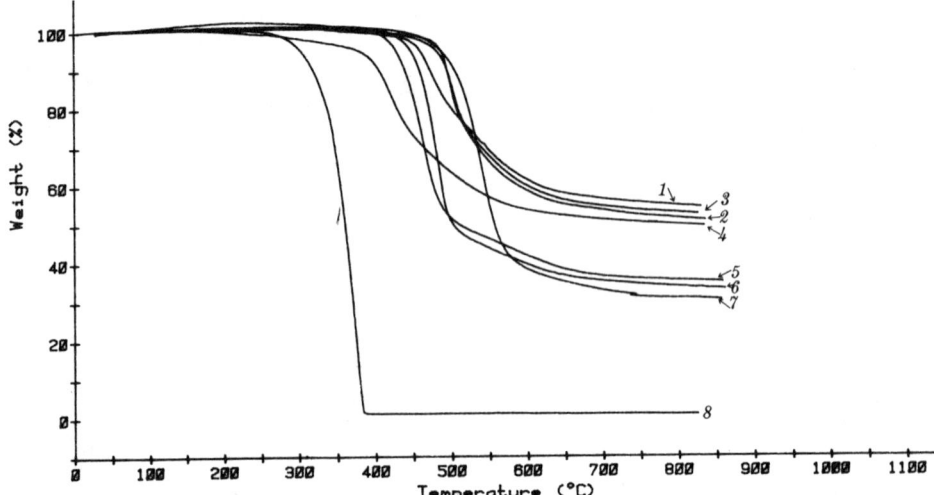

Figure 1. The TGA thermogram of various compounds: (1) poly(2,6-VPDPI), (2) poly(2,6-VPDPT), (3) poly(2,4-VPDPT), (4) poly(2,4-VPDPI), (5) p,p'-BVPDPT, (6) p,p'-BVPDPI, (7) p,p'-2,6-VPDPDB, and (8) p-VPPB; ATM: N_2 (flow rate 0.5 LPM) and heating rate: 20°C/min.

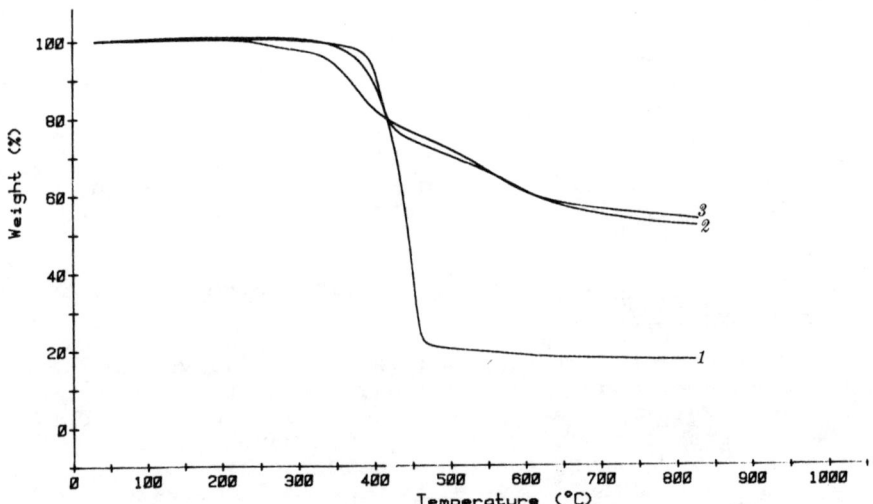

Figure 2. The TGA thermogram of various compounds: (1) p,p'-BVPDPC, (2) poly(2,4-VPDPC) and poly2,6-VPDPC); ATM: N_2 (flow rate 0.5 LPM) and heating rate: 20°C/min.

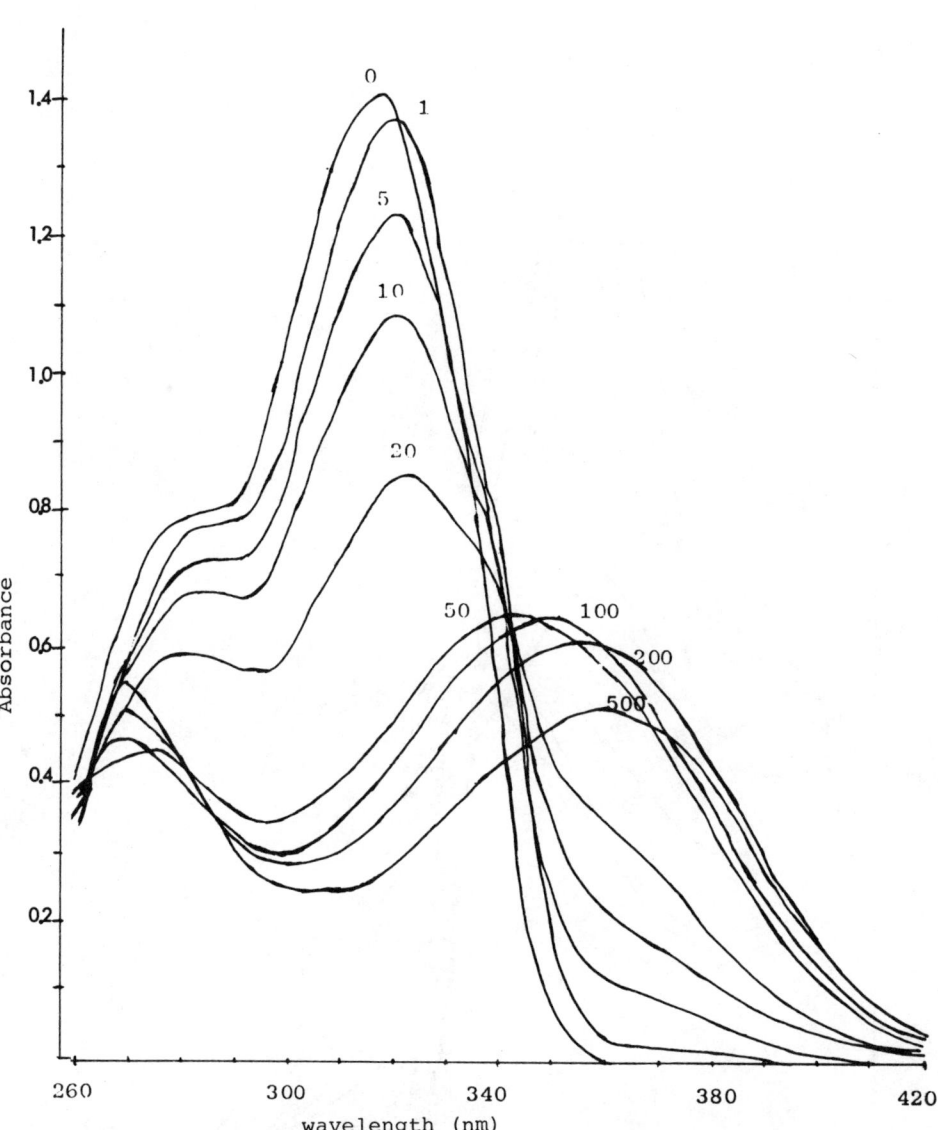

Figure 3. Change in UV spectra of p-VPPB in 1,2-dichlorothane solution before and after irradiation for different periods of time (seconds).

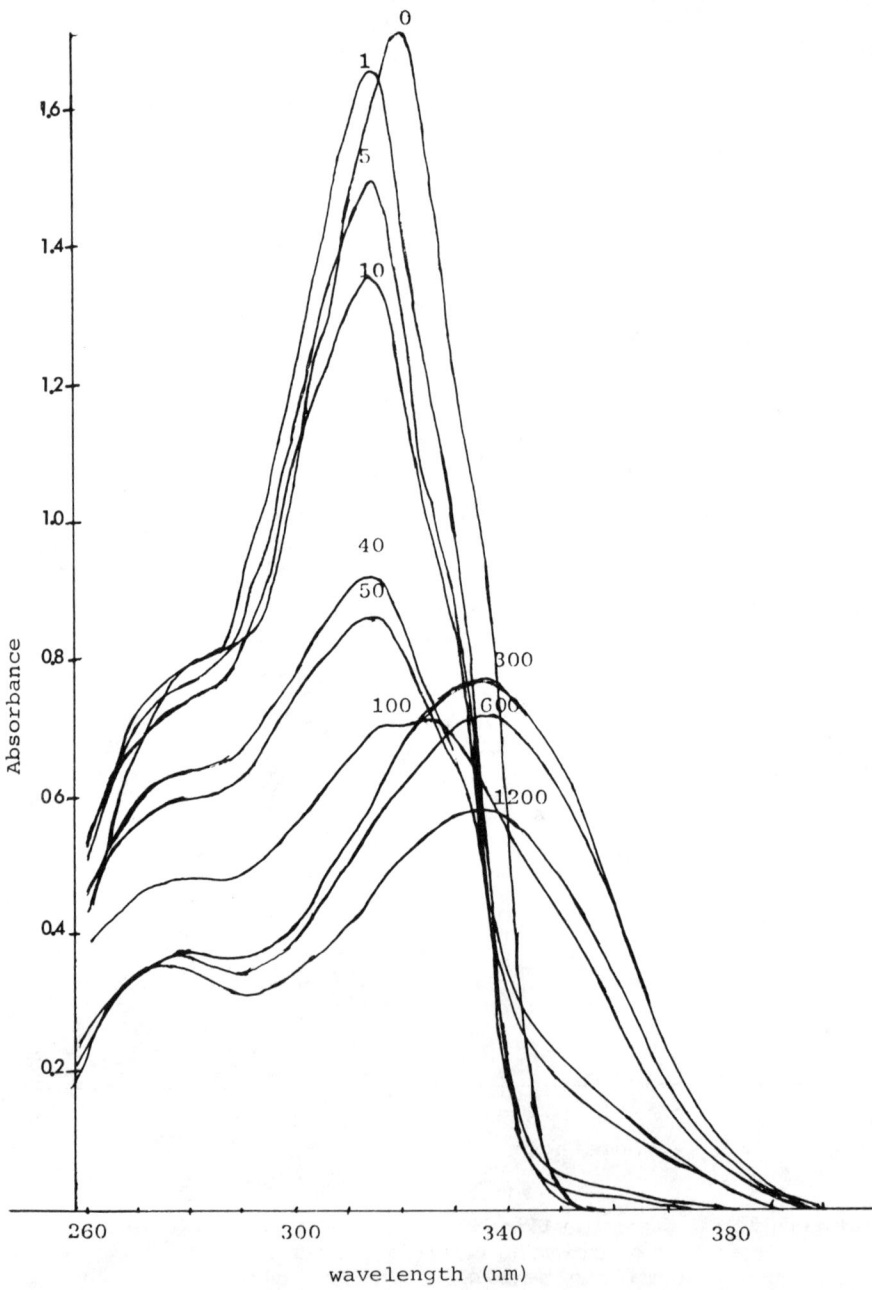

Figure 4. Change in UV spectra of p,p'-BVPDPC in 1,2-dichloroethane solution before and after irradiation for different periods of time (seconds).

340 nm range with a maximum at 350 nm was due to the
formation of the hydroxybenzophenone structure during the
Fries rearrangement (20,22) (see eq. 1). The primary
absorption maximum (max= 319 nm) rapidly diminished in
intensity, and a new absorption peak appeared on
irradiation at 270 nm. This was probably related to
either the dimerization or to isomerization of the C=C
between the phenyl and pyridine groups (23,24,25) (eq.
2). In the case of p,p'-BVPDPC, there was no increased
absorption at 270 nm probably due to steric effects
associated with a large molecule (ponderal effect) and to
increased aplanarity of the styrylpyridine ring to
suppress the dimerization and isomerization of the double
bond preventing cis-isomer formation.

Conclusions

An increase of char yield is generally reflected as an
improvement in oxygen index. In the styrylpyridine based
polyesters and polycarbonate an intermolecular thermally
induced Diels-Alder reaction has occurred through the
double bond, this increased the char yield and decreased
the flammability. The Fries rearrangement, as well as
dimerization and isomerization, occurred simultaneously
during the UV irradiation of p-VPPB, but no dimerization
or isomerization occurred for p,p'-BVPDPC, probably due to
steric effects.

Acknowledgment

We acknowledge the partial support of this research by the NASA-Ames Research Center under Grant No. NAG 2-229

Literature Cited

1. D. W. van Krevelen, Polymer, 16, 615 (1975).
2. E. M. Pearce and R. Liepins, Environ. Health Perspectives, 11, 69 (1975).
3. G. I. Nelson and J. L. Webb, J. Fire Flammability, 4, 325 (1973).
4. R. D. Ritchie, Symposium on "High Temperature Resistance and Thermal Degradation of Polymers" London, September, 1960.
5. (a) Z. Jedlinski and D. Sek, Eur. Polym. J., 7, 827(1971).
   (b) Z. Jedlinski and D. Sek, J. Polym. Sci., Part A-1, 7, 2587(1969).
6. D. W. van Krevelen, Chimia, 28, 504(1974).
7. J. A. Parker, G. M. Fohlen, and P. M. Sawko, Development of Transparent Composites and Their Thermal Response, paper presented at Conference on Transparent Aircraft Enclosures, Las Vegas, Nevada, Feb. 5-8, 1973.
8. M. S. Lin and E. M. Pearce, J. Poly. Sci., Polym. Chem. Ed., 19, 2151, 2659(1981).
9. A. M. Trozzolo, "Stabilization Against Oxidative Photo-degradation in Polymer Stabilization, ed. by W. L. Hawkins, Wiley Interciences, N. Y. 1972.. pp. 189.
10. R. A. Finnegan and J. J. Mattice, Tetrahedron, 21, 1015(1965).
11. H. Kobsa, J. Org. Chem., 27, 2293(1962).
12. G. C. Newland and J. W. Tanblyn, J. Appl. Polym. Sci., 8, 1946(1964).
13. R. B. Fox, Pure Appl. Chem., 30, 87 (1972).
14. V. V. Korshak, S. V. Vinogradova, S. A. Siling, S. R. Rafikov, Z. Ya. Fomina and V. V. Rode, J. Polym. Sci., Part A-1, 7, 157(1969).
15. D. Bellus, P. Slama and L. Durisinova, J. Polym. Sci., Part C, 22, 629 (1969).
16. J. L. Lo, S. N. Lee and E. M. Pearce, J. Appl. Polym. Sci., 29, 35(1984).
17. S. M. Cohen, R. H. Young, and A. H. Markhart, J. Polym. Sci., Part A-1, 9, 3263 (1971).
18. H. J. Yan and Eli M. Pearce, The Div. of Org. Coatings & Plastics Chem., American Chemistry Meeting in Las Vegas, Nevada, March 28-April 2, 1982.
19. D. W. van Krevelen and P. J. Hoftyzer, "Properties of Polymers" Elsevier, N. Y. 1976, Chapter 21, pp.459-465.

20. S. B. Maerov, J. Polym. Sci., Part A, 3, 48(1965).
21. D. Bellus, Z. Manasek, P. Hrdlovic, and P. Slama, J. Polym. Sci., Part C, 16, 267(1967); D. Bellus, P. Slama, P. Hrdkovic, Z. Manasek, and L. Duristnova, Part C, 22, 629(1969); D. Bellus, P. Hrdlovic and Z. Manasek, Part B, 4, 1 (1966).
22. J. S. Humphrey, Jr., A. R. Shultz and D. B. G. Jaquiss, Macromolecules, 6, 305 (1973).
23. N. R. Bertoniere, W. E. Franklin, and S. P. Rowland J. Appl. Polym. Sci., 15, 1743 (1971).
24. C. C. Unruh, J. Polym. Sci., 45, 325 (1960).
25. S. A. Zahir, J. Appl. Polym. Sci., 23, 1355 (1977).

RECEIVED February 19, 1985

19

Glycol Bis(allyl Phthalates) as Cocross-linkers for Diallyl Phthalate Resins

AKIRA MATSUMOTO and MASAYOSHI OIWA

Department of Applied Chemistry, Faculty of Engineering, Kansai University, Suita, Osaka 564, Japan

> First, several glycol bis(allyl phthalate)s (GBAP) as reactive oligomers were prepared and their polymerization behaviors were investigated in terms of cyclopolymerization and gelation as compared with diallyl phthalate (DAP). Second, DAP was cocrosslinked with GBAPs and the resultant crosslinked products were evaluated for the elucidation of correlations between mechanical properties and cocrosslinkers with the intention of improving mechanical properties of commercially important DAP resins. Polyethylene glycol bis(allyl phthalate) having a long-chain glycol unit acted as flexible crosslinked units between microgels to form the macrogel with improved flexibility through microheterogeneous curing process. This type of microheterogeneous curing process was successfully developed to the cocuring of DAP with vinyl monomers having long-chain alkyl groups.

The commercial interest in allyl resins can be attributed to their excellent electrical properties and ability to maintain these under conditions of high temperature and humidity, their excellent dimensional stability, and their chemical resistance. Diallyl phthalate (DAP) and diallyl isophthalate resins are in widest use, being supplied as monomers and B-staged prepolymers for conversion to thermosetting molding compounds, preimpregnated glass cloth and paper, sealants, coating, etc. The prepolymers are dry, free-flowing white powders that are stable on storage and, even when compounded by using catalysts decomposing at high temperatures, do not polymerize at room temperature. Under heat and pressure as in molding and laminating, they soften, flow, and crosslink to yield three-dimensional insoluble thermosets, being used in critical electrical/electronic applications requiring high reliability such as electronic connectors in communications, computer, and aerospace system (1,2).
In spite of these excellent properties, allyl resins are, however, quite brittle and have a rather limited application,

although DAP-ethylene precopolymer was developed for producing a crosslinked copolymer having desirable properties such as resistance to cracks, improved flexibility, and improved drillability (3). In this sense, the development of allyl resins with improved flexibility would be required for versatile applications.

We have extensively investigated the radical polymerization of a variety of symmetrical or unsymmetrical divinyl compounds including diallyl dicarboxylates (4), ethylene glycol dimethacrylate (5), and allyl unsaturated carboxylates (6) in terms cyclopolymerization and gelation. As part of these investigations, we commenced a research program concerned with the polymerization of glycol bis(allyl fumarate)s (7), glycol bis(allyl dicarboxylate)s (8), and polyfunctional oligomers or precursors of three-dimensional polymers (9) as reactive oligomers, the polymerization of which could provide us a useful information for the mechanical elucidation of network formation which is essential for molecular design of three-dimensional polymers with high performance and also, these reactive oligomers are applicable to the modification of thermosetting resins.

Thus, as an extension of our continued investigations concerned with gelation in the radical polymerization of divinyl monomers, this article deals with the application of glycol bis(allyl phthalate)s as cocrosslinkers to the modification of commercially important DAP resins. DAP was cocrosslinked with glycol bis(allyl phthalate)s and the resultant crosslinked products were evaluated for elucidation of correlations between mechanical properties and cocrosslinkers with the intention of improving their mechanical properties (10). The improved flexibility of DAP resins cocrosslinked with glycol bis(allyl phthalate) having a long-chain glycol unit are discussed in connection with the microheterogeneous polymerization process of DAP beyond the gel-point conversion leading to the microgel and, moreover, macrogel formation (11). Finally, the above discussion is developed to the microheterogeneous cocuring of DAP with vinyl monomers having long-chain alkyl groups resulting in the DAP resins with improved flexibility (12).

Preparation and Polymerization of Glycol Bis(allyl phthalate)s

R: $-(CH_2)_m-$ or $-CH_2+CH_2OCH_2\}_n CH_2-$

(m=2,4,6,10)   (n=1,2,3)

Eight glycol bis(allyl phthalate)s were prepared via the reaction path described above (8,10).

The purity of each monomer thus obtained was confirmed from the ^{1}H-NMR spectrum, the saponification value, the bromine value, and the molecular weight.

These monomers were polymerized in bulk using 0.1 mol/L AIBN at 80°C; their polymerization rates were reduced compared to DAP. This may be ascribable to enhanced steric effect on the intermolecular propagation of the uncyclized radical (13).

The cyclization constant K_c, the ratio of the rate constant for the intramolecular cyclization to that for the intermolecular propagation of the uncyclized radical, was estimated to be 1.90(m=2), 1.77(m=4), 1.63(m=6), 1.38(m=10), 1.82(n=1), 1.60(n=2), and 1.24(n=3) mol/L; the K_c value decreased with increasing the cyclic member of the cyclized structural unit.

In our previous article (14) the gelation in the polymerization of diallyl aromatic dicarboxylates has been experimentally examined in detail and discussed according to Gordon's theory (15). The discrepancy between actual and theoretical gel point conversion was quite large. In this connection, the gelation behavior of glycol bis(allyl phthalate)s was explored in detail from both experimental and theoretical standpoints.

Table I. Comparison of Actual and Theoretical Gel Points(GPs)

Monomer	Gel point (%)		Actual GP
	Actual	Theoretical	Theoretical GP
DAP	25	5.5	4.5
EGBAP (m=2)	31	7.3	4.2
TMGBAP (m=4)	33	7.6	4.3
HMGBAP (m=6)	33	7.1	4.6
DMGBAP (m=10)	34	7.5	4.5
2-EGBAP (n=1)	36	12.1	3.0
3-EGBAP (n=2)	40	15.1	2.6
4-EGBAP (n=3)	43	16.7	2.6

Table I summarizes the results obtained; the actual gel-point conversions increased with an increase in the molecular weight of monomer. On the contrary, the discrepancy between actual and theoretical gel-point conversions was notably reduced from DAP to tetraethylene glycol bis(allyl phthalate)(n=3). This may be ascribed to the reduction of excluded volume effects on crosslinking (16,17) with the increase of the molecular weight of monomers, although the details of excluded volume effects on crosslinking will be discussed elsewhere.

Mechanical Properties of DAP Resins Cocrosslinked with Glycol Bis(allyl phthalate)s

Three kinds of glycol bis(allyl phthalate)s, including ethylene glycol bis(allyl phthalate)(m=2; EGBAP), tetraethylene glycol bis(allyl phthalate)(n=3; TEGBAP), and polyethylene glycol bis(allyl phthalate)(n=12.2; PEGBAP) were cocrosslinked with DAP and the resultant crosslinked products were evaluated for the elucidation of correlations between mechanical properties and cocrosslinkers with the intention of improving their mechanical properties (10). Noteworthily, in the direct preparation of nonfilled DAP resin from DAP monomer, an excessively brittle material is obtained. This may be due to the rigidity of structural units, particularly crosslinked units, and also the high crosslinking density, i.e., the low molecular weight between crosslinking points because 56 mol% of DAP units incorporated into the polymer chain obtained in the bulk polymerization have pendant unreacted double bonds that can function as crosslinkers, although DAP is a highly cyclopolymerizable monomer and 44 mol% of DAP units can form cyclized structural units having no functionality of crosslinking (18).

Figures 1 and 2 show the dependences of glass transition temperature, T_g, and the concentration of the network chains, ν, on monomer composition for the DAP resins cocrosslinked with EGBAP, TEGBAP, and PEGBAP, respectively, in which T_g was regarded as the temperature where the elastic modulus decreased drastically and the damping was maximum and ν was also calculated from the equation of rubber elasticity (19). The high value of T_g and ν were estimated for DAP resin to be 206°C and 9.3×10^{-3} mol/cm^3, respectively, as expected. Also, both T_g and ν decreased with an increase in the mole fraction of cocrosslinkers and their decreasing tendencies become remarkable when going from EGBAP to PEGBAP, i.e., with lengthening glycol units of cocrosslinkers.

On the other hand, tensile strength was reduced with an increase in the mole fraction of cocrosslinkers, whereas the elongation and fracture energy showed reverse dependences. In addition, good correlations between the mechanical properties and ν were observed, regardless of the length of glycol units of cocrosslinkers.

Thus, the effect of introducing flexible crosslinked units into DAP resins was reflected in the improved flexibility. In particular, PEGBAP having a long-chain glycol unit showed a remarkable effect: The DAP resin cocrosslinked with only 3 mol% of PEGBAP had a tensile strength of 680 kg/cm^2, a value comparable to that of commercially available glass-fiber-filled DAP resin (1). Whereas the DAP resin cocrosslinked with 20 mol% of EGBAP was quite brittle and, therefore, could not even be subjected to the measurement of mechanical properties. Furthermore, it seemed that the mechanical properties of the crosslinked DAP resins should be, at least apparently, governed by the crosslinking density. The latter fact may suggest that the decrease in the crosslinking density, i.e., the increase in the molecular weight between crosslinking points, leads to the improved flexibility. In this connection, the DAP-allyl benzate(ABz)(90:10-20:80 by mol) copolymerization systems were cured for the measurement

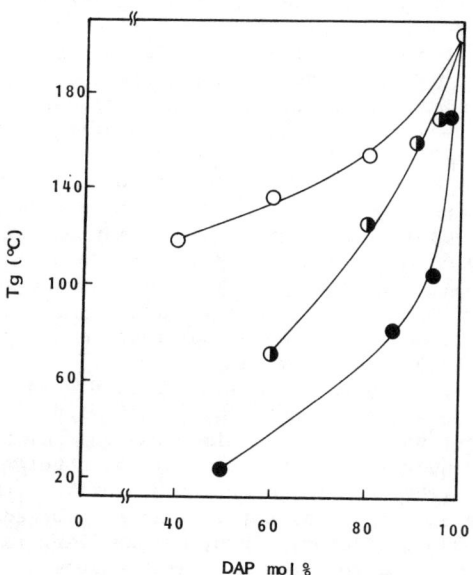

Figure 1. Dependences of T_g on monomer composition for the DAP resins cocrosslinked with (○) EGBAP, (◐) TEGBAP, and (●) PEGBAP. Reproduced with permission from Ref. 10. Copyright 1983, John Wiley & Sons, Inc.

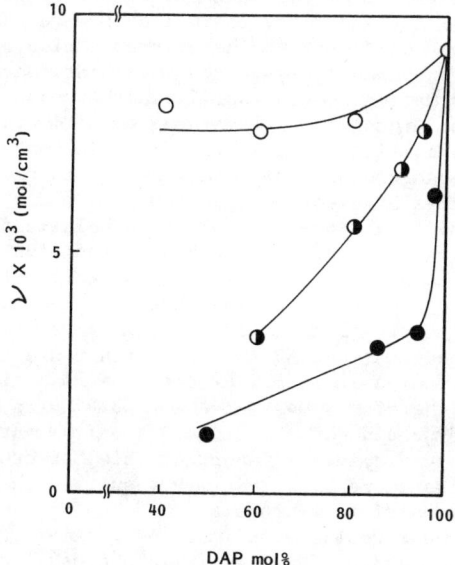

Figure 2. Dependences of ν on monomer composition for the DAP resins cocrosslinked with (○) EGBAP, (◐) TEGBAP, and (●) PEGBAP. Reproduced with permission from Ref. 10. Copyright 1983, John Wiley & Sons, Inc.

of their mechanical properties because the molecular weight between crosslinking points would be increased with an increase in the mole fraction of ABz as a monovinyl monomer in the DAP-ABz mixture. However, the resulting cured resins were quite brittle, regardless of monomer composition. Thus, the DAP-ABz copolymerization result means that the decrease in the crosslinking density of the DAP resins cocrosslinked with EGBAP, TEGBAP, and PEGBAP is not merely relevant to the increase in the molecular weight between crosslinking points. In other words, this suggests the occurrence of heterogeneous network formation such as accompanied by microgel formation, although the direct correlation of the crosslinking density to the molecular weight between crosslinking points might hold for the homogeneous network formation.

In this connection, our recent results concerned with the elucidation of gelation mechanism in the radical polymerization of diallyl dicarboxylates should be noticed (11): DAP was polymerized in bulk at 80°C. In Figure 3, both total and gel polymer conversions were plotted against polymerization time; the gel point of 25% was estimated as the conversion at the time, at which gel starts to form. Noteworthily, beyond the gel-point conversion, the content of gel polymer increased rapidly as polymerization proceeded and reached more than 90% at 50% conversion. Then, the swelling ratio of the gel polymer thus obtained was measured by using benzene as a solvent. Figure 4 shows the dependence of the swelling ratio of gel polymer on conversion; the swelling ratio of the gel polymer obtained just beyond the gel-point conversion was very high, but it decreased rapidly with conversion in the conversion range of 25-50%. In addition to these results, the molecular weight and molecular weight distribution of sol fraction, the residual unsaturation of both sol and gel fractions, and the primary chain length were examined in detail. Finally, we conclude that the polymerization of DAP proceeds microheterogeneously beyond the gel-point conversion, that is, the microgel formation occurs rapidly in the conversion range of 25-50% and then, the microgels agglomerate to the macrogel, although the details will be published elsewhere. In this connection, Erath and Robinson (20) observed directly the macrogel as the agglomerate of colloidal particles by electron microscopy.

Thus, in order to interpret the above correlations between mechanical properties and cocrosslinkers, the function of glycol bis(allyl phthalate)s as cocrosslinkers on the polymerization process of DAP beyond the gel-point conversion should be considered in connection with the microgel and, moreover, macrogel formation. Here it should be recalled that PEGBAP showed a rather drastic effect: The polyethylene glycol unit in PEGBAP can't be compatible with DAP polymer chain and, therefore, most of the PEGBAP units incorporated into the DAP polymer chain may come to exist on the surface of microgel, effectively playing an important role for crosslinking of microgels to form the macrogel. This may result in the formation of flexible units between rigid microgels.

In conclusion, DAP resins with improved flexibility were successfully obtained by introducing flexible crosslinked units on the surface of the microgel, i.e., by crosslinking flexibly microgels to form the macrogel. This finding may become a useful guideline for the modification of three-dimensional polymers: As an extension of

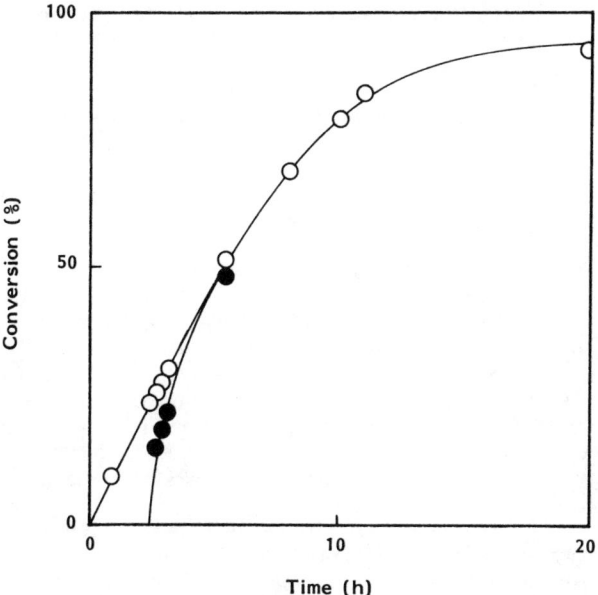

Figure 3. Time-conversion curves in the bulk polymerization of DAP using 0.1 mol/L of BPO at 80°C: (○) total polymer, (●) gel polymer.

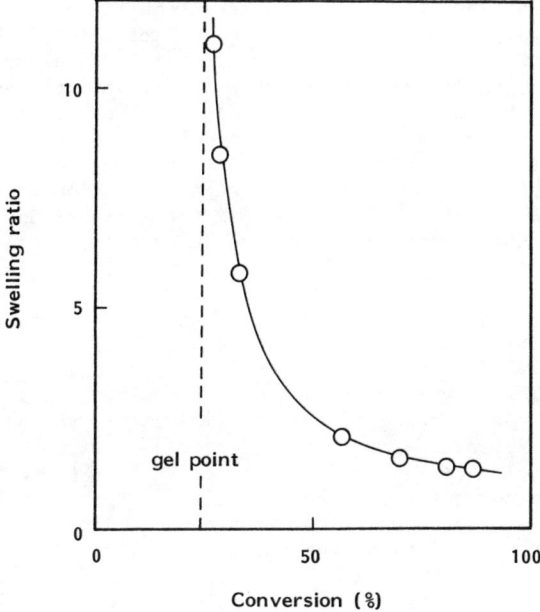

Figure 4. Dependence of the swelling ratio of gel polymer on conversion.

the above discussion, a next section deals with the cocuring of DAP with vinyl monomers having long-chain alkyl groups because the polymer chain having less polar, long-chain alkyl groups which are not compatible with DAP polymer chain can exist predominantly on the surface of the microgel, resulting in the formation of flexible crosslinked units.

Mechanical Properties of DAP Resins Cocured Microheterogeneously with Vinyl Monomers Having Long-Chain Alkyl Groups

DAP was cocured in bulk with vinyl monomers having long-chain alkyl groups, including vinyl laurate (VL), dioctyl fumarate (DOF), lauryl methacrylate (LMA), and stearyl methacrlate (SMA) (12).

Figure 5 shows the dependences of tensile strength of the cocured DAP resins on monomer composition; the tensile strength decreased with an increase in the mole fraction of comonomers and its dependency tended to be remarkable in the order VL < DOF < LMA < SMA. Particularly, the addition effects of LMA and SMA were quite great, being comparable to the result of PEGBAP as the dotted line in Figure 5.

On the other hand, the relationships between elongation or fracture energy and monomer composition showed a reverse tendency to the case of the tensile strength shown in Figure 5; both elongation and fracture energy increased with increasing the mole fraction of comonomers and their extents were enlarged in the order VL < DOF < LMA < SMA.

Thus, the introduction effect of long-chain alkyl groups into the DAP resins was reflected in the improved flexibility. In particular, LMA and SMA as comonomers showed a remarkable effect; for example, the DAP resin cocured with 10 mol% of LMA had the tensile strength of 600 kg/cm^2 and the elongation of 9.0%, although the DAP resin obtained by homopolymerization was quite brittle and, therefore, could not even be subjected to the measurement of mechanical properties as mentioned above.

These results may be interpreted considering the copolymerization of DAP with the comonomers employed and the function of precopolymers in the copolymerization process to form the microgel and, moreover, macrogel. In this connection, we have investigated in detail the copolymerization of DAP (M_1) with vinyl monomers (M_2) (4,21), the copolymerization parameters for which were estimated as follows: M_2, r_1, r_2; vinyl acetate (VAc), 0.82, 0.88 (at 80°C); DOF, 0.02, 0.96 (at 60°C); methyl methacrylate (MMA), 0.057, 35.0 (at 80°C). Here the r_1 and r_2 values correspond to those for one allyl group of DAP, the values which were calculated from the equation derived by assuming the nonoccurrence of the intramolecular cyclization reaction of DAP (21). By considering that the r_1 and r_2 values for VAc or MMA may approximately correspond to those for VL or LMA and SMA, respectively, we can now calculate the composition of the copolymer formed at any instance, the composition which changes with the progress of polymerization, according to the method of Skeist (22); Figure 6 shows the variation of the instantaneous copolymer composition with conversion for a starting monomer composition of 20 mol% of M_2 for the copolymerization of DAP (M_1) with VL, DOF, and LMA (or SMA) (M_2) as an example; in the case of LMA the copolymer composition was

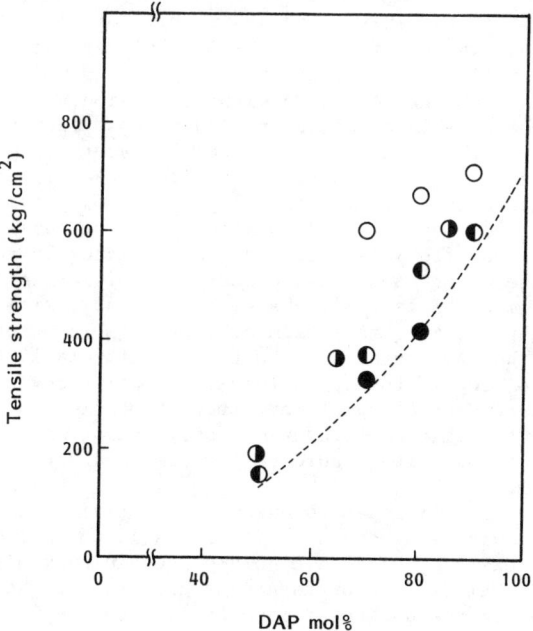

Figure 5. Dependences of tensile strength on monomer composition for the DAP resins cocured with (○) VL, (◐) DOF, (◑) LMA, (●) SMA, and (-----) PEGBAP.

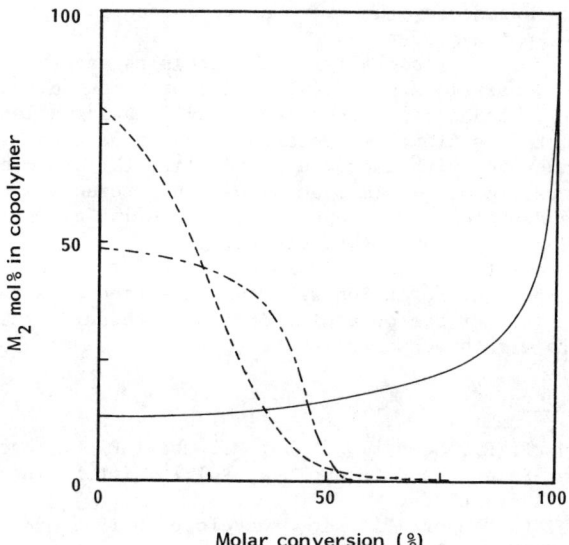

Figure 6. Variation of the instantaneous copolymer composition with conversion for a starting monomer composition of 20 mol% of M_2 in the copolymerization of DAP (M_1) with (———) VL, (— - —) DOF, and (-----) LMA (M_2).

rapidly changed from the initial value of 78.7 mol% of M_2 to 2.3 mol% at 50% conversion, whereas for the DAP-VL copolymerization the slow change was observed as the opposite tendency accompanied by the increase in the mole fraction of M_2 with conversion.

Thus, on the DAP-LMA copolymerization process leading to the microgel formation the initially obtained copolymers or precopolymers of high LMA contents can't be compatible with DAP-enriched polymer chains which increased rapidly with the progress of polymerization, and, therefore, they may exist predominantly at the space among microgels, acting as flexible crosslinkers of microgels to form the macrogel. On the other hand, in the DAP-VL copolymerization system the copolymer compositions don't change considerably with conversion as shown in Figure 6; the long-chain alkyl groups incorporated uniformly into the DAP chain may come to exist favorably on the surface of the microgel, forming a local, non-polar environment where the local concentration of unreacted VL monomer may be increased leading to a microheterogeneous polymerization, although the extent of heterogeneity should be low compared with DAP-LMA copolymerization.

This type of the microheterogeneous copolymerization of DAP with vinyl monomers having long-chain alkyl groups is now being investigated (23): For example, DAP was copolymerized in bulk with 10 mol% of LMA at 80°C. Gelation occurred at 33% conversion. No unreacted LMA was observed in the unreacted monomer mixture recovered at 30% conversion. Once gel started to form, DAP unit was rapidly incorporated into the gel polymer compared to a rather slow incorporation of LMA; thus, the LMA content in the sol obtained just beyond the gel-point conversion was considerably higher than in the precopolymer obtained just before gelation. These results may be strongly support the occurrence of the microheterogeneous polymerization beyond the gel-point conversion.

In summary, glycol bis(allyl phthalate)s as reactive oligomers were used as cocrosslinkers for DAP resins with the intention of improving their mechanical properties; PEGBAP having a long-chain glycol unit acted as flexible crosslinked units between microgels to form the macrogel with improved flexibility through microheterogeneous curing process. This type of microheterogeneous curing process was successfully developed to the cocuring of DAP with vinyl monomers having long-chain alkyl groups.

Further investigation is now in progress for the control of microgel and macrogel formation with the intention of establishing a novel guideline for the molecular design of three-dimensional vinyl polymers with high performance.

Literature Cited

1. Greenspan, F. P.; Beacham, H. H.; McCombie, R. L. "Encyclopedia of Polymer Science and Technology, Vol.1"; Interscience: New York, 1964; p.785.
2. Thomas, J. L. "Modern Plastics Encyclopedia 1980-1981, Vol.57, No.10A"; McGraw-Hill: New York, 1981; p.14.
3. Oiwa, M.; Matsumoto, A.; Yokoyama, K.; Sasaki, H.; Tojima, M.; Matsumoto, K. U. S. Pat. 3,766,149 (1973).

4. Oiwa, M.; Matsumoto, A. "Progress in Polymer Science Japan, Vol.7"; Kodansha Ltd.: Tokyo, 1974; p.107.
5. Matsumoto, A.; Noguchi, A.; Oiwa, M. Polym. Prepr. Jpn. 1982, 31(2), 348.
6. Matsumoto, A.; Ishido, H.; Urushido, K.; Oiwa, M. J. Polym. Sci. Polym. Chem. Ed. 1982, 20, 3207.
7. Matsumoto, A.; Jiang, G. J.; Oiwa, M. J. Polym. Sci. Polym. Chem. Ed. 1983, 21, 3191.
8. Matsumoto, A.; Kukimoto, Y.; Aoki, K.; Oiwa, M. J. Polym. Sci. Polym. Chem. Ed. 1983, 21, 3493.
9. Matsumoto, A.; Araki, Y.; Oiwa, M. J. Polym. Sci. Polym. Lett. Ed. 1983, 21, 771.
10. Matsumoto, A.; Aoki, K.; Kukimoto, Y.; Oiwa, M.; Ochi, M.; Shimbo, M. J. Polym. Sci. Polym. Lett. Ed. 1983, 21, 837.
11. Matsumoto, A.; Nakajima, H.; Oiwa, M. Presented at the 32nd Symposium on Thermosetting Resins, Osaka, October 1982.
12. Matsumoto, A.; Aoki, K.; Oiwa, M.; Ochi, M.; Simbo, M. Polym. Bull. 1983, 10, 438.
13. Matsumoto, A.; Nakane, T.; Oiwa, M. J. Polym. Sci. Polym. Lett. Ed. 1983, 21, 699.
14. Matsumoto, A.; Yokoyama, S.; Khono, T.; Oiwa, M. J. Polym. Sci. Polym. Phys. Ed. 1977, 15, 127.
15. Gordon, M. J. Chem. Phys. 1954, 22, 610.
16. Honda, K.; Matsumoto, A.; Oiwa, M. Polym. Prepr. Jpn. 1981, 30(5), 916.
17. Matsumoto, A.; Noguchi, A.; Oiwa, M. Polym. Prepr. Jpn. 1982, 31(6), 1109.
18. Matsumoto, A.; Asano, K.; Oiwa, M. Nippon Kagaku Zasshi 1969, 90, 290.
19. Smith, T. L. J. Polym. Sci. Symp. 1974, 46, 97.
20. Erath, E. H.; Robinson, M. J. Polym. Sci. 1963, C3, 65.
21. Matsumoto, A.; Kamigaki, H.; Oiwa, M. J. Polym. Sci. Polym. Chem. Ed. 1982, 20, 2611.
22. Skeist, I. J. Am. Chem. Soc. 1946, 68, 1781.
23. Matsumoto, A.; Aoki, K.; Kurokawa, M.; Oiwa, M. Polym. Prepr. Jpn. 1983, 32(7), 1647.

RECEIVED February 19, 1985

20

Special Functional Triglyceride Oils as Reactive Oligomers for Simultaneous Interpenetrating Networks

L. H. SPERLING[1], J. A. MANSON[2], and G. M. JORDHAMO

Polymer Science and Engineering Program, Lehigh University, Bethlehem, PA 18015

[1] Department of Chemical Engineering
[2] Departments of Chemistry and Metallurgy and Materials Engineering

> Castor oil and styrene were polymerized to make simultaneous interpenetrating networks. On reaction, the materials phase separate and sometimes phase invert.

Triglyceride oils are natural products abundant in both plants and animals. Common examples include corn oil and tallow. Triglyceride oils have glycerol as a backbone and three fatty acids as side chains. Most of these triglyceride oils have fatty acids which contain only double bonds as functional groups. However, a few of these oils are endowed with special functionalities (1,2) which include hydroxyl groups and epoxy (or oxirane) groups. These oils can react directly with many chemical reagents yielding polyurethane or polyester polymer networks. The most important of these special functional group oils is castor oil. Castor oil contains three hydroxyl groups, one on each acid residue:

$$\begin{array}{l}CH_2-O-\overset{O}{\overset{\|}{C}}-(CH_2)_7-CH=CH-CH_2-\overset{OH}{\overset{|}{C}H}-(CH_2)_5-CH_3\\[4pt] \overset{|}{C}H-O-\overset{O}{\overset{\|}{C}}-(CH_2)_7-CH=CH-CH_2-\overset{OH}{\overset{|}{C}H}-(CH_2)_5-CH_3\\[4pt] \overset{|}{C}H_2-O-\overset{O}{\overset{\|}{C}}-(CH_2)_7-CH=CH-CH_2-\overset{OH}{\overset{|}{C}H}-(CH_2)_5-CH_3\end{array} \qquad (1)$$

The present research program began in 1974 as an international program with Colombia, South America (3-22). Castor beans grow wild in Colombia, and it should be especially pointed out that castor oil does not come from an oil well. Today, the most important commercial sources of castor oil are India and Brazil.

Beginning in 1978, the program was broadened to include polymers containing oxirane groups. Two types of materials were examined: (a) Vernonia oil, which is a triglyceride oil that naturally contains 80% epoxy groups on an acid residue basis (22), and (b) ordinary triglyceride oils, such as linseed oil, whose double bonds were deliberately epoxidized (21). These epoxy-containing oils can be polymerized with materials such as sebacic acid, which itself is derived from castor oil, to produce polyester networks.

Another oil of interest is Lesquerella oil, which comes from a desert wild flower native to Arizona, and known locally as pop weeds or bladder pods. This oil is similar to castor oil, except that it has two more $-CH_2-$ groups in between the glycerol and hydroxyl groups. Consequently, its polymers tend to have slightly lower glass transition temperatures (20).

In addition to esterification, these oils can also be reacted with isocyanates to make polyurethanes. Some of the most interesting products to be described below consist of mixed ester-urethane compositions, where the ester portion is made before gelation where the water can be evaporated easily, and the urethane component is added as a type of postcure.

Simultaneous Interpenetrating Networks. An interpenetrating polymer network, IPN, can be defined as a combination of two polymers in network form, at least one of which was polymerized or synthesized in the presence of the other (23). These networks are synthesized sequentially in time. A simultaneous interpenetrating network, SIN, is an IPN in which both networks are synthesized simultaneously in time, or both monomers or prepolymers mixed prior to gelation. The two polymerizations are independent and non-interfering in an SIN, so that grafting or internetwork crosslinking is minimized (23-26).

The synthesis of an IPN is illustrated in Figure 1, which shows both types of interpenetrating polymer syntheses. First, the reaction for a sequential IPN is shown, where monomer I is polymerized together with crosslinker I to produce a network. Then monomer II and crosslinker II are swollen in and polymerized in a sequential mode to make the IPN.

In the simultaneous interpenetrating networks (SIN), the two reactions are run simultaneously. This reaction will be emphasized in the present paper. One reaction, for example, can be a polyesterification or a polyurethane stepwise reaction, while the other is an addition reaction using styrene to make polystyrene via free radical chemistry.

Dual Phase Continuity. Dual phase continuity has been shown to be important in numerous polymer blends and IPN's, to achieve special properties. Dual phase continuity is defined as a region of space where two phases maintain some degree of continuity. An example of dual phase continuity is an air filter and the air that flows through it. A Maxwell demon could traverse all space within the air filter phase, as well as within the air phase.

Quantities useful for predicting phase continuity and inversion in a stirred, sheared, or mechanically blended two-phased system include the viscosities of phases 1 and 2, η_1 and η_2, and the volume fractions of phases 1 and 2, ϕ_1 and ϕ_2. (Note: These are phase characteristics, not necessarily polymer characteristics.) A theory was developed predicated on the assumption that the phase with the lower viscosity or higher volume fraction will tend to be the continuous phase and vice versa (23,27). An idealized line or region of dual phase continuity must be crossed if phase inversion occurs. Omitted from this theory are interfacial tension and shear rate. Actually, low shear rates are implicitly assumed.

Taking into account the symmetrical behavior of both the reduced volume fraction axes, a rheological equation expressing the mid-point of the phase inversion region may be written:

$$\frac{\eta_1}{\eta_2} \cdot \frac{\phi_2}{\phi_1} = 1 \qquad (2)$$

If the quantity on the left is greater than unity, phase 2 is likely to be continuous; if the quantity is less than unity, phase 1 is likely to be continuous.

This review will emphasize the appearance of phase inversion during the synthesis of caster oil/polystyrene SIN's (19).

Experimental

Oil-Based SINs. The SINs produced were based on a castor oil polyester-urethane and styrene crosslinked with 1 mole percent of technical grade (55%) divinyl benzene (DVB) (7). This structure may be written poly[(castor oil, sebacic acid, TDI)-SIN-(Styrene, DVB)], poly[(CO,SA,TDI)-SIN-(S,DVB)]. Benzoyl peroxide (BP) (0.48%) was used as the free radical initiator for the styrene and 1,4-tolylene-diisocyanate (TDI) was used as the crosslinker for the polyester prepolymer. A 500 ml resin kettle equipped with a N_2 inlet, condenser, thermometer, and high torque stirrer was used as the polymerization reactor.

A castor oil polyester prepolymer was synthesized by combining DB grade castor oil from Cas Chem Inc. with sebacic acid from Pfaltz and Bauer in a COOH/OH ratio of 0.7 and reacting almost completely. The reaction was carried out at 180°C and followed by titrating the unreacted carboxyl groups with a standard methanolic KOH solution. Typical acid values as a function of time are shown in Table I. Reaction times ranged from 19-22 hours and the prepolymer formed was an amber colored, transparent, viscous fluid. Calculations of the extent of reaction via the Flory-Stockmeyer equation (28) indicated that the extent of reaction was 0.916. The Carothers equation (29), not quite so sensitive in this range, suggested that the system was fully reacted. A plot of the inverse carboxylic acid concentration remaining vs time, not shown, gave a straight line. At this point the material is a highly reactive oligomer.

SIN's were prepared by adding the proper amount of oil prepolymer to the reactor and heating to 80°C. The styrene/DVB/BP mixture was charged to the reactor followed by purging with N_2 gas for 10 minutes. Next, the required amount of TDI needed to react with the remaining hydroxyl groups was added. The TDI crosslinks the reactive oligomers into a three-dimensional network. The reaction was carried out under a flow of N_2 gas of 40 cm^3/min and the temperature maintained at 80°C.

Figure 2 illustrates the following sequence of steps. During the reactions, forty ml samples were removed from the reactor by pipeting the solution into bottles, followed by quenching in an ice bath. Samples were removed every 10 minutes early in the reaction and every 5 minutes close to the phase inversion point.

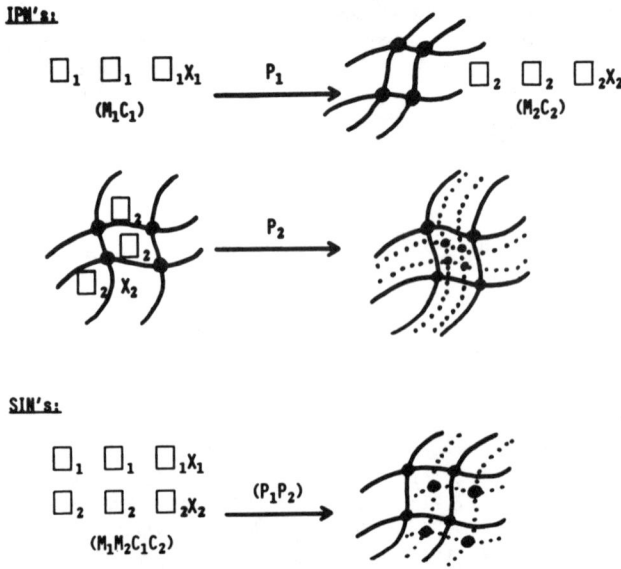

Figure 1. The synthesis of sequential IPN above and simultaneous interpenetrating networks, SIN, below. For the synthesis of SIN, two different reactions operate simultaneously such as condensation polymerization and addition polymerization. Reproduced with permission from Ref. 23. Copyright 1981, Plenum Publishing.

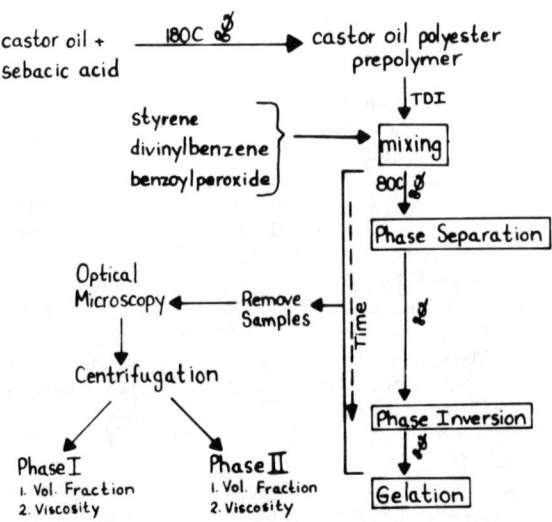

Figure 2. Reaction scheme for castor oil/polystyrene SIN's. Samples were removed for optical microscopy, followed by centrifugation. The volume fraction and viscosity of each phase fraction was determined for values used in equation (2).

Optical microscopy of the samples was done using a Riechert Zetopan-Pol[R] transmission light microscope equipped with a transmitted light polarizer and halogen lamp source. Photographs were taken using a 35 mm camera mounted on the microscope equipped with high speed film. Slide preparation involved placing a small drop of sample onto a clean dry slide and covering with a cover slip. Samples were viewed under 10X and 63X using a 10X eyepiece making the overall magnification 100X and 630X, respectively.

The viscosity and volume fraction of each phase were determined by separate experiments. The two phases were separated by centrifugation. Samples were placed into heavy duty, graduated centrifuge tubes which in turn were placed in a Precision Scientific Vari-Hi Speed Centricone [R]. Samples were spun at 2500 rpm's (665G's) for 4 hours after which the volumes of each phase did not change with further time. Volume fractions were recorded by noting the volume of each phase in the centrifuge tube. The phases were separated by pipeting into separate vials. The viscosity of each phase was measured using a Gilmont Inc. Simplified Falling Ball Viscometer [R]. This viscometer is based on Stokes' rheological theory. All viscosities were measured at 80°C±1°C. Densities of each phase were measured by pyconometry at the same temperature. Since 80°C is the reaction temperature, all measurements were made as rapidly as possible to reduce error. When storage was absolutely necessary, the samples were placed in a refrigerator.

Results

SINs. Two SIN compositions were studied, one having 10% oil prepolymer and the other 20% oil prepolymer, both dissolved in the styrene-DVB monomer solution. During the synthesis several morphological changes occurred in the mixture. Early in the reaction all components formed a mutual clear solution which was slightly yellow due to the original color of the prepolymer. As the free radical polymerization of styrene began, the polystyrene first produced remained soluble. At a critical concentration, phase separation

Table I. Acid Values of Castor Oil Prepolymer Formation at 180°C.

Time, min.	Acid Value
0	93.66
30	92.54
60	78.09
90	71.38
120	70.0
150	69.02
180	64.35
210	62.32
240	61.95
270	59.26
300	59.38
330	54.84
555	46.84
1290	36.85

occurs. The reaction time necessary for phase separation was 30-35 minutes for the 20/80 composition and 45-50 minutes for the 10/90 system. The phase separation point was noted by the onset of turbidity leading to a completely opaque solution over a range of 4-5 minutes. However, at the earliest point of phase separation, the phases are so nearly alike that the mixture is clear. Only on centrifugation was phase separation noted. The results for the poly[(CO,SA,TDI)-SIN-(S,DVB)], 10/90 system is illustrated in Figure 3. Below 70 minutes, the oil is the continuous phase, with growing polystyrene droplets dispersed in it. As the polystyrene-rich phase increases in volume, the oil phase volume decreases due to the loss of the styrene monomer. The polystyrene-rich phase remains discontinuous because the oil-rich phase has a lower viscosity and larger volume. Eventually the polystyrene phase reaches a critical volume fraction at 80 minutes and dual phase continuity is seen to exist. After five more minutes, the polystyrene-rich phase became completely continuous. The phase inversion process is bordered on one side by phase separation and the other side by polystyrene gelation.

Tables II and III provide the viscosity ratios and volume ratios for the poly[(CO,SA,TDI)-SIN-(S,DVB)], 10/90 and 20/80 synthesis, respectively. In Tables II and III, ϕ_2 and η_2 represent the polystyrene-rich phase, and ϕ_1 and η_1 the oil-rich phase.

It can be seen from Table I that the oil phase viscosity increases at a much slower rate than the polystyrene phase due to its lower reactivity at 80°C. Also, the volume of the polystyrene-rich phase is increasing at the expense of the oil-rich phase as styrene and polystyrene migrate to the polystyrene-rich phase.

The reaction path is plotted in Figure 4, and compared with the theoretical line. For the 20/80 reaction system, polystyrene gelation takes place substantially simultaneously with the onset of dual phase continuity. The system was consequently not observed to undergo phase inversion.

It should be emphasized that the reaction paths plotted in Figure 4 do not cross the idealized phase inversion region. This was an unexpected finding, since polymer blend systems (19) exhibit phase continuity and inversion according to the theory illustrated in Figure 4. A few observations on the centrifugation were important. After centrifugation was complete, the oil phase was thought to be relatively pure. However, since $\eta_2 \to \infty$ with the polystyrene gelation, one would expect the oil-rich phase to become continuous again. Thus, these experiments cast a kinetic as well as a morphological aspect on the process. The polystyrene-rich phase appeared to be composed of close-packed droplets of polystyrene, with greater or lesser amounts of oil in the interstitial regions. Thus, the results in Tables II and III may only be approximate.

After polymerization was complete, transmission electron microscopy was carried out on thin sections of the 10/90 and 20/80 compositions. Confirming the optical micrographs, the polystyrene phase was continuous for the fully reacted product. As illustrated in Figure 5 for the 10/90 system, the oil phase (stained dark) contains a considerable amount of occluded polystyrene. For the 20/80 system, data not shown, dual phase continuity was found. The polystyrene phase was relatively pure, but the oil-rich phase had much occluded

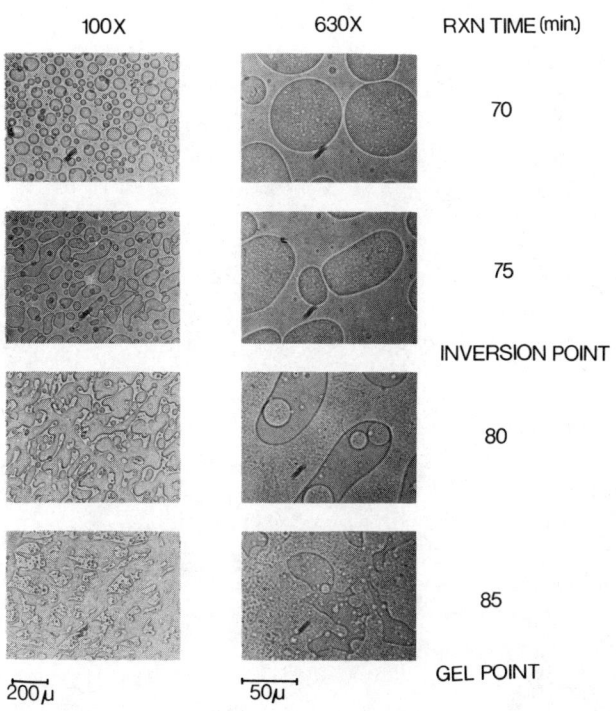

Figure 3. Phase inversion process during the synthesis of a poly[(CO,SA,TDI)-SIN-(S,DVB)] 10/90.

Table II. Viscosity and Volume Fraction Data for Poly[(CO,SA,TDI)-Sin-(S,DVB)], 10/90

Rxn Time (min)	ϕ_1 (ml)	ϕ_2 (ml)	ϕ_2/ϕ_1	η_1 (cps)	η_2 (cps)	η_1/η_2	$\eta_1/\eta_2 \times \phi_2/\phi_1$
*30	---	---	---	---	---	---	---
40	10.3	2.5	0.24	4.06	5.43	0.75	0.180
50	10.2	2.4	0.24	4.84	6.29	0.77	0.185
60	9.80	3.2	0.33	6.02	8.26	0.73	0.241
70	7.30	5.1	0.70	10.75	16.23	0.66	0.462
75	7.00	5.5	0.79	9.62	25.37	0.38	0.300
**80	6.40	6.2	0.97	10.74	45.74	0.23	0.223
85	5.20	7.4	1.42	---	---	---	---

* = phase separation point
** = phase inversion point

phase 1 = castor oil rich
phase 2 = polystyrene rich

Table III. Viscosity and Volume Fraction Data for Poly[(CO,SA,TDI)-Sin-(S,DVB)],20/80

Rxn Time (min)	ϕ_1 (ml)	ϕ_2 (ml)	ϕ_2/ϕ_1	η_1 (cps)	η_2 (cps)	η_1/η_2	$\phi_2/\phi_1 \times \eta_1/\eta_2$
30	----	---	----	----	-----	----	-----
*40	10.7	2.2	0.21	7.01	25.71	0.27	0.057
50	10.4	2.6	0.25	7.27	28.63	0.25	0.063
60	9.2	3.5	0.38	8.89	59.79	0.15	0.057
65	8.6	4.1	0.48	9.24	124.90	0.074	0.036
**70	8.1	4.4	0.54	----	-----	----	-----
75	7.5	5.0	0.67	----	-----	----	-----

* = phase separation point
** = dual phase continuity

phase 1 = castor oil rich
phase 2 = polystyrene rich

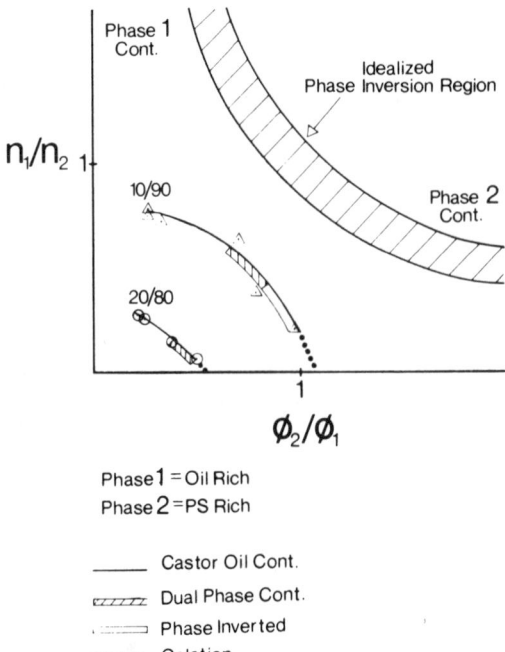

Figure 4. Phase inversion followed by the reaction path of poly[(CO,SA,TDI)-SIN-(S,DVB)].

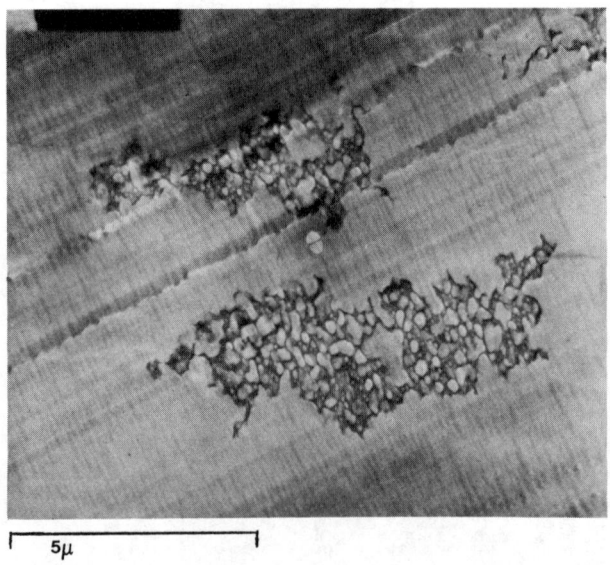

Figure 5. Transmission electron micrograph of poly[(CO,SA,TDI)-SIN-(S,DVB)], 10/90, after being fully polymerized and postcured. Oil phase is stained dark.

polystyrene. This may have influenced the results shown in Figure 4. It should be emphasized that for non-reacting blends of polybutadiene and polystyrene, and other blend systems, equation (2) is followed within experimental error (19). Thus, kinetic restrictions imposed by a system gelling simultaneously with phase inversion dominate this particular experiment.

Physical and Mechanical Behavior of the Oils and SIN's. One of the most important properties of any polymer is its glass transition temperature. This defines its range of use, as well as a host of fundamental properties. This holds for IPN's and SIN's. In particular, for multiphase materials, the rubber phase must have a T_g below about -40°C if significant impact resistance is to be obtained.

Table IV summarizes the glass transition temperature of a number of special functional oil-based homopolymers. Many of the polymers shown indeed have T_g's below the critical value. It is now believed that the synthetically epoxidized oils were fully epoxidized, producing materials with unusually high degrees of crosslinking, which tended to raise their T_g's beyond the desirable range.

Table IV. Glass Transition Temperatures of Single Networks Made from Special Functional Oils

Oil	Crosslinker	T_g, °C	(Ref.)
Not Synthetically Epoxidized			
Castor	Sebacic Acid	-66	(12)
Castor	Sebacic Acid-TDI	-50	(12)
Castor	TDI	- 4	(12)
Lesquerella Palmeri	Sebacic Acid-TDI	-55	(20)
Naturally Epoxidized			
Vernonia anthelmintica	Sebacic Acid	-50	(22)
Synthetically Epoxidized			
Linseed	Dimer Acid	-18	(21)
Lunaria annua	Dimer Acid	-34	(21)
Crambe abyssinica	Dimer Acid	-35	(21)
Lesquerella gracilis	Dimer Acid	-27	(21)
Lesquerella palmeri	Sebacic Acid	-28	(20)

The SIN's from castor oil and the other oils were tough materials, either reinforced elastomers or impact resistant plastics depending on their composition and whether phase inversion had occurred. Impact strengths in the range of 40-60 J/m^2 were obtained. The glass transitions of the rubber phase of the SIN's tended to be a little higher than those shown in Table IV. The polystyrene phase has T_g's near 100°C (9-14).

Conclusions

Castor oil and other special function group triglyceride oils can be made into low T_g elastomers. In SIN form with polystyrene, a range of reinforced elastomers or tough plastics can be made. Points of interest both to academia and industry include phase continuity and inversion during polymerization, glass transition behavior of the phases, and a host of other properties, many of which remain to be characterized.

Acknowledgment

The authors wish to thank the National Science Foundation for support under Grant No. CPE-8109099.

Literature Cited

1. Princen, L. H. J. Coat. Technol. 1977, 49, 88.
2. Princen, L. H. JOACS 1979, 56, 845.
3. Yenwo, G. M.; Manson, J. A.; Pulido, J.; Sperling, L. H.; Conde, A.; Devia, N. J. Appl. Polym. Sci. 1977, 21:1531.
4. Yenwo, G. M.; Sperling, L. H.; Pulido, J.; Manson, J. A. Polym. Eng. Sci. 1977, 17:251.
5. Pulido, J. E.; Yenwo, G. M.; Sperling, L. H.; Manson, J. A. Rev. UIS (Colombia) 1977, 7:35.
6. Yenwo, G. M.; Sperling, L. H.; Manson, J. A.; Conde, A.. In "Chemistry and Properties of Crosslinked Polymers"; Labana, S. S., Ed.; Academic Press: New York, 1977; p. 257.
7. Sperling, L. H.; Manson, J. A.; Yenwo, G. M.; Devia, N.; Pulido, J. E.; Conde, A. In "Polymer Alloys"; Klempner, D.; Frisch, K. C., Eds.; Plenum: New York, 1977.
8. Devia, N.; Conde, A.; Sperling, L. H.; Manson, J. A. Rev. UIS (Colombia), 1977, 7:19.
9. Devia-Manjarres, N.; Conde, A.; Yenwo, G.; Pulido, J., Manson, J. A.; Sperling, L. H. Polym. Eng. Sci., 1977, 17:294.
10. Devia, N.; Manson, J. A.; Sperling, L. H.; Conde, A. Polym. Eng. Sci., 1978, 18:200.
11. Devia, N.; Manson, J. A.; Sperling, L. H.; Conde, A. Macromolecules, 1979, 12:360.
12. Devia, N.; Manson, J. A.; Sperling, L. H., Conde, A. Polym. Eng. Sci., 1979, 19:870.
13. Devia, N.; Manson, J. A.; Sperling, L. H.; Conde, A. Ibid., 1979, 19:878.
14. Devia, N.; Manson, J. A.; Sperling, L. H.; Conde, A. J. Appl. Polym. Sci. 1979, 24:569.
15. Sperling, L. H., Devia, N., Manson, J. A., Conde, A. In "Modification of Polymers; Carraher, C. E.; Tsuda, M., Eds.; ACS Symposium Series No. 121, American Chemical Society, Washington, DC, 1980.
16. Sperling, L. H.; Manson, J. A.; Qureshi, S., Fernandez, A. M. Ind. Eng. Chem. Prod. Res. Dev. 1981, 20:163.
17. Sperling, L. H.; Manson, J. A.; Devia-Manjarres, N.; U.S. Patent 4,254,002, 1981.
18. Sperling, L. H.; Manson, J. A. JOACS, 1983, 60(11).

19. Jordhamo, G. M.; Manson, J. A.; Sperling, L. H. Accepted *Polym. Eng. Sci.*
20. Linne, M. A.; Sperling, L. H.; Fernandez, A. M.; Qureshi, Shahid; Manson, J. A. In "Rubber-Modified Thermoset Resins"; Riew, C. K.; Gillham, J. K., Eds. ACS ADVANCES IN CHEMISTRY SERIES NO. 208, American Chemical Society: Washington, DC, 1984.
21. Qureshi, Shahid; Manson, J. A.; Sperling, L. H.; Murphy, C. J. In "Polymer Applications of Renewable Resource Materials," Carraher, C. E., Jr.; Sperling, L. H., Eds.; Plenum, NY, 1983.
22. Fernandez, A. M.; Manson, J. A.; Sperling, L. H. In "Polymer Applications of Renewable Resource Materials," Carraher, C. E., Jr.; Sperling, L. H., Eds. Plenum: New York, 1983.
23. Sperling, L. H. "Interpenetrating Polymer Networks and Related Materials." Plenum: New York, 1981.
24. Klempner, D. *Angew Chem.* 1978, 90, 104.
25. Frisch, H. L.; Frisch, K. C.; Klempner, D. *Mod. Plast.* 1977, 54, 76, 84.
26. Lipatov, Yu. S.; Sergeeva, L. M. *Russ. Chem. Rev.* 1976, 45(1), 63.
27. Paul, D. R.; Barlow, J. W. *J. Macromol. Sci.-Rev. Macromol. Chem.* 1980, C18, 109.
28. Stockmayer, W. H. *J. Chem. Phys.* 1943, 11, 45.
29. Carothers, W. H. *Trans. Faraday Soc.*, 1936, 32, 29.

RECEIVED February 19, 1985

Author Index

Arnold, F. E., 17,31
Auman, Brian C., 9
Bailey, William J., 147
Chen, Wei-Chih, 139
Cheng, Pao-Luo, 139
Dawson, D. J., 63
Economy, J., 63
Evans, B., 31
Faust, Rudolf, 125
Fehervari, Agota, 125
Feld, W. A., 17
Fleming, W. W., 63
Gapud, Benjamin, 147
Harris, Frank W., 81
Hergenrother, Paul M., 1
Hinkley, Jeffrey A., 43
Holubka, J. W., 117
Jordhamo, G. M., 237
Kane, J. J., 31
Kennedy, Joseph P., 125
Lin, S. C., 105
Lin, Yin-Nian, 147
Lindley, P. M., 31
Loucks, George R., 187
Lyerla, J. R., 63
Maekawa, Toshihiko, 175
Malhotra, Vinay, 53
Manson, J. A., 237
Matsumoto, Akira, 225
McGrath, James E., 161
Ni, Zhende, 147
Nishide, Hiroyuki, 175
Oiwa, Masayoshi, 225
Pearce, Eli M., 209
Percec, Virgil, 91
Picklesimer, L. G., 31
Quirk, Roderic P., 139
Riffle, Judy S., 161
Sperling, L. H., 237
Sridhar, K., 81
Sukenik, Chaim N., 53
Tsuchida, Eishun, 175
Varde, Uday, 53
Wallace, J. S., 17
White, Dwain M., 187
Wu, Shang-Ren, 147
Yan, Hoh-Jiear, 209
Yilgor, Iskender, 161
Yu, Simon H., 199

Subject Index

A

Acetic anhydride, capping, PTC reaction, PPO, 188
Acetone-protected acetylene-terminated products, preparation, 20
Acetone-protecting groups, cleavage of AT-products, 21
Acetylene end-capping, DMSO, bisphenol-based AT-resins, 25
Acetylene-terminated (AT) arylether sulfone oligomers, Ullman ether reaction, 31-44
Acetylene-terminated (AT) imide oligomers
  cured, 11
  thermally curable, 10
Acetylene-terminated (AT) isoimide oligomers, thermally curable, 11
Acetylene-terminated (AT) polymers, synthesis, 25
Acetylene-terminated resins (ATR)
  bisphenol-based, synthesis, 17-29
  thermally curable, 6
Acetylene-terminated sulfone (ATS), thermally curable, 5
Acid values, castor oil prepolymer formation, 241
Addition polymerization
  biodegradable, free radical ring-opening, 151
  SINs, 240f
Adhesive applications, thermosetting epoxy resin based on N-cyanourea-terminated oligomer, 107
Adogen 464, coupling reagent, poly(phenylene oxide), 190
Amination, polymeric organolithium compounds, 139-145
Amine, telechelic, synthesis, 139-145
Amine end-group functionality, primary, synthesis, 140-143

Amine-cured bisphthalonitrile polymers, network structure, 50f
α,ω-Aminopropyl-terminated poly(dimethyl-diphenyl)siloxane oligomers, synthesis, 165,169-73
Aromatic nucleophilic substitution, AT arylether sulfone oligomers, 33,36
Aromatic polyether sulfone (PSU), containing styrene groups, 92
Arylether sulfone oligomers, AT, Ullman ether reaction, 31-44

B

Backbone composition, siloxane oligomers, 163
Base-catalyzed equilibrations of hydroxyl-terminated siloxane oligomers, 169
3,3',4,4'-Benzophenone tetracarboxylic dianhydride (BTDA)
 anhydride-terminated oligomers, 82
 oligomers with diamines, 83-87
Benzoyl peroxide, effect on synthesis of 2-methylene-1,3-dioxolane, 149
Benzyl ketene acetal, addition-elimination with free radicals, 157
Benzyl methyl ketene acetal, mechanism of production of end-capped oligomer, 156
Biodegradable addition polymer, free radical ring-opening polymerization, 151
α,ω-Bis(aminopropyl)poly(dimethyl-diphenyl)siloxane, structures, 167
α,ω-Bis(benzylether) polyether sulfone, DSC curves, 101f
α,ω-Bis(hydroxybutyl)polydimethyl-siloxane oligomers, synthesis and structures, 167
α,ω-Bis(hydroxyphenyl) polyether sulfone
 dependence between the Tg and Mn, 101f
 DSC curves, 97f,101f
 thermal characterization, 99
α,ω-Bis(vinylbenzyl) polyether sulfone
 DSC curves, 97f
 synthesis and characterization, 93-96
 thermal characterization, 99
Bismaleimides, preparation, thermally curable oligomers, 3
Bisphenol-based acetylene-terminated resins, synthesis, 17-29

Bisphthalonitrile polymers, network structure, 43-51
Block copolymers containing poly(phenylene oxide), PTC coupling system, 191

C

Capping, PTEB, determined by DSC, 68
Capping and coupling reactions, PPO, 187-197
Carbomethoxy-substituted norbornenyl imides, intermolecular and intramolecular reactions, 53-61
Carbon dioxide, carbonation of polymeric organolithium compounds, 143
Carbon-13 CPMAS NMR, PTEB, 75f,76f
Carbon-13 NMR
 characterization of cured diacetylene oligomers, 64
 HTE liquid polymers, 203,206f
 PTEB, 75f
Carbonation, solid-state, polymeric organolithium compounds, 139-145
Castor oil-polystyrene, SINs, reaction scheme, 240f
Castor oil-styrene, SINs, 237
Catalysis, effect on urethane-modified epoxy-diol oligomers, 119
Catalyst, PTC reaction, PPO, 188
Catalyst metals removal, effect of acetylene end-capping agent, 27
Catalytic cycle, bisphenol-based AT-resins, 22,24f
Catalyzed Williamson etherification, phase transfer, polyaromatics containing styrene groups, 93
Cationic polymerization technique, novel semicontinuous, telechelic PIB, 125-137
Cationic ring-opening polymerization, ECH, 199
Cationic routes, synthesis of α,ω-hydroxybutyl-terminated polydimethylsiloxane oligomers, 165
Cetyltrimethylammonium bromide, coupling reagent, poly(phenylene oxide), 190
Chain end functionalization, polymeric organolithium compounds, 139-145
Chain extension
 mechanism, urethane-modified epoxy-diol oligomers, effect, 119

INDEX

Chain extension—Continued
  oligourethane oligomers, IR study, 120
  urea-urethane oligomers, 119
Chain flexibility, effect of various synthesis routes, AT matrix resins, 36
Chain-reaction mechanism, electrooxidative polymerization of 2,6-dimethylphenol and its dimer, 178
Chain-transfer agents, ketene acetals, 155
Char yield, styrylpyridine-based polymers, 215t
Chemical structure, influence on thermal stability, 210
Cleavage
  AT-product acetone protecting groups, 21
  effect of benzyl group in ketene acetal, 156
Coating, high solids, chain extendable oligomers, 117-123
Coating applications, thermosetting epoxy resin based on $\underline{N}$-cyanourea-terminated oligomer, 107
Coating materials, $\underline{N}$-cyanourea oligomers, 112
Cocross-linkers, DAP resins, glycol bis(allyl phthalate)s, 225-234
Condensation polymerization, SINs, 240f
Conjugating substituents, effect on high temperature cure of substituted norbornenyl imides, 53-61
Copper(I) iodide, proposed role, bisphenol-based AT-resins, 23f
Coupling and capping reactions, PPO, 187-197
Cross-coupling process, PPO, 191
Cross-linking
  bisphthalonitriles, 48
  $\underline{N}$-cyanourea-terminated oligomers, 113f
  effect on gelation, diallyl phthalate resins, 227
  effect on toughness of AT systems, 32
  network structure, polymers from bisphthalonitriles, 48
  oligomers of BTDA and nonlinear diamines, 87
  urea-urethane oligomers, 119
  urethane-modified epoxy-diol oligomers, effect, 119
Cross-polarization magic-angle spinning techniques (CPMAS), 63

Cure behavior, oligomers of BTDA and linear diamines, 83
Cure mechanism, urea-urethane oligomers, 119,121
Cure reactions, PTEB, 72
Cured acetylene-terminated imide, properties, 11
Cured ethynyl-terminated ester oligomers, preparation, 13
Cured ethynyl-terminated sulfones, properties, 8
Curing agents, based on $\underline{N}$-cyanourea-terminated oligomers, 106,109
Curing study, oligomers of BTDA and linear diamines, 86
Cyanamides, preparation, thermally curable oligomers, 3
Cyanates, preparation, thermally curable oligomers, 2
Cyanourea chemistry, schemes, 114
$\underline{N}$-Cyanourea-terminated resins, curing and polymerization mechanisms, 105
Cyclic voltammogram, oxidation of 2,6-dimethylphenol, 177f,181f

D

Diacetylenic polymers, thermally stable polymers for electronic applications, 64
Diamines
  containing oxyethylene linkages, planarizing coatings, 82
  linear, and oligomers of BTDA, 83-86
  nonlinear, and oligomers of BTDA, 87
α,ω-Diaminopolystyrene
  amination, 141
  molecular weight characterization, 142t
Dibenzoyl derivative, α,ω-diaminopolystyrene, elemental analyses, 142t
Dibromobenzene isomers, effect on synthesis of AT arylether sulfone oligomers, 37
Dibromobenzene structure, effect on chain flexibility, 36
Dielectric insulating layers, model polymer, 64
Dielectric loss peaks, secondary, bisphthalonitrile polymers, 47f
Diels Alder synthetic routes, effect on monomer isomerization, substituted norbornenyl imides, 55

Diethyl ketene acetal, chain transfer agent, 155
Differential scanning calorimetry (DSC)
  bisphenol-based AT-resins, 27
  effect of phenylacetylene capping, PTEB, 71f
  effect of UV radiation, PTEB, 71f
  ethynyl end-capped polyimide oligomers containing oxyethylene linkages, 84f
  nature of PTEB curing process, 68
  oligomers of BTDA
    and linear diamines, 83
    and nonlinear diamines, 87
  polyaromatics with styrene groups, 97f, 101f
  quenching experiment, capped PTEB, 73f
  thermal curing of PSU and PPO containing pendant vinyl groups, 100
Difunctional N-cyanourea compound, polymerization mechanism, 109
Difunctional siloxane oligomers, α,ω-terminated, synthesis and characterization, 161-173
Dihydroxybiphenyl, effect on synthesis of AT arylether sulfone oligomers, 37
α,ω-Dilithiumpolystyrene, amination, 141
Diol structure
  effect on chain flexibility, 36
  effect on synthesis of AT arylether sulfone oligomers, 37
Diol synthesis, 4-methoxyphenol, 38f
2,6-Disubstituted phenols, electrooxidative polymerization, 176-178
Dual phase continuity, SINs, definition, 238
Dynamic mechanical properties, bisphthalonitriles, 44-46, 49

E

Electrochemical features, polymerization, 2,6-disubstituted phenols, 180
Electrodes, electrooxidative polymerization of phenols, 175-185
Electrolysis apparatus, polymerization of phenols, 177f
Electronic applications, thermally stable polymers, 63-79

Electrooxidative polymerization of phenols, synthesis of poly(phenylene oxide)s, 175-185
End groups, functional, variations in the structure and composition of siloxane oligomers, 163
End-capped oligomers, reactive, free radical ring-opening copolymerization, 147-158
End-capping agent, bisphenol-based AT-resins, 18, 21-24, 25
End-groups, semicontinuous polymerization, inifers, 131
End-reactive polyisobutylenes (PIB)
  comparison between conventional batch and semicontinuous inifer techniques, 136
  well-defined, semicontinuous polymerization, inifer technique, 125-137
Epichlorohydrin (ECH), hydroxyl-terminated liquid polymers, characterization, 199-208
Epoxidation, phenyl-substituted norbornenyl imides, 60
Epoxies, preparation, thermally curable oligomers, 2
Epoxy curing agents, N-cyanourea-terminated oligomers, 109
Epoxy oligomer, chain extendable urethane modified, preparation, 118
Epoxy-diol oligomers, urethane modified, discussion, 119
Equilibration route, synthesis of functionally terminated siloxane oligomers, 164
Ether-linked bisphthalonitriles, synthesis, 43
Etherification, PTC, Williamson, polyaromatics containing styrene groups, 93
Ethylene glycol (EG), polymerization of ECH, 200
Ethynyl end-capped polyimide oligomers, synthesis and characterization, 81-90
Ethynyl-terminated ester oligomers, cured, preparation, 13
Ethynyl-terminated oligomers, thermal stability, 63
Ethynyl-terminated sulfone (ETS) cured, properties, 8
  preparation, 7
Exo-endo isomerization, effect on polymerization of norbornenyl imides, 58

INDEX

F

Flame detector mass spectroscopy, HTE liquid polymers, 205,207f
Flammability characterization, styrylpyridine polyesters and polycarbonates, 217
Flexible cross-linked units, DAP resins cocross-linked with glycol bis(allyl phthalate)s, 228
Free radical ring-opening polymerization, synthesis of reactive oligomers, 147-158
FTIR
  HTE liquid polymers, 201,202f
  α,ω-hydroxybutyl-terminated polydimethylsiloxane oligomer, 168f
Functionality, HTE liquid polymers, 201,204t
Functionalization, polymeric organolithium compounds, 139-145
Functionally-terminated oligomers, free radical process, 147

G

Gel permeation chromatography (GPC)
  HTE liquid polymers, 202f
  polymers by batch and semicontinuous inifer technique, 133f
Gelation
  bisphthalonitriles, 46
  glycol bis(allyl phthalate)s, 227
  radical polymerization of diallyl dicarboxylates, 230
Glaser oxidative coupling, TEB to give PTEB, 68
Glass transition temperatures (Tg)
  aminopropyl-terminated poly(dimethyl-diphenyl)-siloxane oligomers, 171
  AT-type systems, 31
  bisphenol-based AT-resins, 28f
  bisphthalonitrile monomers, 44-46
  DAP resins cocross-linked with glycol bis(allyl phthalate)s, 228
  monomeric model of AT matrix resin, 36
  phthalonitrile mixtures, 50f
  polyaromatics containing styrene groups, 98
  single networks made from special functional oils, 247t
Glycol bis(allyl phthalate), cocross-linkers for DAP resins, 225-234

H

Halo-terminated intermediate, bisphenol-based AT-resins, 19
Halo-terminated products, bisphenol-based AT-resins, 25
High temperature polymers
  definition, 2
  reactive oligomer approach, 1
Hydrogen donor, curing of phenyl-substituted norbornenyl imides, 59
Hydrogenation, phenyl-substituted norbornenyl imides, 60
Hydrolysis
  ethylene-2-methylene-1,3-dioxepane copolymer, 152
  hydroxybutyl-terminated siloxane oligomers, 167
  oligomeric polystyrene, 154
  poly(phenylene oxide)s, 189
Hydroquinone, effect on synthesis of AT arylether sulfone oligomers, 36,37
α,ω-Hydroxybutyl-terminated polydimethylsiloxane oligomers, synthesis, 165
Hydroxybutyl-terminated siloxane oligomers, synthesis and characterization, 166-69
Hydroxyl groups, substituent effect on carbon chemical shifts of polyethers, 203
Hydroxyl-terminated epichlorohydrin (HTE) liquid polymers,
  characterization, 199-208
  synthesis by cationic ring-opening polymerization, 199
Hydroxyl-terminated poly(phenylene oxide), electrooxidative polymerization, 182
Hydroxyl-terminated siloxane oligomers, base catalyzed equilibrations, 169

I

Imides
  cured AT, properties, 11
  substituted norbornenyl, intermolecular and intramolecular reactions, 53-61
  thermally curable oligomers, 8
Imidization procedures, general, 89
Initiator-transfer agents (inifers), polyfunctional, new

Initiator-transfer agents—Continued
  telechelic polymers and sequential copolymers, 125-137
Intra-aggregate coupling, carbonation of polymeric organolithium compounds, 143
IR analysis
  chain extension reaction, oligourethane oligomers, 120t
  cure of (tert-butyl carbamate) terminated urea-urethane coating, 121
  Epon 828-di(N-cyanourea) mixture, 110f,113f
  monitoring of thermal cure, capped PTEB, 74f
  oligomers of BTDA
    and linear diamines, 83
    and nonlinear diamines, 87
  poly(1,4-phenylene oxide) obtained by electrolysis of phenols, 184f
  urea-urethane oligomers, 122f
  urethane-modified epoxy-diol oligomer, 120f
Isomerization, monomer, substituted norbornenyl imides, 55
Isothermal aging (ITA), bisphenol-based AT-resins, 27

K

Ketene acetal
  chain transfer agents from, 155
  cyclic
    nitrogen analogs, free radical ring-opening polymerization, 152
    sulfur analog, free radical ring-opening polymerization, 153
    synthesis, free radical ring-opening polymerization, 149
  seven-membered, free radical ring-opening polymerization, 150
Ketones, macromolecular, synthesis, 139-145
Kinetics
  idealized semicontinuous polymerization technique, 128-131
  polymerization of isobutylene by the inifer-BCl3 system, 127

L

Lap shear strength, N-cyanourea-terminated resins, 108f

Liquid polymers, HTE, functionality, 204t
Long-chain alkyl groups, vinyl monomers with, DAP resins cocured microheterogeneously, 232

M

Maleimide-terminated sulfones, preparation, thermally curable oligomers, 6
Mechanical damping of thin films, bisphthalonitrile polymers, 49f
Mechanical properties, DAP resins cocross-linked with glycol bis(allyl phthalate)s, 228
2-Methyl-1,3-dioxolane ring system, introduction of phenyl group, 156
Methylene bromide, coupling reagent, PTC reaction, poly(phenylene oxide), 191
Methylene chloride, coupling reagent, PTC reaction, poly(phenylene oxide), 191
2-Methylene-1-dioxepane, free radical ring-opening polymerization, 150
2-Methylene-1-dioxolane, free radical ring-opening polymerization, 149
Microgel formation, dependence on conversion, DAP resins, 230
Model monomer systems, synthesis, 35f
Model substrate, synthesis and characterization, substituted norbornenyl imides, 54
Molecular weight characterization, α,ω-diaminopolystyrene, 142t
Molecular weight dispersities, semicontinuous polymerization, inifers, 131
Mono(bromophenoxy)phenols, synthesis, 34f
Monomer
  AT, synthesis, 25
  BTDA, preparation, 82
  composition effect on DAP resins cocross-linked with glycol bis(allyl phthalate)s, 229f,233f
  isomerization, substituted norbornenyl imides, 55
  structure
    ether-linked bisphthalonitriles, 44
    polymers from bisphthalonitriles, 46
Monomer-oligomer, acetylene-terminated, 25

INDEX

Monomeric AT arylether sulfones, evaluation, 35f
Monomeric systems, model, AT matrix resins, 33-36

N

Nadimide-terminated polysulfones, preparation, thermally curable oligomers, 6
Network structure, polymers from bisphthalonitriles, 43-51
Nitrogen analogs, free radical ring-opening polymerization, cyclic ketene acetals, 152
NMR instrumentation, characterization of cured diacetylene oligomers, 67
Norbornenyl imides, substituted, intermolecular and intramolecular reactions, 53-61

O

Oil-based homopolymers, special functional, physical and mechanical behavior, 247
Oil-based prepolymers, SIN compositions, 241
Oil-based SINs, preparation, 239
Oil-phase viscosity, SINs with special functional triglycerides, 242
Oligomer
  AT arylether sulfones
    evaluation, 38f
    reaction sequence, 34f
  BTDA and diamines, 83-86,87
  chain extendable urethane-modified epoxy, preparation, 118
  $\underline{N}$-cyanourea-terminated, 106,107
  evaluation, synthesis of AT arylether sulfones, 37
  reactive difunctional siloxane, synthesis and characterization, 161-173
  reactive end-capped, free radical ring-opening copolymerization, 147-158
  thermally curable, synthesis, physical, and mechanical properties, review, 1-14
  thermosetting, based on $\underline{N}$-cyanourea-terminated resins, 105
  urea-urethane, preparation, 118

Oligomer--Continued
  urethane-modified epoxy-diol, discussion, 119
Oligomeric coupling reagents, PTC reaction, poly(phenylene oxide), 191
Oligourethane oligomers
  chain extendable, high solids coating systems, 117-123
  chain extension reaction, IR study, 120
Organolithium compounds, polymeric, functionalization, 139-45
Organosiloxane-based segmented elastomers, discussion, 161
Oxidation, curing of phenyl-substituted norbornenyl imides, 59
Oxyethylene linkages, synthesis, ethynyl end-capped polyimide oligomers, 81
Oxygen index, styrylpyridine-based polymers, 215t

P

Palladium catalyst, bisphenol-based AT-resins, 21
Pendant or terminal styrene groups, polyaromatics, synthetic methods, 91-103
Pendant unreacted double bonds, DAP resins cocross-linked with glycol bis(allyl phthalate)s, 228
Phase inversion process
  castor oil-polystyrene SINs, 243f,246f
  SINs with special functional triglycerides, 242
Phase separation, SINs with special functional triglycerids, 242
Phase transfer catalysis (PTC)
  etherification of polyaromatics with styrene groups, 96
  PPO, 187
  Williamson etherification, polyaromatics containing styrene groups, 93
Phenols, electrooxidative polymerization, synthesis of poly(phenylene oxide)s, 175-185
Phenyl group, introduction into 2-methyl-1,3-dioxolane ring system, 156
Phenyl-substituted norbornenyl imides, intermolecular and intramolecular reactions, 53-61

Phenylacetylene
  effect of capping on DSC, PTEB, 71f
  thermal polymerization, 92
Phenylene R systems, flexible, Ullman
  ether reaction, 32
Phenylquinoxalines-polyphenyl-
  quinoxalines (PPQ), preparation,
  thermally curable oligomers, 12
Photo-Fries rearrangement,
  styrylpyridine polyesters and
  polycarbonates, 217
Photodegradation, polymers, 210
Phthalonitrile mixtures, Tg
  values, 50f
Planarizing coatings
  ethynyl end-capped polyimide
    oligomers, 81
  PTEB systems, 64
Polyamic acid oligomers, general
  procedure, 89
Polyaromatics with terminal or pendant
  styrene groups, synthetic
  methods, 91-103
Polyarylates, thermal and photo
  stability, 209-221
Polybenzimidazole (PBI), reactive
  oligomeric precursor, 12
Polycarbonates, styrylpyridine based,
  synthesis and
  characterization, 209-221
Poly(diethynylbenzene), synthesis, 65
Poly(dimethyldiphenyl)siloxane
  oligomers
  aminopropyl-terminated,
    synthesis, 169-73
  characteristics of aminopropyl
    terminated, 171
Poly(2,6-dimethyl-1,4-
  phenylene oxide) (PPO)
  coupling and capping
    reactions, 187-197
  pendant vinyl groups
    DSC of thermal curing, 100-103
    synthesis, 95f,100
  synthesis of polyaromatics
    containing styrene
    groups, 92
Polyesters, styrylpyridine based,
  synthesis and
  characterization, 209-221
Polyether sulfone (PSU)
  pendant vinyl groups
    DSC of thermal curing, 100-103
    synthesis, 94f,100
Polyfunctional initiator-transfer
  agents (inifers), new telechelic
  polymers and sequential
  copolymers, 125-137

Polyimide (PI) oligomers
  ethynyl end-capped, synthesis and
    characterization, 81-90
  reactive, thermal polymerization, 53
  thermally curable, 8
Polyisobutylenes (PIB)
  end-reactive, comparison between
    conventional batch and semicon-
    tinuous inifer techniques, 136
  well-defined end-reactive, semicon-
    tinuous polymerization, inifer
    technique, 125-137
Polymer
  biodegradable addition, free radical
    ring-opening polymerization, 151
  bisphthalonitriles, network
    structure, 43-51
  high molecular weight, based on
    $\underline{N}$-cyanourea-terminated
    oligomers, 106
  high temperature, reactive oligomer
    approach, 1
  primary amine end-group
    functionality,
    synthesis, 140-143
  structure, phenyl-substituted
    norbornenyl imides, 59-61
  thermally stable, electronic
    applications, 63-79
Polymeric organolithium compounds,
  functionalization, 139-145
Polymerization
  free radical ring-opening, synthesis
    of reactive oligomers, 147-158
  glycol bis(allyl phthalate)s, 226
  mechanism, 2,6-dimethylphenol, 178
  model substrates, substituted
    norbornenyl imides, 56-59
  novel semicontinuous cationic,
    telechelic PIB, 125-137
  onset and peak, bisphenol-based
    AT-resins, 28f
  phenols, electrooxidative, synthesis
    of poly(phenylene
    oxide)s, 175-185
  thermal, reactive polyimide
    oligomers, 53
  variations in the structure and
    composition of siloxane
    oligomers, 163
Polyorganosiloxanes, discussion, 161
Poly(phenylene oxide)s
  bifunctional, coupling and capping
    reactions, 187-197
  synthesis, 175-185
Polystyrene, poly(styryl)amine,
  α,ω-diaminopolystyrene, and
  derivatives, TLC, 144f

INDEX

Poly(styryl)amine, preparation, 141
Poly(styryl)lithium
  amination, 141
  carbonation in benzene, 143
Poly(triethynylbenzene) (PTEB)
  cure reactions, 72
  cured, 13C CPMAS NMR, 76f
  effect of phenylacetylene capping on DSC, 71f
  effect of UV radiation of DSC, 71f
  highly capped
    IR to follow thermal cure, 72
    synthesis, 66
  physical properties, 78f
  planarizing coatings, 64
  synthesis, 65
  TGA in air and helium, 70f
  thermal curing products, 77
  uncured, 13C CPMAS NMR, 75f
Proton NMR
  derivatized HTE, 206f
  HTE liquid polymers, 205
  α,ω-hydroxybutyl-terminated polydimethylsiloxane oligomer, 170f
  polymers by semicontinuous inifer technique, 134f, 135f

Q

Quaternary ammonium halide catalyst, PTC reaction, PPO, 188
Quenching experiment, DSC, capped PTEB, 73f

R

Reaction sequence, AT arylether sulfone oligomers, 34f
Reactive difunctional siloxane oligomers, synthesis and characterization, 161-173
Reactive end-capped oligomers, free radical ring-opening copolymerization, 147-158
Reactive oligomeric precursor, PBI, 12
Reactivity of siloxane oligomers, effect of terminal functional groups, 162
Redistribution reaction, synthesis of functionally-terminated siloxane oligomers, 164
Resin
  acetylene-terminated bisphenol-based, synthesis, 17-29
  cocross-linkers for DAP, glycol bis(allyl phthalate)s, 225-234
  $\underline{N}$-cyanourea-terminated, curing and polymerization mechanisms, 105
Resin properties
  ATS, neat, 5
  resorcinol dicyanate cured, 3
Resorcinol, effect on synthesis of AT arylether sulfone oligomers, 37
Resorcinol dicyanate cured resin properties, 3
Ring-opening polymerization, free radical, discussion, 149

S

Salicylic acid, electrooxidative polymerization, 183
Seven-membered ketene acetal, free radical ring-opening polymerization, 150
Siloxane oligomers, reactive difunctional, synthesis and characterization, 161-173
Simultaneous interpenetrating networks (SIN), special functional triglyceride oils as reactive oligomers, 237-249
Solid-state carbonation, polymeric organolithium compounds, 139-145
Solution thermolysis reactions, substituted norbornenyl imides, 55
Solvents, various, effect on monomer isomerization, substituted norbornenyl imides, 55
Special functional triglyceride oils as reactive oligomers, SINs, 237-249
Stabilizers, polymers containing certain chromophores, 210
Stoichiometric excess of hydroquinone, bisphthalonitrile polymers, 49f
Structures
  bisphenol-based AT-resins, 23f
  HTE liquid polymers, 205
Styrene groups, polyaromatics with terminal or pendant, synthetic methods, 91-103
Styrene-castor oil, SINs, 237
Styrylpyridine-based polymers, thermal and photo stability, 209-221
Substituent effect of hydroxyl groups on carbon chemical shifts of polyethers, 203t

Substituted norbornenyl imides,
  intermolecular and intramolecular
  reactions, 53-61
Substitution
  effect on monomer isomerization of
    norbornenyl imides, 56
  effect on the rate of polymerization
    of norbornenyl imides, 58
Substrate synthesis and
  characterization, substituted
  norbornenyl imides, 54
Sulfone oligomers
  arylether, AT, Ullman ether
    reaction, 31-44
  cured ethynyl-terminated,
    properties, 8
  preparation, thermally curable
    oligomers, 3
Sulfur analog, free radical ring-
  opening polymerization, cyclic
  ketene acetal, 153
Swelling ratio of gel polymer,
  dependence on conversion, DAP
  resins, 230

T

Telechelic amines, synthesis, 139-145
Telechelic polyisobutylenes,
  preparation, comparison of
  semicontinuous and conventional
  polymerization techniques, 132t
Telechelic polymers, definition,
  functionality, 200
Telechelic polymers and sequential
  copolymers, new, polyfunctional
  inifers, 125-137
Temperature
  effect on monomer isomerization,
    substituted norbornenyl
    imides, 55
  use, AT type systems, 31
Temperature dependence, dynamic
  modulus and loss tangent,
  bisphthalonitrile monomers, 47f
Tensile strength
  DAP resins cocross-linked with
    glycol bis(allyl
    phthalate)s, 228
  effect on monomer composition, DAP
    resins cocross-linked with
    glycol bis(allyl
    phthalate)s, 233f
Terminal or pendant styrene groups,
  polyaromatics, synthetic
  methods, 91-103

Tetrabutylammonium hydrogen sulfate
  (TBAH), etherification of
  polyaromatics with styrene
  groups, 96
Thermal characterization,
  α,ω-bis(hydroxyphenyl)PSU and
  α,ω-bis(vinylbenzyl)PSU, 99t
Thermal cross-linking, oligomers of
  BTDA and linear diamines, 86
Thermal curing
  DSC, 96
  possible products, PTEB, 77
Thermal degradation, linear polymer
  synthesized from di-$\underline{N}$-cyanourea
  compound, 111
Thermal polymerization, reactive
  polyimide oligomers, 53
Thermal properties
  oligomers of BTDA and linear
    diamines, 86
  oligomers of BTDA and nonlinear
    diamines, 87
Thermal stabilities, oligomers of BTDA
  and nonlinear diamines, 87
Thermally curable oligomers,
  synthesis, physical, and
  mechanical properties, review, 1-14
Thermally stable polymers, electronic
  applications, 63-79
Thermogravametric analysis (TGA)
  ethynyl end-capped polyimide
    oligomers containing oxyethylene
    linkages, 84f,85f
  PTEB in air and helium, 70f
  styrylpyridine-based polymer, 218f
Thermolysis reactions, solution,
  substituted norbornenyl imides, 55
Thermomechanical analysis (TMA),
  bisphenol-based AT-resins, 27
Thermooxidative stability, bisphenol-
  based AT-resins, 27
Thermosets, fast curing, based on
  $\underline{N}$-cyanourea-terminated
  oligomers, 106
Thermosetting epoxy compounds,
  one-package, 106,107
Thermosetting resin, irreversible
  cross-linking reaction, 105
Thin-layer chromatography (TLC),
  polystyrene, poly(styryl)amine,
  α,ω-diaminopolystyrene, and
  derivatives, 144f
Three-arm star products, semicontinuous reaction condition, 126
Time-conversion curves, bulk
  polymerization DAP, 231f
Transmission electron microscopy (TEM)
  castor oil-polystyrene SINs, 246f

INDEX

Transmission electron microscopy (TEM)—Continued
  SINs with special functional triglycerides, 242
1,3,5-Triethynylbenzene (TEB), synthesis, 66,69
Triglyceride oils, description, 237

U

Ullman ether reaction, AT arylether sulfone oligomers, 31-44
Urea-urethane oligomers, preparation, 118
Urethane-modified epoxy oligomer, chain extendable, preparation, 118
Urethane-modified epoxy-diol oligomers, discussion, 119
UV radiation, effect on DSC, PTEB, 71f
UV spectrum
  poly(dimethyl-diphenyl)siloxane oligomer, 172f
  styrylpyridine-based polymer, 219f,220f

V

Vinyl monomers with long-chain alkyl groups, mechanical properties of DAP resins cocured microheterogeneously, 232

Viscosity, oil-based SINs, 244t
Viscosity increase, bisphthalonitriles, 46
Volume fraction data, oil-based SINs, 244t

W

Williamson etherification, PTC, polyaromatics containing styrene groups, 93

X

X-ray diffraction pattern, oligomers of BTDA and linear diamines, 86

Y

Young's modulus, bisphthalonitrile monomers, 44-46

Production by Hilary Kanter
Indexing by Susan Robinson
Jacket design by Pamela Lewis

Elements typeset by Hot Type Ltd., Washington, D.C.
Printed and bound by Maple Press Co., York, Pa.

RECENT ACS BOOKS

"Organic Phototransformations in Nonhomogeneous Media"
Edited by Marye Anne Fox
ACS SYMPOSIUM SERIES 278; 308 pp.; ISBN 0-8412-0913-8

"Xenobiotic Metabolism: Nutritional Effects"
Edited by John W. Finley and Daniel E. Schwass
ACS SYMPOSIUM SERIES 277; 382 pp.; ISBN 0-8412-0912-X

"Bioregulators for Pest Control"
Edited by Paul A. Hedin
ACS SYMPOSIUM SERIES 276; 540 pp.; ISBN 0-8412-0910-3

"Nutritional Bioavailability of Calcium"
Edited by Constance Kies
ACS SYMPOSIUM SERIES 275; 200 pp.; ISBN 0-8412-0907-3

"Chemical Process Hazard Review"
Edited by John M. Hoffmann and Daniel C. Maser
ACS SYMPOSIUM SERIES 274; 124 pp.; ISBN 0-8412-0902-2

"Dermal Exposure Related to Pesticide Use:
Discussion of Risk Assessment"
Edited by Richard C. Honeycutt, Gunter Zweig,
and Nancy N. Ragsdale
ACS SYMPOSIUM SERIES 273; 530 pp.; ISBN 0-8412-0898-0

"Macro- and Microemulsions: Theory and Applications"
Edited by Dinesh O. Shah
ACS SYMPOSIUM SERIES 272; 502 pp.; ISBN 0-8412-0896-4

"Purification of Fermentation Products:
Applications to Large-Scale Processes"
Edited by Derek LeRoith, Joseph Shiloach, and Timothy J. Leahy
ACS SYMPOSIUM SERIES 271; 200 pp.; ISBN 0-8412-0890-5

"Reaction Injection Molding: Polymer Chemistry and Engineering"
Edited by Jiri E. Kresta
ACS SYMPOSIUM SERIES 270; 302 pp.; ISBN 0-8412-0888-3

"Materials Science of Synthetic Membranes"
Edited by Douglas R. Lloyd
ACS SYMPOSIUM SERIES 269; 496 pp.; ISBN 0-8412-0887-5

"The Chemistry of Allelopathy: Biochemical Interactions Among Plants"
Edited by A. C. Thompson
ACS SYMPOSIUM SERIES 268; 466 pp.; ISBN 0-8412-0886-7

"Rubber-Modified Thermoset Resins"
Edited by Keith Riew and J. K. Gillham
ADVANCES IN CHEMISTRY SERIES 208; 370 pp.; ISBN 0-8412-0828-X

"The Chemistry of Solid Wood"
Edited by Roger M. Rowell
ADVANCES IN CHEMISTRY SERIES 207; 588 pp.; ISBN 0-8412-0796-8

NACAN PRODUCTS LIMITED